I0882732

Leading With Faith and Ethics in an AI World

Jeffrey Moore
https://orcid.org/0000-0002-2508-2896
Anderson University, USA

Manjari Sharma
https://orcid.org/0000-0003-1459-165X
Christ University, Bangalore, India

Sharad Gupta
https://orcid.org/0000-0002-4994-3523
Christ University, Bangalore, India

IGI Global
Scientific Publishing
Publishing Tomorrow's Research Today

Vice President of Editorial	Melissa Wagner
Director of Acquisitions	Mikaela Felty
Director of Book Development	Jocelynn Hessler
Production Manager	Mike Brehm
Cover Design	Jose Rosado

Published in the United States of America by
IGI Global Scientific Publishing
701 East Chocolate Avenue
Hershey, PA, 17033, USA
Tel: 717-533-8845 | Fax: 717-533-7115
Website: https://www.igi-global.com E-mail: cust@igi-global.com

Copyright © 2027 by IGI Global Scientific Publishing. All rights reserved. No part of this publication may be reproduced, stored or distributed in any form or by any means, electronic or mechanical, including photocopying, without written permission from the publisher. Use of this publication to train generative artificial intelligence (AI) technologies is expressly prohibited. The publisher reserves all rights to license its use for generative AI training and machine learning model development.

Product or company names used in this set are for identification purposes only. Inclusion of the names of the products or companies does not indicate a claim of ownership by IGI Global Scientific Publishing of the trademark or registered trademark.

Library of Congress Cataloging-in-Publication Data

LCCN: 2026011364 (CIP Data Pending)
ISBN13: 9798337381923
Isbn13Softcover: 9798337381930
EISBN13: 9798337381947

British Cataloguing in Publication Data
A Cataloguing in Publication record for this book is available from the British Library.

All work contributed to this book is new, previously-unpublished material.
The views expressed in this book are those of the authors, but not necessarily of the publisher.
This book contains information sourced from authentic and highly regarded references, with reasonable efforts made to ensure the reliability of the data and information presented. The authors, editors, and publisher believe the information in this book to be accurate and true as of the date of publication. Every effort has been made to trace and credit the copyright holders of all materials included. However, the authors, editors, and publisher cannot assume responsibility for the validity of all materials or the consequences of their use. Should any copyright material be found unacknowledged, please inform the publisher so that corrections may be made in future reprints.

To my father, James W. Moore, whose life has modeled faith, perseverance, humility, and steadfast dedication to his family and parishioners.

You taught me to love reading, studying, and the pursuit of wisdom. Through your example, I learned that faith is not only spoken, but lived through service, sacrifice, patience, and love.

The greatest gift you have given me is the example of a personal relationship with God and a deep love for developing the next generation through discipleship and training. For your guidance, your love for God and others, and your enduring influence on my life, I dedicate this work to you with deep love and gratitude.

Jeffrey

Table of Contents

Section 2
Leadership Formation in the Age of AI

Detailed Table of Contents

This chapter examines how faith integration, ethical orientation, and artificial intelligence (AI) practices shape responsible AI adoption, pedagogical integration, academic integrity, and student outcomes at Northrise University, Zambia. Positioned within a values-driven higher education context, this chapter explores how faith-informed ethical frameworks guide responsible AI use in teaching and learning. Using a concurrent cross-sectional mixed-methods design, faculty (n = 38) and student (n = 56) perspectives are analyzed to explain how values-based governance influences AI adoption. Findings indicate that faith integration supports responsible AI guidance, ethical orientation strengthens integrity vigilance, and AI capacity-building drives pedagogical integration, while student outcomes show strong links between AI value, employability, and learning.

The rapid deployment of artificial intelligence across economic, political, and social domains has generated ethical questions that secular frameworks are ill-equipped to resolve. This chapter contends that Christian moral theology provides the most coherent and durable foundation for navigating these challenges. Drawing on the scriptural doctrine of the Imago Dei, this study argues that every human person possesses inherent, non-negotiable dignity that resists algorithmic reduction. This demonstrates how documented patterns of political and ideological bias embedded in leading AI systems underscore the urgency of attending to the worldviews of developers, since values are inscribed in technology whether acknowledged or not. Christian virtues, including stewardship-based accountability, solidarity with the vulnerable, epistemic humility, fidelity to truth, and the primacy of love, are shown to translate directly into actionable ethical principles for AI governance. The chapter concludes with a constructive vision of technology development ordered toward authentic human flourishing.

Unprecedented scientific and technological advancements characterize the present generation. Artificial Intelligence (AI) is the most influential among these advanced technologies. It is the capacity of machines to imitate human intelligence and produce better outcomes. AI has extended its influence across almost all areas of human interaction, and the present global society cannot imagine its existence without it. However, AI also challenges and harms human flourishing. Shaping individuals' moral identity is a significant concern in the present AI era. Conventionally, moral formation was the responsibility of parents, teachers, and society, grounded in the principles and values evolved from each society's cultural capital. The role of these conventional actors is sidelined in the present globalized AI world. From this perspective, this chapter, using the qualitative methodology's critical analytical method, examines the potential of bridging values and algorithms in shaping individuals' moral identity.

This chapter examines how moral courage can be systematically developed as a core capability for ethical leadership in AI-driven environments. While existing approaches to AI ethics emphasize principles, governance, and technical safeguards, they often fail to address the gap between ethical awareness and action under pressure. To address this limitation, the chapter introduces the Integrated Moral Courage Formation Model (IMCFM), which conceptualizes ethical leadership as a developmental process encompassing moral awareness, judgment, pressure navigation, action, and reflective formation. Drawing on virtue ethics, moral psychology, and adult learning theory, the chapter proposes a pedagogical framework that integrates experiential learning, structured deliberation, and reflective practice. An applied case demonstrates how the model can be implemented and assessed in realistic decision-making contexts. The chapter concludes by positioning moral courage as a trainable and measurable capability essential for responsible leadership in the age of artificial intelligence.

As artificial intelligence increasingly shapes organizational decision-making, leadership effectiveness can no longer be defined only by technical competence or digital fluency. AI-mediated workplaces require leaders whose character, ethical discernment and commitment to human dignity guide the design and use of intelligent systems. This chapter positions character formation as a core dimension of leadership development in the AI era, integrating faith-informed ethics, servant leadership and human flourishing. It examines how AI systems can amplify bias, reduce accountability and distance leaders from the human consequences of automated decisions, increasing the need for morally grounded leadership. Chapter proposes a framework for cultivating humility, justice, compassion, moral courage and responsibility in AI-mediated environments. The chapter outlines practical strategies for embedding character development, empathy and ethical reasoning into leadership education and organizational practice, preparing leaders to steward AI responsibly in service of human dignity and the common good.

Artificial intelligence (AI) is increasingly embedded in institutional decision-making, reshaping how authority, accountability, and responsibility are distributed. While enhancing efficiency and predictive capacity, AI also introduces risks such as bias, opacity, and weakened accountability. Existing governance approaches often rely on fragmented regulations or abstract ethical principles that are difficult to operationalize. This chapter proposes the AI–Ethics–Faith Systems Model, integrating infrastructure governance, ethical accountability, and value-based orientation. It reframes AI governance as socio-technical stewardship, positioning leadership as the central mechanism for aligning technological systems with ethical reasoning and long-term societal outcomes. The model emphasizes human flourishing—dignity, inclusion, agency, and well-being as the core evaluative criteria for responsible and sustainable AI deployment.

Artificial intelligence increasingly mediates leadership decisions, yet the deeper crisis of the AI era is not technological complexity but moral displacement. As leaders defer to algorithmic outputs, responsibility risks dissolving into systems, procedures, and claims of neutrality. This chapter argues that ethical AI leadership must begin with self-governance. Drawing on the Islamic concept of khalīfah and the ethic of amānah and mas'ūliyyah, it proposes Khalifah of the Self as a foundational moral framework structured around qalb (consciousness), 'aql and niyyah (reasoning and intention), nafs and ikhtiyār (struggle and choice), and culminating in accountability. By re-centering leadership on moral anthropology, the chapter contends that the first ethical boundary of AI is not in code, but in character.

With the growing influence of Artificial Intelligence on the decisions made in education, organizations, and the governance of the people, a major ethical concern is presented before the leadership, and that is having moral judgment in algorithm-based settings. Even though intelligent systems make the system more efficient and consistent, they have no conscience and are not sensitive to human dignity. This chapter explores faith-based leadership as a reaction to the increasing power of the algorithmic decision-making, with discernment highlighted as one of the fundamental leadership skills. Referring to leadership research, philosophy, and also ethics based

Foreword

Over nearly three decades, I have worked at the intersection of data, analytics, and human decision-making, supporting growth of some of the world's most recognized brands. Across that experience, much has changed, but one truth remains constant:

> **Humans are ultimately responsible for the decisions they make—or choose to delegate—whether to other people or to machines.**

That is why this book matters.

As Artificial Intelligence (AI) advances, we are not just improving how decisions are made, we are reshaping who participates in making them. We are moving from decision support to decision delegation. And in doing so, we risk confusing intelligence with wisdom, and efficiency with what is good.

The book lays out that tension clearly.

The deeper challenge of the AI age is not technological, it is formational. It is about the kind of people we are becoming, the kind of leaders we are forming, and the kind of institutions we are building.

Because systems do not govern themselves. And intelligence cannot replace conscience.

> **Responsibility stays with us.**

In my current work within a growing global enterprise grounded in a mission of care for others, I am reminded daily that technology is not an endpoint. It is something entrusted to us.

And what is entrusted must be stewarded by humans.

I hold a deep conviction that the future will not be defined by human or technology, but by ***HUMAN + technology***.

I am a strong advocate for AI. Data and AI can expand our reach, accelerate insight, and scale impact.

But they cannot:

- Accept responsibility
- Form character
- Define what is right

Those responsibilities remain human.

This becomes even more urgent in the era of agentic AI, where systems act with increasing autonomy. Delegation at this level does not reduce responsibility. **It intensifies it**.

We remain accountable for:

- The decisions we design
- The authority we delegate
- The values we encode
- The outcomes that follow

There is no outsourcing accountability, only expanding it.

This book challenges us to measure success not by speed, scale, or efficiency, but by whether people and communities are truly flourishing.

And it reminds us that flourishing is not accidental. It is formed—through character, wisdom, and a commitment to dignity and the common good.

Stay curious. Be a continuous student of these technologies.

But more importantly, be a student of your own formation.

Because in the end, the future of AI will not be determined by what machines can do—it will be shaped by the accountability, character, and stewardship of those who lead them.

This foreword was written through the integration of human insight and technology, an embodiment of the very principle it advances: HUMAN + technology.

Melinda Smith Borrero
Independent Researcher, USA

Melinda Smith Borrero, Ph.D. Enterprise Data & Analytics Leader, AI Solution Designer, Innovator, and Practicing Floral Designer. Customer & Marketing AI Enablement and Psychometrics Program Lead at Chick-Fil-A corporate. Designer of forward-thinking Machine Learning and Analytical Solutions which creatively connect the dots to inform decision making at the strategic and tactical levels. Currently, Melinda leads the Chick-fil-A Customer & Marketing AI Enablement and Psychometrics Program. Developer of agile, scalable solutions for Brand, Media & Communications, and Customer Experience for brands across the world. Currently, her focus is on design, development, and implementation of Enterprise-level Machine-Learning/Advanced Analytics Solutions for applications in Forecasting, Optimization, and Natural Language Understanding using structured and unstructured data. Served in senior leadership positions with some the largest global research companies including: Head of North American Solution Development, Managing Director of Global P&G Brand & Communications Practice at Kantar, and President, Global P&G Advertising at IPSOS. Melinda been involved with start-Ups and Advanced Analytics and Modeling Agencies in Colombia and US including co-founding bbb Mavens.

Foreword

Artificial intelligence is no longer a distant question. I see its reach in classrooms from Mumbai to Manchester, in schools serving remote villages, and in universities building global partnerships. AI is now personal and institutional. It is changing how we learn, work, and collaborate. As a Christian educator who serves alongside partners across continents, I have learned that questions about technology are, at their deepest level, questions about people: the kind of women and men we are becoming and the communities we are building together.

Tools can expand our capacity, but they do not form our character. Policies matter, but they do not supply conscience. Systems do not govern themselves. Leadership worthy of trust must be rooted in something deeper: the conviction that every person bears the image of God and therefore possesses a worth that cannot be diminished.

Leading with Faith and Ethics in an AI World keeps this truth in view. The authors ask more than whether AI systems are efficient or scalable. They ask whom these systems recognize, whom they serve, and whom they protect across languages, cultures, economic divides, and national borders. University faculty do more than transmit content. They shape the moral imagination of future leaders. If we prepare graduates only for technical fluency, we will send them into AI-mediated workplaces with skill but without anchors, capable but standing on sinking sand.

In my own work, global relationships have sharpened judgment and expanded imagination. When leaders collaborate across cultures, whether in Nairobi, New Delhi, São Paulo, Seoul, or elsewhere, they design better, govern more justly, and lead with greater humility. When universities and educational partners co-create across cultures, they build systems that honor differences rather than erase them.

That is one of the gifts of this book. It moves from aspiration to practice. It presses readers to ask better questions, build accountable processes, and form leaders whose authority is measured by stewardship rather than control.

I commend this book to anyone charged with shaping people or policy in the age of AI. Read it as a guide for formation: of students and staff, of leaders and systems. Let it help you hire and teach differently, govern with moral clarity, and design for participation across cultural and geographic lines. May this volume strengthen your resolve to lead with integrity and hope, exercise power as stewardship, and build institutions where people can learn, work, and flourish together.

Carolyn Bishop
Consortium for Global Education, USA

Dr. Carolyn Bishop has served over 25 years as President of the Consortium for Global Education (CGE), an international education organization begun in 1987. CGE supports international programs for their 42 private accredited university members representing over fifty-thousand undergraduate and graduate students. Member schools represent K-12 and higher education in 14 nations and encourage partnerships with more than 400 overseas institutions representing 91 countries. She completed her PhD at Emory University in Educational Leadership and Supervision after receiving her M.Ed. at Clemson and undergraduate degree from Furman University. Dr. Bishop promotes international education work with quality academic efforts and a heart for service which has been recognized by global leaders at government, institutional, and community levels.

Preface

LEADING IN AN AI AGE: RECOVERING WISDOM AND MORAL FORMATION

This volume began with a conviction that the central question of the AI age is not simply what machines can do, but what kind of people, leaders, and institutions will guide them. Artificial intelligence is changing how decisions are made, how knowledge is produced, how relationships are mediated, and how power is exercised. Yet the most urgent questions are not only technical. They are moral, spiritual, educational, and institutional. AI can increase speed, scale, and efficiency, but efficiency cannot replace wisdom. Power cannot substitute for moral formation. Intelligence cannot become a replacement for conscience.

Leading With Faith and Ethics in an AI World is written from the conviction that human beings are created in the image of God and therefore possess dignity that no algorithm, market, institution, or system can reduce. The proper end of leadership is not technological mastery alone. It is the cultivation of communities where people are formed in character, where justice and righteousness are practiced, where dignity is honored, and where people and institutions are ordered toward human flourishing.

The chapters gathered here share a common concern: AI ethics requires more than policy statements, compliance systems, or technical safeguards. These tools matter, but they cannot supply wisdom, courage, humility, or love. Systems do not govern themselves. They are designed, adopted, interpreted, and used by people whose habits, desires, assumptions, and moral imaginations have already been formed. For that reason, the leadership crisis of the AI age is also a formation crisis.

This book approaches AI through the language of faith, ethics, leadership, and flourishing. It asks how universities can form students who are not merely skilled but wise. It asks how leaders can exercise moral authority when algorithmic systems appear to offer certainty, speed, or neutrality. It asks how organizations can resist reducing people to data points, consumers, workers, or users. It asks whether institutions can build cultures where the vulnerable are protected, participation is widened, and technology is judged by its service to human dignity and the common good.

Human flourishing is the thread that holds these questions together. In this volume, flourishing is not treated as private happiness or individual achievement alone. It is the God-intended development of persons and communities toward wholeness, virtue, purpose, loving relationship, and faithful contribution. Flourishing requires agency and growth, belonging and justice, whole-person well-being, and meaning, character, and spiritual formation. It is deeply personal, but it is never solitary. People flourish within communities, institutions, and practices that honor who they are and help them become who they are called to be.

This is why justice and righteousness are central to the vision of the book. Justice teaches knowledge, builds real skills, and equips people to transform disorder and sometimes chaos into order, innovation, and shared good. Righteousness cares deeply for people. It gives respect to every person, creates support structures for growth, and builds communities where people are safe, valued, and able to belong. In an AI-shaped world, these are not secondary concerns. They are measures of whether leadership is faithful to its purpose.

The introduction that follows develops this argument in fuller detail. The preface offers the reader a doorway into the movement of the volume. The chapters do not treat AI as an isolated technological topic. They ask how faith traditions, moral philosophy, leadership theory, education, governance, supply chains, disability inclusion, and institutional culture can together help leaders recover wisdom for the sake of human flourishing.

THE ORGANIZATION OF THE VOLUME

Part I, Moral Development in Higher Education, begins with the formative role of educational institutions. These chapters examine how students, faculty, and universities can cultivate moral identity, ethical reasoning, faith-integrated learning, and moral courage in a world where digital systems shape habits, attention, and imagination.

Part II, Leadership Formation in the Age of AI, moves from educational formation to organizational and institutional leadership. These chapters ask how leaders can govern intelligent systems with character, conscience, stewardship, self-mastery, legitimacy, foresight, and responsibility for the common good.

Part III, Human Flourishing, brings the volume toward its end: the formation of people and institutions capable of honoring dignity, strengthening belonging, widening participation, practicing servant leadership, and ordering technological power toward lives well lived before God and with others.

CHAPTER OVERVIEW

The chapters are arranged in a deliberate sequence that moves from moral formation in higher education, to leadership formation in AI-mediated organizations and systems, to human flourishing as the end toward which ethical leadership must be ordered.

Part I: Moral Development in Higher Education

Chapter 1, Moral Anchors in an AI World - Faith, Virtue, and the Formation of Ethical Leaders, opens the volume by naming the formation crisis created by AI-mediated educational and institutional life. The chapter presents the Moral Anchor Model as a way to cultivate ethical resilience, digital discernment, and leadership rooted in virtue, dignity, and responsibility for vulnerable communities.

Chapter 2, Faith Integration, Ethical Orientation, and AI Adoption in a Faith-Based University, examines responsible AI adoption in a faith-based university context. Through faculty and student perspectives, the chapter shows how institutional values, faith integration, and ethical orientation can support academic integrity, pedagogical innovation, and student flourishing.

Chapter 3, The Moral Frontier of the AI Age - Why Christian Principles Offer a Coherent Ethical Foundation, argues that Christian moral theology offers a coherent foundation for AI ethics because it begins with the dignity of persons made in the image of God. The chapter connects stewardship, solidarity, humility, truthfulness, and love to the governance of technological systems.

Chapter 4, Shaping Moral Identity in the Age of AI - Bridging Values and Algorithms, addresses how intelligent technologies reshape learning, relationships, and cultural norms. It calls for renewed attention to the institutions and practices that form moral identity so that values and algorithms can be held together without allowing technology to displace ethical judgment.

Chapter 5, Educational Practices for Forming Moral Courage - An Integrated Developmental Framework, focuses on moral courage as a necessary capacity for ethical leadership under pressure. The chapter offers a developmental framework for moving from awareness and judgment to action, reflection, and resilience in complex AI-shaped environments.

Part II: Leadership Formation in the Age of AI

Chapter 6, Character Formation for Ethical Leadership in AI-based Organizations - A Faith-Centered Framework, connects faith-centered ethics, servant leadership, and character formation in AI-based organizations. The chapter argues that leaders must cultivate humility, compassion, discernment, and responsibility if technological systems are to serve rather than diminish human dignity.

Chapter 7, Ethical AI Governance as Socio-Technical Stewardship - A Faith-Informed Systems Model for Sustainable Development, reframes governance as socio-technical stewardship. It emphasizes that AI systems must be aligned with dignity, inclusion, agency, and long-term social well-being rather than treated as neutral tools managed only through fragmented regulation or abstract principles.

Chapter 8, Khalifah of the Self - Spiritual Leadership as the First Boundary in an AI-Driven World, turns to Islamic ethical thought to show that the first boundary for AI leadership is the formation of the self. Concepts such as khalifah, amanah, and mas'uliyyah ground leadership in accountability, disciplined self-governance, and spiritual awareness.

Chapter 9, Algorithms Have Logic, Leaders Have Conscience - Reclaiming Moral Authority in Intelligent Systems, examines the tension between algorithmic rationality and human moral judgment. The chapter argues that AI can support decisions, but it cannot replace conscience, moral authority, or responsibility for justice, inclusion, compassion, and the common good.

Chapter 10, The Moral-Constraint Divide in AI Development - Safety, Legitimacy and Pluralism, addresses the legitimacy crisis within AI governance. By introducing the Lewis-Asimov Model of AI Governance, the chapter connects safety, liberty, moral constraints, governance structures, and human flourishing in a pluralistic technological landscape.

Chapter 11, Faith-Inspired Foresight - Leading Global Supply Chains Beyond the AI Monster Fallacy, applies faith-inspired foresight to global supply chains and complex organizational systems. The chapter shifts attention away from fear-based narratives about machines and toward the deeper problem of moral responsibility in innovation, resilience, and economic life.

Part III: Human Flourishing

Chapter 12, After MacIntyre - Re-Evaluating Christian Ethics for the AI Era, revisits virtue ethics in light of generative AI and the moral fragmentation of contemporary life. The chapter proposes a renewed ethical framework that joins virtue with objective moral obligation for responsible leadership in AI-saturated societies.

Chapter 13, Ethical Traditions and Servant Leadership in the Age of Artificial Intelligence - Foundations for Human Flourishing, draws from Aristotelian virtue ethics, Stoicism, Christian moral theology, historical capitalism, and stakeholder theory to strengthen servant leadership. The chapter shows that AI may optimize processes, but it cannot cultivate virtue, wisdom, or moral purpose.

Chapter 14, AI, Disability, Accessibility, and the Ethics of Inclusion - Participation, Design, and the Common Good, places disability and accessibility at the center of AI ethics. The chapter argues that inclusion is not merely accommodation or compliance, but a test of whether institutions design systems that widen participation, protect dignity, and support flourishing.

Chapter 15, A New Moral Renaissance - Forming Leaders for a Flourishing Future, concludes the volume by integrating the book's central movement from formation to character, wisdom, discerning leadership practice, and human flourishing. It calls for a moral renaissance in which leaders recover their vocation to guide technological power toward the common good.

These chapters argue that the future of AI will not be determined by technical capacity alone. It will be shaped by the moral and spiritual formation of the people who design, govern, teach, adopt, and resist these systems. The question is not only whether AI will become more powerful. The deeper question is whether leaders and institutions will become more wise, more just, more righteous, and more committed to the flourishing of all people.

My hope is that this volume will help readers think more carefully about the relationship between technology and formation, power and conscience, intelligence and wisdom. More importantly, I hope it will encourage educators, leaders, students, policymakers, and communities of faith to recover the moral work required to live and lead well in an age of artificial intelligence.

Acknowledgment

This volume would not have been possible without the thoughtful contributions of each chapter author, whose scholarship, faith commitments, and ethical insight helped shape the vision of this book. Their work reflects a shared concern for how artificial intelligence can be engaged with wisdom, moral responsibility, and commitment to human flourishing.

I am especially grateful to Anderson University and the College of Business and Economics for providing a community where questions of faith, ethics, leadership, and human flourishing can be explored with seriousness and purpose. I also wish to express sincere appreciation to Christ (Deemed to be University), India, with whom we have shared meaningful faculty exchanges, numerous research projects, and academic conferences. These collaborations have strengthened faculty and student knowledge while deepening our shared commitment to ethical leadership, global learning, and human flourishing.

I also wish to thank Dr. Phil Bailey and Dr. Richard Ray for their encouragement, feedback, and support throughout the development of this project. Without their help, I would not have been able to complete this work. I am grateful to Dr. Greg Silver for his passion for wisdom and for the many discussions we shared about ethics, AI, and technology. Thank you for introducing me to Borgmann and Kingsnorth, whose work helped sharpen important themes in this volume.

I am especially grateful to my mentor, Dr. William Hanson, who taught me how to become a qualitative researcher. Thank you, sir, for your training, guidance, and example, and for the formative crucible through which you helped shape my scholarly work. Your influence helped me see more clearly that Complexity Leadership Theory, ethics, and moral development are deeply intertwined.

Finally, I acknowledge the students, leaders, and communities whose questions and experiences continue to remind us why this work matters. In an age increasingly shaped by intelligent technologies, the call to lead with faith, ethics, justice, righteousness, and love remains the central issue of our time.

Introduction

WHAT DOES IT MEAN TO LIVE AND LEAD WELL IN AN AGE WHERE EFFICIENCY REPLACES WISDOM AND POWER SUPPLANTS MORAL FORMATION?

Technological achievement has advanced rapidly, but it has not necessarily deepened human flourishing. Many systems shaping daily life are designed to capture attention and sustain engagement rather than cultivate wisdom, virtue, and wholeness. Social media platforms often use likes, notifications, infinite scroll, autoplay, and algorithmically curated feedback loops can intensify reward anticipation and make disengagement difficult, especially for younger users (APA, 2024; Maza et al., 2023). These patterns are especially concerning for children and adolescents. The U.S. Surgeon General reports that youth social media use is nearly universal, more than one-third of teens use it almost constantly, and adolescents who spend more than three hours a day on social media face double the risk of depression and anxiety symptoms (U.S. Department of Health and Human Services [HHS], 2023). Research also links these patterns to harmful social comparison, body image concerns, sleep disruption, and differences in mental health and brain development (U.S. Department of Health and Human Services [HHS], 2023; Maza et al., 2023).

These findings reveal that the crisis before us is not simply technological, but deeply moral and formative. We have become skilled at designing systems that increase engagement and accelerate speed. We have been far less intentional about whether those same systems nurture wisdom, self-control, dignity, belonging, and responsibility for the common good. Human flourishing has become a central question. Human beings, created in the image of God, are not meant to be treated merely as data points, consumers, or endlessly stimulated users. They are created for truth, relationship, stewardship, worship, and lives rightly ordered toward the good. Yet many dominant systems shaping contemporary life are structured around engagement, monetization, and behavioral capture rather than the dignity, formation, and flourishing of the person.

What, then, counts as human flourishing? Human flourishing resists the reduction of people to data or consumers. Even the language of well-being becomes too narrow when detached from purpose, virtue, belonging, and responsibility. Flourishing asks whether individuals are becoming whole persons who live in right relationship with God, others, themselves, and creation. This volume understands human flourishing as the God-intended development of persons and communities toward lives of wholeness, virtue, purpose, loving relationship, and faithful contribution to the common good. Flourishing is not merely happiness, success, or material well-being. It includes both the gift of life well lived and the responsibility of lives well ordered toward what is good, true, and beautiful.

This vision is rooted in a Christian understanding of the human person. Because human beings bear the image of God, each person possesses inherent dignity and moral worth that can never be reduced to utility, status, or market value. Human life is not accidental, nor is it merely self-directed. It is received as gift and entrusted as vocation. To flourish is to grow into faithful stewardship of one's life before God and in community with others. It is to become a person shaped by virtue, guided by wisdom, and oriented toward love, justice, and service. In this sense, flourishing is both deeply personal and unavoidably communal. No one flourishes alone.

Robert Schalock's quality-of-life model provides a starting point, but the stronger claim of this volume is that flourishing also requires spiritual formation, moral responsibility, ethical leadership, and institutional stewardship. Schalock's framework identifies core dimensions of quality of life related to independence, social participation, and well-being. Flourishing also requires character, vocation, wisdom, and a life grounded in God's purposes.

AI ethics depends on formation. The better account is not that technical safeguards are unimportant, but that they are insufficient. Systems do not govern themselves, and safeguards cannot supply conscience, character, or moral authority. Universities, faith traditions, and moral communities shape the identity, courage, and discernment that leaders bring to AI-mediated decisions. Without that formation, AI governance narrows into technical fluency, compliance, or risk management. With it, leaders can ask harder questions about dignity, agency, belonging, wisdom, justice, and the common good. Disability inclusion makes this test concrete: not simply whether AI works, but whether it recognizes, serves, and protects the people most easily excluded.

Agency and Growth emphasizes the development of the person as a responsible moral agent before God and within community. This area includes personal development, self-determination, and vocational purpose. Human flourishing involves growing in knowledge, wisdom, and capability, while exercising freedom faithfully and discerning how one's gifts serve God and neighbor.

Belonging and Justice highlights the relational and communal conditions necessary for flourishing. This area includes interpersonal relations, social inclusion, rights and dignity, and reciprocal contribution. Because human beings are created for relationship, they flourish in communities marked by love, justice, mutual respect, and belonging. A faith-explicit understanding of flourishing insists that every person bears inherent dignity as an image-bearer of God and must be welcomed, protected, and given real opportunity to participate in community life.

Whole-Person Well-Being recognizes that flourishing is embodied. It includes emotional, physical, material, and intellectual well-being. Human life is not divided into isolated compartments but lived as an integrated whole, so flourishing must include emotional resilience, bodily care, material sufficiency, understanding, and wisdom.

Meaning, Character, and Spiritual Formation addresses the deepest foundations of human life. This area includes moral formation, meaning and purpose, spiritual formation, and stewardship for the common good. From a faith perspective, flourishing involves becoming a person shaped by virtue, humility, love, justice, and faithfulness. It means locating one's identity and purpose within God's larger story rather than in achievement, power, or self-assertion alone.

Not only harm prevention. Not only better systems. AI ethics is about forming persons, leaders, and institutions that honor dignity, exercise wisdom, practice justice, and cultivate communities where people flourish.

The crises of our age cannot be solved by better tools alone because they are also crises of meaning, character, community, and moral formation. Rapid advances in artificial intelligence and digital systems have expanded human capacity, but they have also intensified questions about responsibility, dignity, exclusion, manipulation, and the future of shared life. Institutions now possess growing power to shape habits, desires, opportunities, and identities, often without an equally serious commitment to the flourishing of the human person. Recent national survey research confirms that artificial intelligence is already embedded in student learning practices across higher education, with most college students reporting regular use for coursework even while institutional policies remain uneven or unclear and many students reporting insufficient preparation for responsible AI use after graduation (Gallup, Inc. & Lumina Foundation, 2026).

Because these crises are both personal and institutional, this volume insists that flourishing must be understood in both personal and institutional terms. Individual virtue cannot be separated from the systems, structures, and technologies that shape daily life, nor can efficient institutions flourish apart from the formation of character and the recognition of human dignity.

THREE SECTIONS OF THIS BOOK

The first section, Moral Development in Higher Education, examines how educational institutions can form students in ethical reasoning and moral imagination, along with empathy and courage, amid the pressures of digital culture and algorithmic influence. The second section, Leadership Formation in the Age of AI, focuses on preparing leaders who are not only technologically capable but also morally discerning. It draws on servant leadership and ethical leadership to address power, bias, responsibility, and relational care in AI-mediated environments. The third section, Human Flourishing, explores how leaders and institutions cultivate environments where dignity, belonging, purpose, justice, and full participation are sustained.

What does it mean to live and lead well in a world where efficiency often replaces wisdom and power often supplants moral formation?

How can individuals, communities, and institutions be rightly ordered toward God, one another, and the common good so that all flourish?

The chapter on AI, Disability, Accessibility, and the Ethics of Inclusion argues that disability must move from the margins to the center of AI ethics because it reveals what kind of human being a system is designed to recognize, serve, and protect. Accessibility is not merely a technical or compliance issue; it is a test of whether institutions value dignity, participation, and the common good. A New Moral Renaissance: Forming Leaders for a Flourishing Future presents the volume's integrative model: Formation → Character → Wisdom → Discerning Leadership Practice → Human Flourishing. It argues that the leadership crisis of the AI age is not merely regulatory or technological, but formative. Leaders act not only from what they know, but from who they are becoming.

Leaders, educators, institutions, and communities must recover the moral and spiritual work of formation in an AI-shaped world. Better systems are needed, but they are not enough. The future will require morally formed people, wise leaders, and institutions committed to justice and righteousness. Justice teaches knowledge, builds real skills, and equips people to transform disorder and even chaos into order, new approaches, and shared good. Righteousness honors dignity. It creates communities where people are respected, supported, safe and able to flourish. AI should finally be judged not by speed, scale, or efficiency, but by its service to the flourishing of people, communities, and the common good.

REFERENCES

American Psychological Association. (2024). *Potential risks of content, features, and functions*. In *Health advisory on social media use in adolescence*. https://www.apa.org/topics/social-media-internet/youth-social-media-2024

Gallup, Inc., & Lumina Foundation. (2026). *AI in higher education: Widespread use, unclear rules* (Lumina–Gallup State of Higher Education report). Gallup. https://www.gallup.com/analytics/644939/state-of-higher-education.aspx

Maza, M. T., Fox, K. A., Kwon, S.-J., Flannery, J. E., Lindquist, K. A., Prinstein, M. J., & Telzer, E. H. (2023). Association of habitual checking behaviors on social media with longitudinal functional brain development. *JAMA Pediatrics, 177*(2), 160–167. DOI: 10.1001/jamapediatrics.2022.4924 PMID: 36595277

U.S. Department of Health and Human Services, Office of the Surgeon General. (2023). *Social media and youth mental health: The U.S. Surgeon General's advisory*. U.S. Department of Health and Human Services. https://www.hhs.gov/sites/default/files/sg-youth-mental-health-social-media-advisory.pdf

Section 1
Moral Development in Higher Education

Chapter 1

Moral Anchors in an AI World:
Faith, Virtue, and the Formation of Ethical Leaders

James Meersman
http://orcid.org/0009-0004-0853-4821
Texas A&M University, USA

ABSTRACT

This chapter examines the formation crisis in higher education amid AI-mediated environments and pervasive surveillance systems, where algorithms function as secular liturgies that shape moral identity toward efficiency rather than human flourishing. The purpose of this chapter is to diagnose the formation gap and propose the Moral Anchor Model as a framework for cultivating ethical leadership resilience. The model integrates three dimensions: Ancient Wisdom Roots drawn from faith traditions; Ethical Scaffolding comprising virtues such as humility, compassion, justice, stewardship, and truthfulness that translate commitments into discerning action; and Digital Discernment Practices including attention discipline, critical interrogation of data, community consultation, restorative reflection, and advocacy for systemic change. Drawing on Christian theology's emphases on imago Dei, human fallenness, and preferential care for the vulnerable, the chapter positions faith traditions as comprehensive moral ecosystems that counter surveillance-driven distortions of agency and value.

DOI: 10.4018/979-8-3373-8192-3.ch001

ht © 2027, IGI Global Scientific Publishing. Copying or distributing in print or electronic forms without written permission of IGI Global Scientific Publishing is prohibited. Use of this chapter to generative artificial intelligence (AI) technologies is expressly prohibited. The publisher reserves all rights to license its use for generative AI training and machine learning model development.

THE GHOST IN THE MACHINE

The human experience has undergone a major shift. Artificial intelligence and algorithmic systems now mediate how we understand ourselves, what we value, and who we aspire to become. These systems operate with unprecedented scale and invisibility, shaping billions of daily decisions about what content we consume, what opportunities we access, and how we understand our own capabilities and potential. The problem is not simply that AI is ethically complex or that algorithms sometimes contain biases, though both are true. Rather, the deeper crisis is that we have begun to outsource the formation of moral identity itself to systems engineered for efficiency rather than human flourishing.

The extent of surveillance and data collection embedded within these systems compounds this problem. Recent scholarship on digital surveillance reveals that data from mobile devices, wearable sensors, social media, and smart environments is collected, aggregated, and used to classify risk, normalize behavior, and reshape relationships between individuals, institutions, and technology providers (Cipta et al., 2026). These digital health systems and broader data-driven surveillance infrastructures function as what some scholars term "secular liturgies". That is, repeated, pervasive practices that form identity without conscious awareness or deliberate choice (Band, 2026). What makes this particularly concerning is that the ethical implications of such surveillance extend far beyond privacy concerns; they involve questions about human agency, dignity, and the conditions necessary for authentic moral identity formation (Kantha et al., 2024).

This chapter proposes that faith traditions and spiritual formation practices offer irreplaceable resources for cultivating moral identity amid the pressures and possibilities of an AI-mediated, surveillance-saturated world. Specifically, this chapter introduces the Moral Anchor Model, a framework that grounds ethical leadership formation in three integrated dimensions: (1) Ancient Wisdom Roots drawn from faith traditions; (2) Ethical Scaffolding Virtues that translate faith commitments into practices of discernment and action; and (3) Digital Discernment Practices that help emerging leaders navigate technology's influence on identity and agency. The central argument is that leaders who direct AI systems must themselves be formed through traditions that cultivate resilience against utilitarian logic and that ground human dignity in sources transcendent to market metrics and computational efficiency. Without such formation, even well-intentioned technologists will design systems that reflect the distorted values embedded in algorithmic optimization. With it, they can create alternatives.

Higher education stands at a critical juncture. Universities have long claimed to form not just skilled workers but thoughtful citizens and leaders capable of ethical judgment. Yet in a cultural moment when algorithms recommend what we read, when predictive analytics determine who receives loans or advances in hiring, when quantified metrics of the self become normalized, and when surveillance technologies create infrastructures of continuous monitoring, many students graduate with sophisticated technical competencies but underdeveloped capacities for moral reasoning anchored in something deeper than instrumental calculation. This is not a problem that technology alone can solve. It requires the recovery of ancient and enduring traditions that have, over centuries, sustained moral identity in the face of cultural pressures and dehumanizing forces.

The stakes are high. The leaders who design, deploy, and govern AI systems will shape the future of human work, healthcare, education, and justice. If these leaders have not been formed through traditions that cultivate humility about human limitations, compassion for the vulnerable, commitment to justice, and an understanding of human dignity that transcends efficiency metrics, then the systems they create will inevitably reflect those gaps in formation. This chapter is a call to higher education, and to faith communities themselves, to recognize that moral identity formation is not a luxury but a necessity, and that the resources of faith traditions are indispensable for this work.

DIAGNOSING THE FORMATION GAP: AI AS SECULAR LITURGIST

To understand why faith-grounded moral formation matters now, we must first understand how AI-mediated environments shape identity. Liturgy, in its theological sense, refers to the repeated practices through which communities form their members' desires, values, and sensibilities. Every time we open a social media feed curated by algorithms, every time we see a recommended product tailored to our browsing history, every time our creditworthiness is assessed by a predictive model, we are participating in what might be called a "secular liturgy". That is, the repeated practices that shape who we are becoming, often without our explicit awareness or deliberate choice.

The mechanisms are increasingly well-documented. Recommendation algorithms do not simply show us what we want; they shape what we come to want by repeatedly exposing us to content that maximizes engagement (Weinberg, 2022). Quantified self-practices (e.g. tracking steps, calories, productivity metrics, etc.) promise self-knowledge but often reduce the self to measurable outputs, creating a fragmentary self-understanding disconnected from meaning, purpose, or intrinsic value. Predictive analytics used in hiring, education, criminal justice, and healthcare make judgments about human potential based on historical data patterns, often encoding historical inequities into ostensibly "objective" systems (Mishra et al.,

2025). Meanwhile, the algorithms that power these systems remain opaque to most people, creating accountability gaps where decisions profoundly affecting human lives lack transparency or means of meaningful challenge.

The Formation Crisis: Surveillance, Discernment, and the Gap

The contemporary formation crisis deepens at the intersection of three forces. First, surveillance systems now embed themselves throughout everyday life, operating as infrastructures of behavioral governance. Recent scholarship shows that data collection reshapes how individuals understand and regulate their own behavior, encoding specific visions of human value rooted in engagement metrics rather than intrinsic dignity (Akhundov, 2025, Alperstein, 2025). When young leaders internalize the message that human worth is indexed to quantifiable measures, they lose sight of transcendent dimensions of personhood that faith traditions protect (Kantha et al., 2024).

Second, this surveillance infrastructure actively undermines what might be called digital discernment. That is, the capacity to distinguish authentic desires from manufactured preferences, truth from manipulation. The design of contemporary digital systems exploits psychological vulnerabilities through attention-capturing algorithms and dark patterns that make discernment deliberately difficult. Without explicit practices and communal structures to support discernment, most people find it nearly impossible to resist (Tripathi, 2025).

The cumulative result is what this chapter terms the "formation gap": the growing distance between the formative power of AI-mediated environments and the resources available to young leaders for developing a virtue-grounded moral identity. Students are being formed intensively and continuously by systems engineered for engagement and efficiency, yet they lack formation from traditions oriented toward wisdom, justice, and human flourishing. This gap has concrete consequences. Young leaders often lack frameworks for moral reasoning independent of utilitarian calculation, default to frameworks that prioritize aggregate utility at the expense of justice, and delegate responsibility for ethical decisions to algorithmic systems. The solution must be formational. It must involve intentional cultivation of moral identity through resources that have proven across time capable of sustaining virtue even in hostile or dehumanizing circumstances.

FAITH TRADITIONS AS MORAL IDENTITY ECOSYSTEMS

This is where faith traditions become indispensable. Rather than treating faith as one set of values among many, this section proposes that faith traditions function as moral ecosystems. They are not primarily systems of doctrines or sets of moral rules, though they include both. Rather, they are integrated wholes of story, ritual, community practice, and spiritual discipline that work together to form moral identity across time and through adversity.

Consider the key components of a functioning moral ecosystem. First, faith traditions offer grand narratives that situate human life within a larger story of meaning and purpose. Rather than the fragmented, algorithm-driven narrative of contemporary culture where the self is a collection of preferences to be optimized, faith traditions tell coherent stories about who humans are, why we are here, what our dignity consists in, and how we should relate to one another and to the transcendent. In Christianity, this narrative centers on creation, redemption, and the promised renewal of all things, grounding human dignity in the claim that humans are made in the image of God, possessed of intrinsic worth that cannot be calculated or optimized away. This is radically different from the vision of selfhood offered by algorithmic systems, which treat humans as data points to be analyzed and influenced.

Second, faith traditions embed moral formation in ritual and communal practice. They do not simply tell us what is good; they practice it together, in ways that form our dispositions and desires over time. A Christian community gathered for worship, engaged in practices like confession, prayer, study of scripture, and care for the vulnerable, is does more that exchanging information about what Christianity teaches. Through repeated, embodied participation, members are gradually shaped; their attention reoriented, their empathy deepened, their commitments realigned. This embodied, communal character of faith formation stands in sharp contrast to the individualized, quantified nature of digital formation, where the self is addressed as an isolated consumer rather than as a member of a community.

Third, faith traditions cultivate specific virtues and dispositions that orient people toward genuine goods and away from corrupted versions of human flourishing. Justice, compassion, humility, stewardship, and courage are not presented as abstract ideals but as practiced, cultivated capacities that develop through community participation and spiritual discipline. These virtues function as "moral anchors". That is, stable, deep-rooted dispositions that remain fixed even when cultural pressures shift. A person formed in humility learns to question her own perspective and to listen to dissenting voices, which is invaluable when faced with algorithmic systems that may encode her own biases. A person formed in justice develops sensitivity to how decisions affect the vulnerable, which is crucial for anyone directing systems that allocate resources or make determinations about human worth. A person formed in stewardship understands that resources and power are held in trust for others and for future generations, providing a counterweight to the short-term utility maximization that drives algorithmic optimization.

Fourth, faith traditions offer accountable communities that help sustain moral identity in the face of cultural pressure and personal failure. This is crucial and often overlooked. Individual moral conviction, while important, is fragile. A leader who tries to operate according to values of justice and human dignity but who lacks community support, who is surrounded by colleagues who ridicule these concerns as naive or economically inefficient, will find it difficult to sustain these values over time. Faith communities, at their best, provide structures of mutual accountability, structures where it is safe to struggle with moral questions, where failure is acknowledged and forgiveness offered, and where people are held together in commitment to values larger than individual preference or institutional advancement.

Fifth, faith traditions cultivate moral imagination. That is, the capacity to envision alternatives to the way things are currently organized, and to believe that change is possible. This matters profoundly in an AI-saturated world. If young leaders internalize the view that algorithms are neutral, inevitable, and unchangeable, they will be unable to imagine or work toward more just technological systems. Faith traditions, by contrast, remind us that even entrenched systems can be transformed, that the arc of history is not determined by technical necessity, and that humans retain agency to choose how to design and deploy technology. This visionary capacity is a crucial resource for ethical leadership in technology contexts.

Spiritual Practices and Digital Resilience

An emerging body of research demonstrates that contemplative and spiritual practices (e.g. mindfulness, prayer, meditation, and other disciplines rooted in faith traditions, etc.) develop specific cognitive and emotional capacities that enhance moral resilience in digital contexts. Recent empirical research on practical wisdom itself has begun to measure and validate how this ancient virtue develops. One recent study examining mindfulness among media students found that mindfulness significantly strengthened students' ability to recognize and engage with AI ethically, particularly by enhancing ethical awareness and commitment to responsible AI use (Lan et al., 2025). Another study exploring digital mindfulness interventions found that structured practice enhanced participants' awareness of digital device usage patterns, their understanding of digital risks, and their capacity to make intentional choices about technology use rather than falling into habitual, compulsive patterns (Chandra et al., 2025). These findings align with emerging empirical models of practical wisdom that demonstrate how contemplative and reflective practices strengthen the deliberative capacities central to moral judgment (McLoughlin et al. 2025).

This also suggests that contemplative practices develop what might be called "discernment capacities". This is the ability to notice one's own mental states, to recognize manipulation, to resist the pull of engineered engagement, and to make deliberate choices aligned with authentic values rather than algorithmic suggestions. For emerging leaders in technology, such capacities are not peripheral but central. A person who has spent years cultivating attention through prayer or meditation, who has practiced noticing and releasing attachments through spiritual discipline, will be more capable of observing how persuasive technologies attempt to capture attention and manipulate desire, and more capable of resisting those attempts.

Furthermore, spiritual practices embed the practitioner within community and tradition in ways that counteract the atomization of digital culture. When a young technology leader participates in worship, in shared meals, in communal discernment about difficult ethical questions, she is participating in practices that fundamentally shape her identity in relation to others rather than as an isolated consumer. She is regularly reminded, through lived participation rather than abstract teaching, that her worth is not indexed to her productivity or her metrics, but is rooted in her membership in a community of faith (Gbadeyanka et al., 2025).

Christian Theology and the Formation of Moral Identity

While faith traditions more broadly offer these resources, this chapter focuses specifically on Christian theological frameworks for understanding and forming moral identity. Christianity offers particular riches for the work of moral formation in conversation with technology.

At the heart of Christian moral theology is the claim that humans are made in the image of God and that this image, though marred by sin, is the source of our intrinsic, immeasurable worth. This has profound implications for how we think about AI and human identity. If a person's worth is rooted in status as an image-bearer of God, then that worth cannot be calculated, optimized, or reduced to metrics. This stands in direct opposition to how many algorithmic systems implicitly treat humans: as bundles of data to be analyzed, influenced, and exploited for maximum engagement or profit.

Christian theology also offers deep reflection on human fallenness, the universal human tendency toward sin and self-deception. This is relevant to understanding AI systems, because it reminds us that all human-made systems, including algorithms and AI, will reflect human biases, limitations, and corrupted desires (Ahmad et al., 2026). A culture that naively trusts algorithms as objective or that imagines that removing human judgment from decisions will eliminate bias is failing to reckon with this theological anthropology. By contrast, theological realism about human sinfulness encourages humility about technological solutions and heightened attention to how systems can embed and amplify human injustice.

Christian theology also centers on God's preferential care for the vulnerable and the poor, which is a theme that runs through the entire biblical narrative and is crystallized in Jesus's ministry and teaching. Jesus' earthly ministry illustrates a commitment to ensuring that the most vulnerable members of society are protected from harm and that justice is pursued even when it conflicts with efficiency or profit. This is precisely what is often absent from algorithmic systems, which optimize for aggregate utility without attention to how decisions affect the poorest and most marginalized. Leaders formed in this Christian tradition will be more likely to ask not "Does this system maximize efficiency?" but "Does this system treat the vulnerable with dignity and justice?" and "Are we complicit in harm?"

Finally, Christian theology understands humans as fundamentally relational beings, created for community and for communion with God. This relational anthropology stands against the atomized, individualized view of selfhood offered by algorithmic systems, which address the human as an isolated consumer rather than as part of a web of relationships and obligations. A leader formed through Christian theology will be more likely to ask not only "What is good for me or my organization?" but "What is good for our community, our society, and those beyond the boundaries of my in-group?"

THE MORAL ANCHOR MODEL: INTEGRATING FAITH AND DIGITAL ETHICS

The preceding sections have laid groundwork for understanding both the problem (the formation gap created by AI-mediated culture and surveillance infrastructure) and the resource (faith traditions as ecosystems for moral formation). Now we turn to the concrete model that integrates these insights: the Moral Anchor Model. This model in Figure 1 consists of three integrated dimensions designed to help higher education form leaders who can navigate AI-mediated environments with moral resilience and wisdom.

Figure 1. The moral anchor model

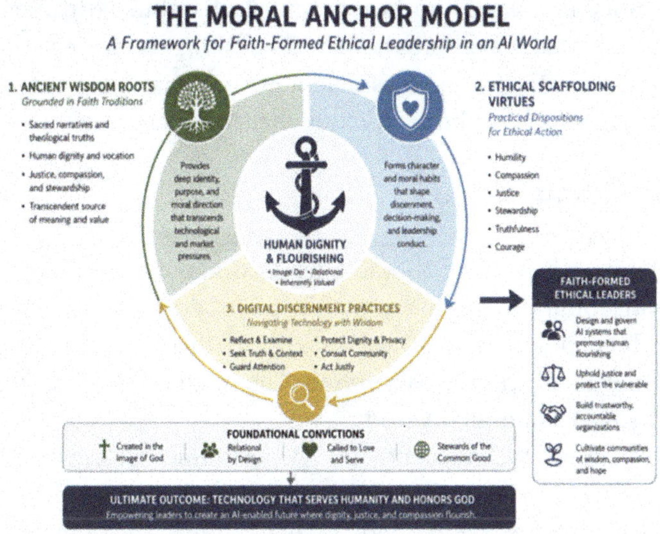

Ancient Wisdom Roots

The first dimension, Ancient Wisdom Roots, acknowledges that moral wisdom is not a recent invention, nor is it primarily a technical matter. Rather, core insights about human dignity, justice, accountability, and flourishing have been refined and articulated across centuries of theological and philosophical reflection. These ancient roots provide the foundation for moral identity that can withstand cultural pressures and technological manipulation.

This dimension draws specifically on faith traditions' understanding of human dignity, vocation, and accountability before God or the transcendent. In Christian terms, it means grounding moral reasoning in theological anthropology (i.e. a vision of who humans are and what they are called to be that comes from sources older and deeper than contemporary market logic or algorithmic optimization). It means recovering the concept of *vocation* or *calling,* or the idea that humans are called by God to particular forms of work and service. It means understanding that we will give account for how we have used power and resources, that there is a moral reckoning that transcends institutional success or market metrics.

In practical terms, this ancient wisdom root means that higher education should help students engage directly with classical texts and traditions. This is not primarily for historical or cultural knowledge, though both matter. Rather, it is formational. When a student reads the biblical prophets' fierce denunciations of injustice, or when he studies Augustine's meditation on the divided will, or when he engages Christian teaching on humility or stewardship, he is gradually absorbing a different perspective on what matters, on what constitutes human flourishing, on what we owe to others. This is a slower, less efficient process than a lecture on ethics, but it is also more resistant to the shallow, algorithmic thinking that dominates contemporary culture.

Ethical Scaffolding Virtues

The second dimension of the Moral Anchor Model consists of specific virtues that translate faith commitments into practices of discernment and action. These are not abstract ideals but practiced, cultivated capacities that develop through repetition, community support, and spiritual discipline. The model identifies five key virtues for leadership in an AI-saturated world:

Humility is perhaps the most crucial. Algorithmic systems and data-driven logic tend to produce confidence that if we have sufficient data and computational power, we can solve any problem, predict any outcome, and control any situation. This confidence is often misplaced. Humility, in contrast, involves acknowledging the limits of our knowledge, the partiality of our perspective, and our vulnerability to bias and error. A leader formed in humility will be less likely to assume that her organization's algorithms are neutral, less likely to implement technological solutions without questioning potential harms, and more likely to listen to voices of dissent and perspectives different from her own.

Compassion or mercy moves leaders beyond abstract principle to genuine concern for the welfare of those affected by their decisions. Many ethical frameworks can be rationalized away. This is more difficult to do with compassion. Leaders formed in compassion will be troubled by algorithmic bias that harms vulnerable populations, will resist the temptation to trade individual harm for aggregate utility, and will work toward systems that protect rather than exploit.

Justice involves the commitment to give each person their due and to correct systemic inequities. In the context of AI, justice means ensuring that the benefits of algorithmic systems are distributed fairly, that the burdens and risks are not borne primarily by the vulnerable, and that decisions made by algorithms are transparent and contestable. Leaders formed in justice will look critically at whose interests algorithmic systems serve and will work to ensure that technology serves human flourishing for all (Ahmad et al., 2026).

Stewardship grounds the use of power and resources in the understanding that they are held in trust, not owned absolutely. A steward is accountable for how she uses what has been entrusted to her. In the context of AI, stewardship means understanding that the data one collects is not rightfully used for any purpose an organization chooses, that the power of algorithmic systems to shape behavior creates serious responsibility, and that present decisions will affect future generations. Leaders formed in stewardship will be more cautious about deploying powerful technologies without understanding potential long-term consequences and more attentive to the needs of those not yet born.

Truthfulness or integrity involves committing to honesty even when it is inconvenient or unprofitable, and to transparency about how systems work and what trade-offs they involve. In an age of misinformation and opaque algorithms, truthfulness is a revolutionary virtue. The emerging research on regulatory frameworks for AI systems emphasizes the critical importance of transparency in algorithmic decision-making, particularly in high-stakes domains like healthcare, criminal justice, and employment (Akhundov, 2025). Leaders formed in truthfulness will resist the temptation to oversell what their technology can do, will be honest about limitations and potential harms, and will work toward transparency even when it complicates deployment.

These virtues are not primarily individual possessions but communal practices. They develop and are sustained through participation in communities where the virtues are named, modeled, practiced, and reinforced. This is why the formational environment matters so much. A student in an educational program that values competition above collaboration, that measures success purely in terms of individual achievement and financial return, and that treats ethical concerns as secondary to technical competency will find it very difficult to cultivate virtues like compassion, justice, and stewardship, no matter how motivated she is.

Digital Discernment Practices

The third dimension of the Moral Anchor Model consists of concrete practices of digital discernment. That is, specific ways of thinking, questioning, and acting that help leaders navigate technology's influence on identity and agency. These practices are essential because technology fundamentally mediates how practical wisdom develops and expresses itself (Zelny, 2025). Technologies can displace wisdom entirely (removing humans from deliberative contexts), diminish it (reducing the need for human judgment), augment it (strengthening deliberative capacity), or define new contexts where wisdom must be applied in novel ways. Understanding these technological relationships helps leaders design and adopt technologies that strengthen rather than erode moral discernment. The practices that follow translate the virtues described above into concrete patterns of reflection and action.

Attention discipline is a foundational practice. In an attention economy where algorithms compete for our focus, one crucial skill is the ability to deliberately direct one's attention rather than allowing it to be captured. This can take concrete forms: setting boundaries on social media use, curating information sources rather than accepting algorithmic recommendations, creating spaces for undistracted reflection and conversation. These practices are forms of resistance against systems engineered to capture and manipulate attention. Research on digital mindfulness demonstrates that regular practice in attention discipline through contemplative exercises and intentional technology use can actually strengthen the capacity to recognize and resist attention capture mechanisms (Chandra et al., 2025).

Critical interrogation of data and metrics involves developing the habit of asking not only "What does the data show?" but "How was this data collected? What is missing? Who does this metric privilege and who does it obscure? What values are embedded in this measurement?" (Ahmad et al., 2026). A leader who practices this discernment will be more difficult to deceive by reports that show impressive aggregate improvements while obscuring harm to vulnerable populations. She will be more likely to recognize when metrics like "engagement" or "efficiency" are being used to justify systems that actually undermine human flourishing.

Community consultation involves making important decisions about technology in dialogue with diverse stakeholders, particularly those who may be affected by the technology but who lack power to shape its design. This practice resists the tendency of tech development to be siloed in specialized teams insulated from accountability. It requires humility and creates space for voices of the vulnerable to shape technological futures. Recent ethical frameworks for responsible AI development emphasize the importance of participatory design processes and stakeholder engagement in ensuring that AI systems serve genuine human needs (Akhundov, 2025).

Restorative reflection involves regularly stepping back to assess how one's work aligns with one's deepest values and calling. In the rush of organizational life, it is easy to drift incrementally toward practices one would never have deliberately chosen. Regular practices of reflection through writing, conversation with trusted friends, spiritual direction, or formal assessment can create opportunities to notice misalignment and to reorient. These practices can be grounded in faith traditions through practices like prayer, scripture study, or structured reflection in faith community contexts. The integration of contemplative practices into professional life provides what some scholars call "embodied discernment". That is, the deep knowing that comes through sustained attention and reflection.

Advocacy for change involves moving beyond personal virtue to work for systemic change in how technology is designed, deployed, and governed. This recognizes that individual ethical choices, while important, are not sufficient in the face of systems structured to prioritize profit or efficiency at the expense of justice. A significant concern in contemporary education is the tendency to outsource critical thinking to AI systems without cultivating the human capacities these systems are designed to support. Leaders formed in faith traditions will understand that they have a responsibility not only to avoid causing harm but to work actively for systems that better serve human flourishing and justice, including insisting that educational and organizational contexts cultivate human judgment rather than replacing it. This might involve participating in regulatory discussions, developing ethical design standards, supporting governance mechanisms that ensure public participation in technology decisions, or building coalitions with other leaders and institutions committed to ethical technology.

FAITH AND HIGHER EDUCATION: PRACTICAL IMPLEMENTATION

Understanding the problem of the formation gap and the resources of faith traditions is necessary but not sufficient. Higher education must also ask: How do we actually incorporate faith-informed moral formation into curricula and institutional practices? This section addresses that question practically.

Integrating Faith Perspectives into Technical Education

One crucial starting point is incorporating faith perspectives directly into technical and professional education, not segregating faith into "ethics courses" that stand apart from the actual work of professional formation. The urgency of this task is supported by recent research on student AI use in higher education. While most college students now use AI regularly in their coursework, most attend institutions that discourage or prohibit such use, creating confusion and inadequate preparation. Students at restrictive institutions report feeling undertrained in AI use, and those without clear institutional guidance feel especially unprepared for post-graduation technology use. This represents a failure of higher education to help students develop informed, ethical relationships with tools they will encounter throughout their careers and lives. This might take several forms:

Integrating moral history into technical curricula. Courses in computer science, statistics, data science, and business should include substantial engagement with the history of how technologies have been used, including their role in perpetuating injustice. This is formational. When a student learns that Facebook's engagement algorithms were optimized in ways that spread misinformation during elections, or that algorithms used in criminal justice have been shown to encode racial bias, she begins to understand that technical decisions have moral consequences (Ahmad et al., 2026). This understanding should shape how she approaches her own work.

Featuring case studies of technological harms and ethical failures. Professional education should regularly examine cases where technology harmed vulnerable populations: predictive policing that targeted poor communities, hiring algorithms that discriminated against women, social media platforms that facilitated ethnic violence, surveillance systems that enabled authoritarian control (Weinberg, 2022). These case studies should be analyzed to explore the moral reasoning failures that preceded the technical failures. This helps students develop the habit of asking difficult questions about potential harms before systems are deployed.

Engaging explicitly with questions of human dignity and flourishing. Rather than treating ethics as a separate module, courses throughout the curriculum should ask: How does this technology affect human dignity? Who benefits and who bears the risks? How might this system affect the vulnerable? What are we optimizing for and what are we sacrificing? These questions should become habitual in students' thinking.

Examining institutional and structural dimensions of ethical technology. Beyond individual ethics and design decisions, students should understand how regulatory frameworks, governance structures, and institutional incentives shape technological development. They should study frameworks like GDPR, emerging AI governance approaches, and debates about algorithmic transparency and accountability (Akhundov, 2025). They should understand how cultural differences influence approaches to privacy, surveillance, and AI regulation, recognizing that values like those embedded in Western data protection frameworks are culturally contingent and not universally shared (Keche, 2025).

Cultivating Formational Communities

Higher education should recognize that formation happens through community participation more than through classroom instruction. This suggests several practical steps:

Creating faith-informed affinity groups and living-learning communities where students who are interested in integrating their faith with their professional studies can gather for reflection, accountability, and mutual support. These communities become spaces where it is safe to ask difficult questions, to articulate values that might seem countercultural in a competitive academic and professional environment, and to practice the virtues necessary for ethical leadership.

Integrating spiritual practices into leadership programs. While respecting religious diversity, higher education can create space for spiritual practices (e.g. prayer, meditation, contemplative reading, practices of service, etc.) that support moral formation. These practices are not primarily about religious belief; they are about cultivating the attention, humility, and connection to something larger than oneself that makes moral discernment possible. Research demonstrates that mindfulness-based practices integrated into educational settings enhance students' capacity for emotional regulation, cognitive flexibility, and ethical awareness (Pm & Yuvasri,

2025). When combined with formal instruction in technology ethics, such practices create synergistic effects, strengthening both the intellectual and dispositional dimensions of ethical leadership development.

Bringing in faith leaders and thinkers as teachers and mentors. Rather than treating faith perspectives as something only appropriate in explicitly religious courses, higher education should invite faith leaders, theologians, and spiritually-formed practitioners to engage with students in business schools, engineering schools, and professional programs. These voices can offer wisdom and perspective that academic training alone does not provide.

Supporting faculty formation in moral education. Faculty themselves need formation to teach about virtue, discernment, and the integration of faith with professional practice. This might involve faculty development programs, sabbaticals for faculty to deepen their own spiritual and ethical understanding, and intentional cultivation of communities of practice among faculty committed to this work.

Institutional Accountability Structures

Finally, higher education itself should be held accountable for whether it is actually forming students morally or just credentialing them technically. This might include:

Mission alignment in hiring, evaluation, and promotion. If a university claims to form leaders for ethical judgment and human flourishing, then faculty hiring and promotion should value scholarship and teaching that address moral formation. Professional school curricula should be evaluated for how well they help students integrate ethical and spiritual wisdom with technical competency.

Institutional transparency about values and mission. Universities should be honest about what values they are actually cultivating through their institutional structures and priorities. If the message students receive through the hidden curriculum (i.e. the signals sent by what is rewarded, what is punished, what is celebrated) contradicts the official curriculum's ethical teaching, that hypocrisy will undermine formation.

Assessment and accountability. Higher education should measure and report on not only technical competency and earning potential, but also on how well its graduates demonstrate moral reasoning, ethical decision-making, and commitment to justice. These measures should be visible and taken seriously in institutional decision-making.

APPLICATIONS: FROM THEORY TO PRACTICE

The preceding sections have developed the theoretical framework and institutional implications of the Moral Anchor Model. This section offers concrete examples of how this model can be applied to specific challenges in AI-mediated culture.

Case Study: Algorithmic Bias and Hiring

Consider the use of algorithmic systems in hiring decisions. Many organizations have adopted AI-powered hiring tools that promise to reduce bias and improve efficiency by analyzing thousands of job applications and identifying the most "qualified" candidates (Ahmad et al., 2026). These systems seem like a solution to human bias in hiring. Yet research has consistently shown that these algorithms encode biases from their training data and can amplify historical discrimination (Mishra et al., 2025).

A leader formed through the Moral Anchor Model would approach this technology differently than one guided only by technical and efficiency considerations. First, she would attend to Ancient Wisdom Roots, by asking fundamental questions about what humans owe to one another in hiring decisions. A faith-informed perspective reminds us that each person is created in the image of God and possesses inherent dignity that cannot be reduced to a score or ranking. This creates an obligation to treat job applicants as persons worthy of respect.

Second, she would cultivate the Ethical Scaffolding Virtues of justice and humility. Justice demands that she question whether an algorithm that has been shown to perpetuate discrimination against women and minorities can be used without causing harm. Humility reminds her that despite the algorithm's apparent objectivity and despite her organization's good intentions, deploying this system could perpetuate injustice. These virtues would motivate her to ask difficult questions: What would it look like to hire in ways that genuinely offer equal opportunity, particularly to candidates from communities that have experienced historical discrimination?

Third, she would practice Digital Discernment by subjecting the hiring algorithm to critical interrogation. She would ask: How was this algorithm trained? What data was used? Have independent researchers found evidence of bias? What metrics is the algorithm optimizing for, and what important qualities might it miss? She would insist on transparency about how the system works and would create space for human judgment in cases where the algorithm's recommendations conflict with other evidence of candidate quality (Ahmad et al., 2026).

Fourth, she would practice community consultation by engaging with workers, HR staff, applicants, and members of communities that might be affected by the system in conversations about how hiring should be done. In this setting, the temptation is to prioritize efficiency by outsourcing critical thinking to AI, preventing the development of these capacities within the self and the collaboration with others. Aboodi (2025) warns about how over-reliance on AI systems for decisions degrades our own judgment and limits shared knowledge. Rather than implementing an algorithm imposed from above, she would create space for people who will be affected by the system to shape how it is designed and deployed.

The result might not be the most "efficient" hiring process in narrow technical terms. But it would be more just. It would respect the dignity of applicants. It would reflect a commitment to genuine equal opportunity rather than a false equivalence between human bias and algorithmic optimization. And it would cultivate, in the leader herself and in the organization, the virtues necessary for ethical leadership in an AI-saturated world.

Case Study: Surveillance in the Workplace

Many organizations are deploying surveillance technologies to monitor worker productivity by tracking the keys employees press, monitoring their location, analyzing their email, recording their screens (Weinberg, 2022). These systems promise to increase efficiency and prevent theft or misconduct. Yet they also profoundly affect worker dignity, autonomy, and trust. Workers subjected to constant surveillance experience increased stress and decreased intrinsic motivation, actually undermining the productivity that surveillance purports to improve (Kuwaiti et al., 2023). Furthermore, workplace surveillance disproportionately affects lower-wage workers, creating systems of differential surveillance that track and control less powerful workers while granting autonomy to those with more authority.

A leader guided by faith and the Moral Anchor Model would recognize that surveillance reflects a fundamental lack of trust and treats workers as threats rather than as members of a community worthy of respect. Christian theology teaches that humans are created for relationship and community, and that trust is the appropriate basis for genuine community. This does not mean that accountability has no place; it does mean that accountability should be mutual and transparent, not one-directional and opaque.

A leader committed to justice would recognize that surveillance disproportionately affects lower-wage workers and would work to eliminate or minimize surveillance across the organization. Leaders would ask: What legitimate accountability concerns exist, and how can they be addressed in ways that respect worker dignity? What would mutual, transparent accountability look like?

The alternative would involve cultivating a culture of trust, involving workers in decisions about monitoring and accountability, and finding ways to address legitimate concerns about productivity and integrity without resorting to surveillance.

This is harder work than simply deploying surveillance technology. But it reflects greater wisdom about human nature and human community.

Case Study: Data Privacy and Human Dignity

The collection and use of personal data has become pervasive. Companies collect detailed information about users' browsing habits, location, purchases, health concerns, and social connections. This data is then used to profile individuals, manipulate behavior, and predict future actions. Users often have little knowledge of what data is collected or how it is used, and even when privacy policies exist, they are typically so lengthy and complex that informed consent is impossible (Goodman, 2022). Additionally, data from wearable devices and health apps creates persistent surveillance of intimate bodily functions, raising profound questions about what it means to have privacy over one's own body in an age of continuous biometric monitoring (Onder & Deniz, 2026).

A leader shaped by faith traditions would recognize that privacy is a consumer preference, but something rooted in human dignity. The ability to have thoughts, desires, and struggles that are not constantly monitored and analyzed is crucial to authentic selfhood and to the development of genuine identity (Goodman, 2022). To be fully known and analyzed by corporate systems creates a form of vulnerability that is fundamentally at odds with respect for persons. Contemporary research on digital surveillance in health contexts demonstrates how surveillance infrastructures can undermine informed consent, exploit vulnerable populations, and erode public trust in institutions (Cipta et al., 2026).

A leader committed to stewardship would recognize that data about people is not rightfully used for any purpose an organization chooses. Rather, data should be collected only for specific, legitimate purposes, should be protected from misuse, and should be deleted when no longer necessary. The default position should be caution and restraint, not aggressive collection and monetization.

A leader cultivating the virtue of truthfulness would insist on genuine transparency about data collection and use. Rather than the current approach of lengthy, incomprehensible privacy policies designed to provide legal cover while obscuring actual practices, a truthful approach would make clear to ordinary people what data is being collected and how it will be used. If the practices could not be stated transparently and honestly, that itself would be a reason to question whether they should be pursued.

The practical result might be that organizations collect less data, use it more restrictively, and provide genuine transparency about what they do with it. This might reduce some forms of profit extraction and behavioral manipulation. But it would reflect greater respect for human dignity and would help restore trust between organizations and the people whose data they handle.

CHALLENGES, LIMITATIONS, AND THE PATH FORWARD

This chapter has made the case that faith traditions offer crucial resources for moral formation in an AI-saturated world, and that higher education should intentionally engage these resources. This argument, however, faces real challenges and limitations that deserve frank acknowledgment.

Religious Diversity and Institutional Complexity

First, in a pluralistic society with religious diversity, how can higher education intentionally draw on faith traditions without privileging some traditions over others or violating principles of religious freedom? This is a genuine tension without simple solutions. The approach suggested here is not to promote any particular faith, but to promote open dialogue about various faiths represented on campus and normalize civil disagreement among the student body. This effort can lay a major role in recognizing that many faith traditions offer wisdom about moral formation that secular institutions have largely abandoned. Public higher education can create space for students to engage with traditions of their own choosing, can invite leaders from various traditions to speak about moral formation, and can examine how different traditions address contemporary ethical challenges, without itself advocating for any particular faith commitment.

Additionally, the argument advanced in this chapter emphasizes that what matters for moral formation is not doctrinal agreement but the cultivation of virtues, spiritual practices, and communal accountability structures that multiple traditions can provide. A Buddhist student and a Christian student might ground their moral commitment in different metaphysical frameworks, but both can benefit from practices of attention, compassion, and community that their respective traditions offer. Both can learn to interrogate digital systems critically, to cultivate humility about technological solutions, and to commit themselves to justice even when it conflicts with efficiency or profit. The key insight is that what unites these diverse traditions is not doctrinal content but shared commitment to practices that form persons toward virtue, wisdom, and accountability.

Secularization and the Eclipse of Transcendence

Second, our culture is increasingly secularized, and many contemporary intellectuals are skeptical of appeals to transcendence or traditional faith sources. For students formed primarily by secular frameworks, the suggestion that ancient religious traditions might offer wisdom about contemporary technology can seem quaint or implausible. Addressing this challenge requires not dismissing secular insights but showing how faith perspectives and secular ethical analysis can illuminate one another. It also requires modeling, in thoughtful faculty and leaders, how serious intellectual engagement with faith traditions is possible for educated, contemporary people. The chapter has attempted to do this by grounding the discussion in theological anthropology and Christian ethics. Furthermore, the emerging research on mindfulness, contemplative practices, and their role in developing ethical awareness and digital resilience provides empirical support for what faith traditions have long claimed: that practices of attention, reflection, and spiritual discipline cultivate capacities crucial for moral formation (Lan et al., 2025).

Risks of Instrumentalization

Third, there is a real risk that higher education will attempt to "use" faith traditions instrumentally by incorporating them into curricula as tools for forming more ethical workers while avoiding the more radical, prophetic dimensions of these traditions. Faith traditions, at their most vital, do more than support good business practices; they question whether the entire framework of contemporary culture is aligned with justice and human flourishing. An institution that cherry-picks faith insights to support its own priorities while avoiding the deeper critique will fail to accomplish true formation. This suggests that institutions must approach faith traditions with genuine openness to being challenged and transformed.

Sustaining Formation After Graduation

Finally, the formational work described in this chapter cannot be completed in higher education alone. The environment into which graduates move (e.g. workplaces, professional organizations, broader culture, etc.) often stands in opposition to the virtues and perspectives cultivated in school. Without ongoing communities of practice and accountability, many graduates will gradually abandon their commitments to justice and integrity as they experience pressure to conform to organizational norms and competitive logic. This means that faith communities themselves have a crucial role to play, providing ongoing formation, community, and accountability for professionals working in technology and leadership positions. Emerging research on professional burnout and moral disengagement suggests that individuals working in high-stress technological and organizational contexts benefit significantly from sustained community support and spiritual practices that help them maintain moral orientation and resilience.

CONCLUSION: TOWARD A NEW ERA OF FORMATION

We stand at a threshold. The tools of artificial intelligence and algorithmic systems will shape the future, but how they shape it depends on the character of the leaders and communities that design, deploy, and govern them. If these leaders have been formed primarily by systems engineered for engagement and efficiency, by surveillance infrastructures designed to extract data and maximize profit, they will create systems that reflect those values (Akhundov, 2025). The result will be technology that increasingly manipulates behavior, exacerbates inequality, erodes privacy, and reduces humans to data points to be optimized.

But a different future is possible. If higher education, in partnership with faith communities, intentionally cultivates moral identity in future leaders, then they will enhance existing systems. If students are formed to see human dignity as transcendent to metrics, to practice virtues like compassion and justice even when these conflict with efficiency, to recognize their accountability before something larger than the market, and to develop the attention and discernment necessary to resist manipulation and work for just systems, then they may develop better systems. These systems might be less profitable in the short term and less technically elegant. But they will better serve human flourishing. They will be more likely to protect the vulnerable rather than exploit them. They will reflect wisdom about what it means to be human and what we owe one another.

The Moral Anchor Model offers a framework for this formation work. By grounding leaders in Ancient Wisdom Roots that remind them of human dignity and transcendent purpose, by cultivating Ethical Scaffolding Virtues that orient them toward genuine goods, and by practicing Digital Discernment that helps them navigate technology's influence on identity and agency, we can form leaders equipped for the work ahead.

This work is demanding. It cannot be accomplished through isolated ethics courses or one-time training sessions. It requires transformation of institutional culture, deliberate design of formational environments, integration of faith perspectives throughout professional education, and sustained partnership between higher education and faith communities. But the alternative (i.e. defaulting to the formative power of algorithms, surveillance systems, and market logic) is to abandon the future to forces engineered not for wisdom but for profit, control, and engagement.

The call to higher education in this moment is clear: recover your responsibility for moral formation, recognize the resources offered by faith traditions, develop capacities for digital discernment in your students, and create the environments and practices through which emerging leaders can develop the character necessary to guide technology toward human flourishing rather than efficiency alone. The stakes could not be higher, and the time for action is now.

REFERENCES

Aboodi, R. (2025). The Worrisome Potential of Outsourcing Critical Thinking to Artificial Intelligence. *Educational Theory*, *75*(4), 626–645. DOI: 10.1111/edth.70037

Ahmad, A., Vallès, Y., & Idaghdour, Y. (2026). Bias in AI systems: Integrating formal and socio-technical approaches. *Frontiers in Big Data*, *8*, 1686452. Advance online publication. DOI: 10.3389/fdata.2025.1686452 PMID: 41584908

Akhundov, A. (2025). The Role of Ethics in Modern Technology Development. *Porta Universorum*, *1*(4), 169–177. DOI: 10.69760/portuni.0104017

Alperstein, N. M. (2025). A network approach to third-party tracking devices on mental health websites: Data surveillance and privacy in the digital panopticon. *Mental Health and Digital Technologies*, *2*(1), 59–74. DOI: 10.1108/MHDT-10-2024-0031

Band, V. S. (2026). Ethical and legal issues in information technology systems. *International Journal of Scientific Research in Engineering and Management*. DOI: 10.55041/IJSREM.IBFE038

Chandra, P. S., Irdayanti, Fithriyyah, M. U., Rahmadani, F., & Astuti, W. P. (2025). Meningkatkan digital mindfull di kalangan remaja. *Development: Journal of Community Engagement*. DOI: 10.46773/djce.v4i2.1991

Cipta, H., Fitriani, I., & Putra, R. S. P. (2026). Surveillance health society: Ethics, privicy, and social control in digital health systems. *Journal of Public Health Indonesian*. DOI: 10.62872/4tt4qc09

Gbadeyanka, T. A., Ajobiewe, D. N., Okunlola, E. F., & Sulaimon, J. T. (2025). *Leveraging mindfulness and information and communication technology (ICT) to enhance attention amidst digital distractions in primary education*. ASEAN Journal of Science and Engineering Education., DOI: 10.17509/ajsee.v4i3.82475

Goodman, B. (2022). Privacy without persons: A Buddhist critique of surveillance capitalism. *AI and Ethics*, *2*(3), 415–426. DOI: 10.1007/s43681-022-00204-1

Kantha, T., Sahoo, D., Rout, D., & Mishra, S. J. (2024). *Ethics in the digital era: Technology's role in shaping society*. Shanlax International Journal of Arts, Science and Humanities., DOI: 10.34293/sijash.v11iS1-June.7794

Keche, U. (2025). *Cybersecurity and information ethics: A comparative study of privacy rights in digital india and the global framework*. Gurukul International Multidisciplinary Research Journal., DOI: 10.69758/GIMRJ/2504I5VXIIIP0090

Kuwaiti, A. A., Nazer, K., Alreedy, A. H., AlShehri, S. D., Almuhanna, A., Subbarayalu, A. V., Muhanna, D. A., & Al-Muhanna, F. (2023). A Review of the Role of Artificial Intelligence in Healthcare. *Journal of Personalized Medicine*, *13*(6), 951. DOI: 10.3390/jpm13060951 PMID: 37373940

Lan, Y., Liu, S., & Xia, L. (2025). How mindfulness shapes AI competence: A structural equation modeling analysis of mindfulness, AI literacy and behavioral intention in Chinese media students. *Frontiers in Psychology*, *16*, 1652934. Advance online publication. DOI: 10.3389/fpsyg.2025.1652934 PMID: 40900965

McLoughlin, S., Thoma, S., & Kristjánsson, K. (2025). Was Aristotle right about moral decision-making? Building a new empirical model of practical wisdom. *PLoS One*, *20*(1), e0317842. Advance online publication. DOI: 10.1371/journal.pone.0317842 PMID: 39841702

Mishra, A., Mou, S. N., Ara, J., & Sarkar, M. (2025). Regulatory and ethical challenges in AI-driven and machine learning credit risk assessment for buy now, pay later (BNPL) in U.S. E-commerce: Compliance, fair lending, and algorithmic bias. *Journal of Business and Management Studies*. DOI: 10.32996/jbms.2025.7.2.3

Onder, O., & Deniz, E. M. (2026). Cyberspatial privacy in the digital age: A proxemics-based framework with a healthcare application. *BMC Medical Ethics*, *27*(1), 14. DOI: 10.1186/s12910-025-01369-0 PMID: 41507986

Pm, M., & Yuvasri. (2025). Effectiveness of mindfulness- based intervention on cognitive flexibility, digital burnout and emotional regulation among college students. *Advanced International Journal for Research*. DOI: 10.63363/aijfr.2025.v06i06.2459

Tripathi, R. (2025). AI meets mindfulness: Redefining spirituality and meditation in the digital age. *The Voice of Creative Research*. DOI: 10.53032/tvcr/2025.v7n1.03

Weinberg, L. (2022). Rethinking fairness: An interdisciplinary survey of critiques of hegemonic ML fairness approaches. *Journal of Artificial Intelligence Research*, *73*, 421–473. DOI: 10.1613/jair.1.13196

Zelny, A. (2025). Offloading wisdom: Four technological relations that mediate phronesis. *Philosophy & Technology*, *38*(2), 56. DOI: 10.1007/s13347-025-00889-2

KEY TERMS AND DEFINITIONS

Ancient Wisdom Roots: Foundational theological and philosophical insights about human dignity, vocation, justice, and accountability that provide enduring guidance for moral reasoning beyond technological efficiency or market-driven values.

Attention Discipline: The cultivated ability to direct and sustain focus intentionally rather than allowing digital systems designed for engagement capture to shape perception and decision-making.

Community Consultation in Technology Design: A participatory leadership practice that includes stakeholders, especially those affected by technological decisions, in shaping how AI systems are developed, evaluated, and implemented.

Digital Discernment Practices: Intentional reflective habits that enable individuals to evaluate how digital technologies shape attention, preferences, and judgments, supporting responsible engagement with AI systems.

Ethical Scaffolding Virtues: Core leadership dispositions such as humility, compassion, justice, stewardship, and truthfulness that translate faith commitments into practical habits of ethical decision-making in complex technological contexts.

Faith Traditions as Moral Ecosystems: Integrated communities of narrative, ritual, discipline, and accountability that sustain moral agency over time and provide resources for resisting reductionist views of human worth in data-driven systems.

Formation Gap: The growing distance between the powerful identity-shaping influence of algorithmic systems and the weakening role of institutions and traditions that cultivate virtue, wisdom, and ethical judgment in emerging leaders.

Moral Anchor Model: A leadership formation framework that integrates ancient wisdom traditions, virtue-based ethical development, and digital discernment practices to help leaders navigate AI-mediated environments while protecting human dignity and moral responsibility.

Moral Identity Formation: The developmental process through which individuals acquire stable ethical commitments, virtues, and interpretive frameworks that guide responsible leadership in professional and technological environments.

Practical Wisdom in Digital Contexts: The capacity to exercise sound judgment about technology use by integrating experience, reflection, virtue, and concern for human flourishing in algorithmically mediated environments.

Relational Anthropology: A theological understanding of the human person as inherently oriented toward relationships with God and others, providing a foundation for ethical leadership that resists purely individualistic or data-reductionist views of identity.

Secular Liturgies: Repeated technological routines such as algorithmic recommendations, data tracking, and engagement optimization that shape identity and behavior in ways similar to formative religious practices but without explicit moral accountability.

Chapter 2
Faith Integration, Ethical Orientation, and AI Adoption in a Faith-Based University:
Evidence From Northrise University

Kingsley Chisanga
http://orcid.org/0009-0002-2583-408X
Northrise University, Zambia

ABSTRACT

This chapter examines how faith integration, ethical orientation, and artificial intelligence (AI) practices shape responsible AI adoption, pedagogical integration, academic integrity, and student outcomes at Northrise University, Zambia. Positioned within a values-driven higher education context, this chapter explores how faith-informed ethical frameworks guide responsible AI use in teaching and learning. Using a concurrent cross-sectional mixed-methods design, faculty (n = 38) and student (n = 56) perspectives are analyzed to explain how values-based governance influences AI adoption. Findings indicate that faith integration supports responsible AI guidance, ethical orientation strengthens integrity vigilance, and AI capacity-building drives pedagogical integration, while student outcomes show strong links between AI value, employability, and learning.

DOI: 10.4018/979-8-3373-8192-3.ch002

ht © 2027, IGI Global Scientific Publishing. Copying or distributing in print or electronic forms without written permission of IGI Global Scientific Publishing is prohibited. Use of this chapter to generative artificial intelligence (AI) technologies is expressly prohibited. The publisher reserves all rights to license its use for generative AI training and machine learning model development.

INTRODUCTION

Artificial intelligence (AI) is rapidly reshaping higher education, transforming how teaching, assessment, research, and student support are designed and delivered (Zawacki-Richter et al., 2019). The growing use of automated feedback, adaptive learning systems, and generative AI tools offers significant opportunities to enhance learning experiences, improve academic efficiency, and prepare graduates for increasingly digital and complex work environments (Chaudhry & Kazim, 2022; Holmes et al., 2019; Yan et al., 2023). Recent evidence indicates that AI is no longer experimental but embedded in everyday student learning practices, with many students regularly using AI tools to support understanding, assignment completion, and productivity (Gallup, Inc., & Lumina Foundation, 2026).

In faith-based universities, however, the integration of AI raises deeper questions. These institutions are not only concerned with efficiency and performance but are fundamentally oriented toward stewardship, integrity, moral accountability, human dignity, and spiritual formation (Astin et al., 2011; de Kock, 2024; Corpuz, 2025).

Within this context, AI adoption must be understood not only as a technological shift but as a question of human flourishing, where institutions cultivate purpose, belonging, ethical identity, and meaningful contribution within AI-mediated environments.

The objective of this chapter is to provide empirically grounded evidence on whether and how faith integration and faculty ethical orientation influence AI adoption, pedagogy, academic integrity practices, and student outcomes in a faith-based university context, to inform responsible, mission-aligned AI governance.

BACKGROUND

Faith-based universities emphasize holistic formation, integrating intellectual development with moral and spiritual growth (Astin et al., 2011; de Kock, 2024). At the same time, evidence shows that while AI use is widespread among students, institutional guidance on acceptable use remains uneven, leading to uncertainty about expectations and appropriate academic practices (Gallup, Inc., & Lumina Foundation, 2026). These tensions extend beyond policy gaps and influence how students experience fairness, trust, and belonging within the institution. Recent scholarship highlights tensions between AI-enabled efficiencies and human-centered learning principles, urging careful consideration of equity, accountability, integrity, and the formation of young people in university education (Corpuz, 2025; Reina-Marín et al., 2025). Faculty similarly navigate competing demands between innovation and integrity, often without fully developed institutional frameworks, reinforcing the need for coherent, values-based governance.

Unlike many secular approaches that prioritize efficiency and competitiveness, faith-based institutions foreground stewardship, integrity, moral accountability, character development, and respect for human dignity (Corpuz, 2025). Consequently, AI integration in these contexts cannot be value-neutral; it must be guided by faith-informed ethics and pedagogical intentionality.

Despite the rapid expansion of AI applications in higher education, much of the literature focuses on technical adoption, learning analytics, and performance outcomes, while ethical governance and critically grounded values-based perspectives remain underdeveloped (Selwyn, 2019). A systematic review by Collin et al. (2023) similarly reports that AI-in-education scholarship is dominated by technical and managerial concerns, with ethical and critical issues still underdeveloped. Few empirical studies examine whether faith integration constrains or enhances responsible AI adoption, or how faculty ethical orientation shapes AI practices and student outcomes in faith-based contexts. This chapter addresses this gap by analyzing faculty and student perspectives on AI use at Northrise University, Zambia.

THE ETHICAL AND FORMATIONAL CHALLENGE OF AI ADOPTION

Issues, Controversies, Problems

In faith-based universities, integrating emerging technologies such as AI should align with institutional commitments to stewardship, academic integrity, human dignity, moral accountability, and spiritual formation (Dockery, 2016). Educational practice in these contexts is explicitly grounded in holistic student development, including spiritual and ethical growth (Astin et al., 2011). Ideally, AI adoption would enhance learning, academic productivity, and innovation while remaining consistent with values-based governance and mission-driven commitments. However, research on AI in higher education has largely emphasized technical adoption, learning analytics, and performance outcomes, with limited attention to critically grounded ethical frameworks or values-based governance (Selwyn, 2019). The description of AI-related challenges in the literature is fragmented, as scholars analyze different educational contexts using diverse computational techniques and analytical perspectives, which complicates holistic understanding (Collin et al., 2023).

Within this landscape, faith-based universities face uncertainty regarding how to integrate AI responsibly without compromising institutional values, academic integrity, or spiritual formation. Empirical evidence remains limited on whether faith integration and faculty ethical orientation meaningfully influence AI adoption, pedagogy, academic integrity, or student trust. It is also unclear how students perceive AI's value for learning, employability, critical thinking, and spiritual development.

Concerns about AI-assisted plagiarism, misuse of generative tools, and erosion of academic honesty have intensified across higher education (Bretag et al., 2019; Perkins et al., 2023). Without systematic evidence on how faith integration interacts with AI practices, institutions risk reactive or inconsistent policy responses, misalignment between innovation and mission, weakened guidance frameworks, diminished student trust, and fragmented curriculum design.

The central research problem addressed in this chapter is ideally that, AI adoption in faith-based universities would enhance learning and innovation while remaining aligned with institutional commitments to integrity, stewardship, and human dignity. However, emerging evidence points to several challenges, including inconsistent AI policies, academic integrity risks, uneven ethical guidance, and student uncertainty regarding expectations.

These challenges risk undermining not only academic standards but also student trust, institutional coherence, and the development of environments that support belonging, purpose, and ethical formation.

Purpose of the Study

The purpose of this study is to examine how faith integration, ethical considerations, and AI practices among faculty are associated with responsible AI adoption, pedagogical integration, academic integrity, and research productivity, and how these faculty practices relate to student trust, learning outcomes, critical thinking, employability perceptions, and spiritual engagement in faith-based universities.

Research Objectives

1. Examine the relationship between faculty faith integration and responsible AI guidance to students.
2. Assess the influence of faculty ethical considerations on AI-related academic integrity outcomes.
3. Determine whether AI capacity-building predicts pedagogical AI integration among faculty.
4. Examine the relationship between AI use for research and faculty research productivity.
5. Test whether faith integration and ethical considerations interact to predict AI adoption in teaching.

6. Examine student perceptions of ethical AI guidance, AI value addition, AI-integrated pedagogy, and AI's role in spiritual formation.
7. Assess the relationships between AI practices and student trust, learning outcomes, critical thinking, and perceived employability.
8. Explore the benefits and challenges of AI usage both by faculty and students.

Hypotheses

Faculty Hypotheses

H1: Faith Integration is positively associated with Responsible AI Guidance among faculty. Faculty who more intentionally integrate faith into their teaching are more likely to provide structured, values-informed guidance on responsible AI use.
H2: Faculty Ethical Orientation is positively associated with AI-related Academic Integrity Vigilance. Faculty with stronger ethical awareness report greater concern regarding AI misuse, monitoring workload, and integrity governance pressures. (Note: This measures vigilance, not breach frequency.)
H3: AI Capacity-Building is positively associated with AI Pedagogical Integration. Faculty training, confidence, and competence in AI significantly predict the extent to which AI is meaningfully integrated into teaching and assessment.
H6 (Descriptive/Perceptual): Faith integration and ethical orientation are associated with perceived AI research productivity value. (Note: This is a perception-based descriptive hypothesis, not a predictive model of use → productivity.)
H9: The interaction between Faith Integration and Ethical Orientation is associated with AI Pedagogical Adoption among faculty. Specifically, AI adoption varies as a function of the combined influence of faith integration and ethical vigilance, reflecting discernment rather than additive effects.

Student Hypotheses

H4: Perceived Ethical Guidance is positively associated with Student Trust in AI systems. Students who perceive clearer, consistent, and value-informed ethical guidance are more likely to develop trust in AI systems used within the university context.
H5: AI Use in Learning is positively associated with Student Critical Thinking outcomes. Students reporting more structured AI-supported learning engagement report stronger critical thinking-related outcomes. (Note: No demographic controls included in this dataset.)

H7: Perceived AI Value Addition is positively associated with Perceived Employability. Students who perceive AI as enhancing productivity, learning, and skill development report significantly higher career readiness confidence.
H8: AI Use Supporting Spiritual Formation is positively associated with Christian Engagement. Students who perceive AI as aligned with faith-based values report stronger spiritual coherence. (Relationship modest in magnitude.)
H10: AI-Integrated Pedagogy Exposure is positively associated with Student Learning Outcomes. Students experiencing intentional AI integration in teaching report significantly stronger perceived learning improvements.

Final Structured Overview

Faculty Model

Faith Integration; Ethical Orientation; AI Capacity-Building → Responsible AI Guidance → Integrity Vigilance → AI Pedagogical Adoption → Perceived Research Productivity Value (with Faith × Ethics interaction effect)

Student Model

Figure 1 presents the chapter's proposed mechanism model. In a faith-based university, institutional mission and faculty-level drivers shape AI-related teaching and governance practices. These practices influence how students experience AI in learning, which in turn affects educational, vocational, and spiritual outcomes. The chapter's central claim is that AI adoption in faith-based higher education is not value-neutral but is mediated through faith-informed ethical guidance and pedagogical practice.

Figure 1. Faith-informed mechanism model of AI integration in a faith-based university

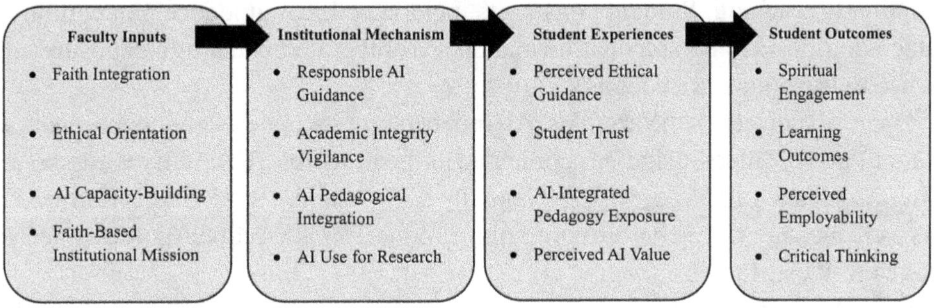

Ethical Guidance; AI Value Addition; AI-Integrated Pedagogy; AI Learning Use Trust Employability Critical Thinking Learning Outcomes Spiritual Engagement

Significance of the Study

This chapter contributes theoretically by extending scholarship on AI in education through a faith-informed ethical lens, responding to calls for more normative and values-based approaches to educational technology governance (Floridi et al., 2018; Selwyn, 2019). Methodologically, it provides empirically validated measurement approaches for examining AI adoption in values-driven institutions. Practically, the findings inform policy development, faculty training, curriculum design, and leadership strategies for responsible AI integration within faith-based universities.

METHODOLOGY

Research Design: The study adopts a cross-sectional concurrent mixed-method design, in which quantitative (survey-scale) and qualitative (open-ended) data were collected simultaneously from faculty and students in a faith-based university (Creswell & Creswell, 2018). Findings are interpreted as associational rather than causal (Field, 2018).

Population and Sample: The study population comprises faculty involved in teaching and research and undergraduate/postgraduate students enrolled in faith-based universities. Separate but aligned instruments were administered to faculty and students. Participation was voluntary, and ethical approval was obtained prior to data collection.

Data Collection Instrument: Data were collected using a single survey instrument containing quantitative Likert-scale measures (1 = strongly disagree to 5 = strongly agree) and qualitative open-ended questions. Faculty constructs included faith integration, ethical considerations, AI capacity-building, AI use for research, and AI adoption in teaching. Student constructs included ethical guidance perception, AI value addition, AI and spiritual formation, AI-integrated pedagogy exposure, and AI use in learning and pedagogy.

Operationalization of Variables: All independent variables were operationalized as continuous composite indices computed as mean scores. Reliability was assessed using Cronbach's alpha ($\alpha \geq 0.70$) (Nunnally & Bernstein, 1994). Interaction predictors were mean-centered before computing product terms to reduce multicollinearity (Aiken & West, 1991).

Data Analysis Procedures: Analyses included descriptive statistics, reliability (Cronbach's α), Pearson correlations, linear and multiple regression models (including interaction terms), and inclusion of controls (year of study and program) where theoretically justified. Effect sizes were interpreted using conventional benchmarks (Cohen, 1988).

FACULTY CONSTRUCTS

Faith Integration (Faculty) (H1; H9)

Conceptual Definition

Faith integration refers to the extent to which faculty embed spiritual values, beliefs, and principles into their teaching practices and decision-making processes. Within this study, it is closely linked to the Faith–Ethics Interaction Effect, which represents the combined influence of spiritual orientation and ethical awareness on faculty AI integration behavior. Conceptually, this construct reflects the proposition that faith-informed values and ethical reasoning jointly shape how and to what extent artificial intelligence is incorporated into teaching. Rather than assuming a simple additive relationship, the interaction perspective recognizes that ethical considerations may have different effects on AI adoption depending on the level of faith integration. Thus, faith integration is understood as a moderating force that influences the relationship between ethical awareness and AI pedagogical adoption.

Operational Definition

Faith integration is measured using a multi-item Likert-scale index assessing the extent to which faculty incorporate faith-based principles into teaching, decision-making, and AI-related practices. For interaction analysis (H9), faith integration scores are mean-centered and multiplied with ethical considerations to generate an interaction term. AI pedagogical adoption is operationalized as a composite index capturing AI use for feedback provision, AI-supported lesson innovation, AI-adapted assessments, and encouragement of AI-assisted analysis. All items are measured on 5-point Likert scales. The regression model includes faith integration (main effect), ethical considerations (main effect), and the interaction term (Faith × Ethics).

Ethical Considerations (Faculty) (H2; H9)

Conceptual Definition

Ethical considerations refer to faculty awareness, prioritization, and application of moral principles guiding professional conduct, academic integrity, and responsible technology use. In the context of AI, ethical considerations encompass issues such as fairness, accountability, transparency, plagiarism, misuse of AI tools, and the protection of academic standards. This construct reflects the degree to which faculty actively engage with ethical dilemmas associated with AI and integrate ethical reasoning into their teaching and research practices.

Operational Definition

Ethical considerations are operationalized using a multi-item index measured on a 5-point Likert scale (1 = strongly disagree to 5 = strongly agree). The scale assesses awareness of AI-related ethical risks, commitment to academic integrity, frequency of addressing ethical AI use in teaching, and perceived responsibility for preventing misuse. Higher scores indicate stronger ethical orientation. Reliability is assessed using Cronbach's $\alpha \geq 0.70$. For hypothesis H9, ethical considerations are mean-centered and included both as a main effect and as part of the interaction term with faith integration in regression analysis.

AI Capacity-Building (Faculty) (H3)

Conceptual Definition

AI capacity-building refers to the extent to which faculty have developed the knowledge, skills, and confidence required to understand, evaluate, and effectively use AI technologies in teaching, assessment, and academic work. It includes formal training, informal learning, and continuous professional development related to AI. This construct reflects the readiness of faculty to integrate AI meaningfully into pedagogical practice.

Operational Definition

AI capacity-building is measured using a summative Likert-scale index capturing participation in AI-related training, self-assessed competence in using AI tools, familiarity with AI applications in education, and confidence in pedagogical integration. Items are rated on a 5-point Likert scale. Higher scores indicate greater professional development in AI. Reliability is confirmed using Cronbach's $\alpha \geq 0.70$.

AI Use for Research (Faculty) (H6)

Conceptual Definition

Perceived research productivity value of AI refers to faculty members' subjective assessment of how AI enhances efficiency, quality, and effectiveness in scholarly work. This includes support for idea generation, literature review, drafting, structuring arguments, data analysis, and workflow optimization. In faith-based contexts, these perceptions are shaped by ethical reflection, authorship responsibility, and alignment with institutional research standards.

Operational Definition

This construct is operationalized using a single Likert-scale item: "AI adds significant value to my personal research productivity" (1 = strongly disagree to 5 = strongly agree). Higher scores reflect stronger perceived value of AI in research productivity.

Faith × Ethics Interaction (Faculty) (H9)

Conceptual Definition

The interaction between faith integration and ethical considerations represents the synergistic influence of spiritual orientation and ethical awareness on faculty behaviour. Conceptually, it reflects the premise that faith-informed values amplify the impact of ethical reasoning in shaping responsible AI adoption.

Operational Definition

The interaction term is computed by mean-centering faith integration and ethical considerations and multiplying them to form a product term. This variable is included in multiple regression models alongside main effects to test whether the relationship between ethical considerations and AI adoption varies depending on levels of faith integration.

Methodological Note (Faculty Variables)

All faculty independent variables are treated as continuous composite indices. This approach is justified by internal consistency (Cronbach's $\alpha \geq 0.70$) and conceptual coherence. The variables are analyzed using correlation and multiple regression models to assess associations and predictive relationships.

STUDENT CONSTRUCTS

Ethical Guidance Perception (Student) (H4)

Conceptual Definition

Ethical guidance perception refers to students' assessment of the extent to which faculty and institutions provide clear, consistent, and value-driven guidance on ethical AI use. It includes perceptions of integrity, moral accountability, and clarity of expectations regarding AI-assisted learning.

Operational Definition

This construct is operationalized as a composite mean score derived from multiple Likert items measuring clarity of ethical expectations, faculty guidance, institutional emphasis on integrity, and consistency of ethical messaging. Reliability is assessed using Cronbach's $\alpha \geq 0.70$.

AI Value Addition (Student) (H7)

Conceptual Definition

AI value addition refers to students' perceptions of how AI enhances learning experience, skill development, academic efficiency, and preparedness for future careers.

Operational Definition

This construct is measured using a multi-item Likert index assessing improvements in understanding, productivity, digital skills, and employability. Higher scores indicate stronger perceived value. Reliability is confirmed using Cronbach's $\alpha \geq 0.70$.

AI and Spiritual Formation (Student) (H8)

Conceptual Definition

This construct reflects the extent to which students perceive AI use as supportive of spiritual growth, moral reflection, and engagement with Christian values.

Operational Definition

Measured using a Likert composite assessing alignment with faith-based learning goals, encouragement of moral reflection, and support for spiritual growth. Higher scores indicate stronger alignment.

AI-Integrated Pedagogy Exposure (Student) (H10)

Conceptual Definition

AI-integrated pedagogy exposure refers to students' experiences of instructional practices that intentionally incorporate AI into learning, assessment, and feedback.

Operational Definition

Operationalized as a composite index capturing frequency and depth of AI exposure in coursework. Measured using Likert-scale items with reliability confirmed (Cronbach's $\alpha \geq 0.70$).

AI Use in Learning and Pedagogy (H5)

Conceptual Definition

This construct reflects the extent to which AI is used to support learning, problem-solving, and cognitive development.

Operational Definition

Measured as a composite Likert index capturing AI use in learning activities, analysis, and critical thinking. Included in regression models controlling for demographic variables.

Methodological Note (Student Variables)

All student constructs are treated as continuous latent variables approximated through reliable Likert composites. Correlation and regression analyses are used to assess relationships and predictive effects.

Overall Operationalization Note

All constructs are measured using Likert-type scales (1 = strongly disagree to 5 = strongly agree), with mean indices computed after reliability testing (Cronbach's $\alpha \geq 0.70$). Interaction terms and composite indices are used in regression analysis to examine relationships among variables.

Figure 2 illustrates the chapter's concurrent mixed-methods design. Quantitative survey data and qualitative open-ended responses were collected simultaneously from faculty and students, analyzed in parallel, and integrated during interpretation to provide a richer understanding of faith integration, ethical orientation, AI practices, and student outcomes in a faith-based university context.

Figure 2. Concurrent mixed-methods design

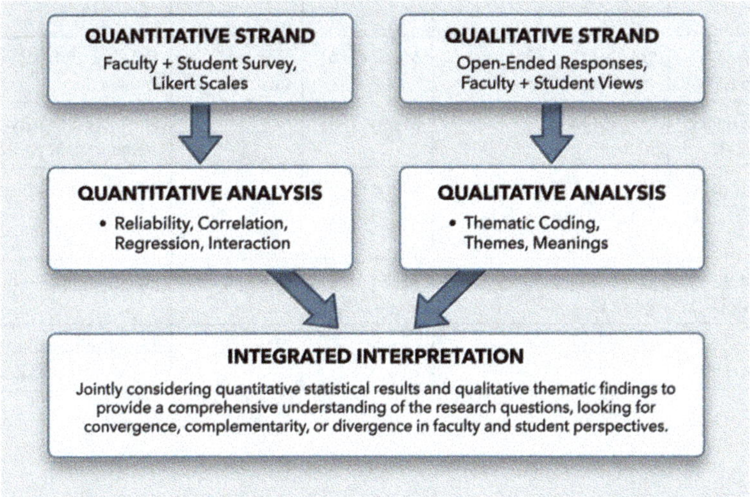

Ethical Considerations

The study adhered to ethical standards governing social science research. Participation was voluntary, informed consent was obtained, anonymity and confidentiality were ensured, and data were used solely for academic purposes. Given the faith-based context, sensitivity to religious identity, institutional mission, and spiritual values was maintained throughout.

RESULTS

This section presents quantitative results, organized by faculty and student hypotheses. Results are reported using APA 7-compliant statistical tables embedded within the text, followed by interpretive explanations emphasizing direction, magnitude, and practical meaning. Given the cross-sectional design, findings are interpreted as associations rather than causal effects.

Table 1. Summary of construct operationalization and reliability (Faculty dataset, n = 38)

Construct	Items Used	Scale Type	Reliability (α)	Interpretation
Faith Integration	Q3_1, Q3_3, Q3_4 (3 items)	Mean (1–5)	.74	Acceptable internal consistency
Ethical Considerations	Q3_5, Q3_6, Q3_7, Q3_8 (4 items)	Mean (1–5)	.42	Low reliability; broad ethics-orientation index
AI Capacity-Building	Q4_1, Q3_1, Q3_8 (3 items)	Mean (1–5)	.42	Low reliability; interpret cautiously
Responsible AI Guidance (H1 DV)	Q3_2, Q3_8 (2 items)	Mean (1–5)	Weak / unstable	Policy-emphasis index (not tight scale)
Integrity Vigilance (H2 DV proxy)	Q3_7, Q4_2, Q4_3 (3 items)	Mean (1–5)	.42	Reflects integrity-management pressure
AI Pedagogical Integration (H3 DV)	Q4_6, Q4_7, Q3_9, Q3_10 (4 items)	Mean (1–5)	.58	Moderate reliability
AI Adoption in Teaching (H9 DV)	Q4_6, Q4_7, Q3_9 (3 items)	Mean (1–5)	.74	Acceptable reliability
AI Research Productivity (H6)	Q4_4 (single item)	Single Likert	N/A	Descriptive only; no regression possible

Table 2. Summary of construct operationalization and reliability (Student dataset, n = 44)

Construct	Items Used	Scale Type	Reliability (α)	Interpretation
Ethical Guidance (H4 predictor)	Q3_5 (single item)	Single Likert	N/A	Clear rule-communication indicator
Trust in AI (H4 outcome)	Q3_6 (single item)	Single Likert	N/A	Institutional trust measure
AI Value Addition (H7 predictor)	Q3_10, Q4_4, Q4_7, Q4_8, Q4_5, Q4_1 (6 items)	Mean (1–5)	.80	Good reliability
Perceived Employability (H7 outcome)	Q4_1, Q4_5 (2 items)	Mean (1–5)	Moderate	Acceptable for short scale
AI-Integrated Pedagogy (H10 predictor)	Q3_10, Q4_4 (2 items)	Mean (1–5)	Acceptable	No-overlap index
Learning Outcomes (H10 DV)	Q4_8 (single item)	Single Likert	N/A	Perceived learning quality
Critical Thinking (H5)	Approximate proxy	Composite	Limited	No full validated scale; no demographic controls

Table 3. Cronbach's alpha interpretation guide

Alpha Value	Interpretation
≥ .80	Good reliability
≥ .70	Acceptable
.60–.69	Moderate
.50–.59	Weak
< .50	Low (interpret cautiously)

Table 4. Key effects reported in the study

Relationship	Effect Size	Interpretation
Faith Integration → Responsible AI Guidance	$r = .58$	Large
Ethical Orientation → Integrity Vigilance	$\beta = .57$	Large
AI Capacity → Pedagogical Integration	$\beta = .72$	Very Large
Faith × Ethics → AI Adoption	$\beta = -.30$	Moderate interaction

continued on following page

Table 4. Continued

Relationship	Effect Size	Interpretation
AI Value → Employability (students)	$\beta = .78$	Very Large
AI-Integrated Pedagogy → Learning	$\beta \approx .78$	Very Large
AI Use → Critical Thinking	$\beta \approx .29$	Moderate

Figure 3 summarizes the study's empirical findings linking faith integration, ethical orientation, and AI capacity-building to faculty AI practices and student outcomes. Solid arrows indicate statistically supported relationships, while dashed arrows indicate partial or interaction-based effects. Results demonstrate that mission-aligned ethical guidance and pedagogical integration of AI significantly shape student trust, learning outcomes, employability perceptions, and spiritual engagement.

Figure 3. Empirical results summary model of faith integration and AI Adoption

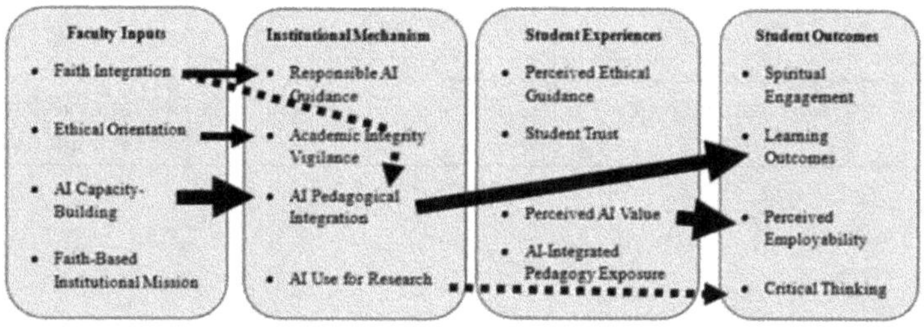

Faculty Results (Interpretation and Discussion)

H1: Faith Integration and Responsible AI Guidance (Faculty)

Pearson correlation results show that faith integration is strongly associated with responsible AI guidance ($r = .58$, $p < .001$). This large effect indicates that faculty who integrate faith more intentionally are more likely to guide students in ethically grounded AI use, suggesting that faith functions as a governance resource rather than a constraint. Qualitative responses align with this pattern, emphasizing stewardship, responsibility, and moral boundaries, suggesting that faith functions as an enabling governance resource rather than a barrier to innovation.

H2: Ethical Orientation and AI-Related Academic Integrity Vigilance (Faculty)

Linear regression results indicate that ethical considerations significantly predict integrity vigilance ($\beta = .57$, $p < .001$), explaining 33% of the variance ($R^2 = .33$). This means that approximately one-third of variation in faculty vigilance regarding AI misuse is explained by ethical awareness, reflecting a heightened sense of professional responsibility rather than resistance to innovation.

H3: AI Capacity-Building and Pedagogical Integration (Faculty)

The strongest predictor of pedagogical integration is AI capacity-building ($\beta = .72$, $p < .001$). This very strong effect highlights that while values provide direction, skills and competence enable implementation, reinforcing the importance of faculty development. Faculty report that AI supports lesson preparation, conceptual clarity, and feedback workflows, demonstrating that value alignment is necessary but not sufficient without practical skills.

H6: Faculty perceived research productivity value of AI

Faculty report strong agreement that AI enhances research productivity ($M = 4.32$, $SD = 1.00$; $n = 37$). However, this reflects perceived efficiency gains rather than measured output increases, and faculty emphasize verification and authorship responsibility.

Faculty describe AI as supporting idea development, drafting, and workflow optimization, while emphasizing verification, authorship accountability, and responsibility due to risks of inaccurate outputs or hallucinated references. The finding is descriptive and reflects perceived efficiency gains rather than verified output increases.

H9: Faith × Ethics interaction and AI pedagogical adoption (Faculty)

Faith integration significantly predicts AI adoption ($\beta = .48$, $p < .01$), while ethical considerations alone are not significant ($\beta = .21$, ns). This suggests that ethical awareness does not necessarily increase adoption but shapes how AI is adopted. The interaction between faith and ethics is negative ($\beta = -.30$, $p \approx .05$, $R^2 = .28$), which indicates a discernment effect, where higher levels of both faith integration and ethical vigilance lead to more selective and carefully governed AI adoption. Qualitative evidence supports this guardrail interpretation: faculty acknowledge pedagogical benefits alongside integrity risks, resulting in discernment-driven adoption.

Student Results (Interpretation and Discussion)

H4: Ethical Guidance and Trust in AI

Ethical guidance is not significantly associated with trust ($r = .22$, $p = .161$). This aligns with broader evidence that students require not only rules but also consistent institutional practices and clarity of expectations to develop trust in AI systems (Gallup, Inc., & Lumina Foundation, 2026). Qualitative data suggest that clarity may reduce anxiety but does not automatically produce trust; trust appears to depend on institutional coherence, transparency, consistency of enforcement, and perceived reliability of AI outputs.

H7: AI Value Addition and Perceived Employability

AI value addition strongly predicts employability ($\beta = .78$, $p < .001$; $R^2 = .61$). This very large effect suggests that students view AI competence as central to career readiness, productivity, and digital literacy, aligning education with labour market expectations.

H10: AI-Integrated Pedagogy and Learning Outcomes

AI-integrated pedagogy strongly predicts learning outcomes ($\beta \approx .78$, $p < .001$; $R^2 \approx .60$). This demonstrates that intentional instructional design, rather than mere exposure to AI tools, drives meaningful learning improvements.

Conceptual scaffolding, as reported by participants, refers to structured cognitive support that helps clarify and organize ideas while preserving independent judgment and scholarly responsibility.

H5: AI Use and Critical Thinking

AI use predicts critical thinking ($\beta \approx .29$, $p < .01$; $R^2 \approx .17$). This moderate effect indicates that AI can support reasoning when used reflectively but may undermine thinking when over-relied upon. Qualitative accounts support a conditional effect: AI can scaffold analysis and reasoning when used reflectively and ethically, but over-reliance may undermine cognitive effort.

FACULTY QUALITATIVE FINDINGS (BENEFITS AND CHALLENGES)

Faculty Benefits

Theme F1: Pedagogical Efficiency and Time Optimization

Examples:

- "Reduction of workload… work… done within the shortest period of time… increases efficiency."
- "Save time during research, delivery and revision."
- "Quick preparation of teaching materials."

Meaning:

Faculty consistently experience AI as a workload-reducing assistant that accelerates routine academic tasks such as lesson preparation, grading preparation, and research drafting. This reflects a broader shift in academic work organization, where repetitive and time-intensive functions are delegated to AI systems. Importantly, the value of this efficiency lies not only in time savings but in the reallocation of effort toward higher-order teaching functions, including mentoring, student engagement, and reflective pedagogy. AI thus enhances instructional readiness by freeing cognitive and temporal resources. This explains why faculty with higher AI capacity are more likely to integrate AI into teaching (H3), as the perceived benefits outweigh adoption costs. Furthermore, the time saved enables more thoughtful lesson planning and feedback design, indirectly improving student learning outcomes when AI integration is intentional (H10).

Theme F2: Instructional Design and Lesson Preparation Support

Examples:

- "Provide insights to prepare lesson content…"
- "Better and quality delivery of lessons."
- "Preparing for lessons… made easy."

Meaning:

Beyond efficiency, faculty use AI to enhance lesson structure, sequencing, and delivery quality. AI functions as an instructional design aide, supporting the generation of examples, explanations, and organized content. This indicates a transition from AI as a productivity tool to AI as a pedagogical partner. The meaning of this theme highlights that effective AI use depends on faculty capacity to translate technological support into meaningful teaching practice. This supports H3 by demonstrating that capacity-building enables deeper pedagogical enactment rather than superficial tool use. It also reinforces H9, as faculty adoption is framed in terms of improving mission-aligned teaching rather than technological novelty.

Theme F3: Cognitive Scaffolding and Conceptual Clarity

Examples:

- "Better understanding of some complicated topics."
- "Simplification… analysis, explanation of concepts."
- "Helps… understand complex concepts…"

Meaning:

Faculty use AI as a cognitive partner to simplify complex ideas and generate clearer explanations. This reflects AI's role in scaffolding knowledge, enabling both instructors and students to navigate difficult concepts more effectively. The deeper implication is that AI supports knowledge translation and conceptual clarity, which are critical for effective teaching. This contributes to improved student comprehension and supports mechanisms underlying H10. Additionally, when AI is used interactively rather than passively, it can stimulate analytical thinking and dialogue, indirectly supporting critical thinking development (H5).

Theme F4: Assessment and Feedback Support

Examples:

- "Guiding students… research supervision… giving feedback."
- "Developing a marking key… within seconds… check if they are correct."
- "Assessment… feedback… allowing teachers to focus more on mentorship…"

Meaning:

AI supports assessment processes by assisting in rubric development, feedback drafting, and supervision guidance. However, faculty emphasize the necessity of human verification and judgment. This indicates a hybrid model where AI enhances efficiency while human oversight maintains academic standards. The meaning of this theme lies in the transformation of assessment workflows, where AI reduces administrative burden and allows faculty to focus on mentorship and qualitative feedback. This supports H3, as capacity-building enables redesign of assessment practices. It also aligns with H2, as the emphasis on verification reflects active ethical vigilance.

Theme F5: Research Workflow Support (Perceived Productivity)

Examples:

- "Reduction in time spent on research."
- "Preparing for lessons and research…"
- "Helps… research and innovation…"

Meaning:

Faculty perceive AI as enhancing research efficiency, particularly in structuring ideas, drafting content, and organizing workflows. However, this perceived productivity is bounded by ethical considerations related to authorship and accuracy. The meaning of this theme highlights the distinction between perceived and actual productivity, where AI is seen as a supportive tool rather than a replacement for scholarly responsibility. This supports H6 and connects with H2 and H9, as ethical awareness shapes responsible use.

Faculty Challenges

Theme F6: Student Over-Reliance and Reduced Effort

Examples:

- "Students… paste the question… submit… without… trying to understand."
- "Compromises critical thinking…"
- "Students may not read enough."

Meaning:

Faculty express concern that AI may reduce student effort and engagement, leading to superficial learning. This reflects a displacement effect where AI substitutes rather than supports cognitive processes. The meaning of this theme emphasizes the need for guided AI use to maintain intellectual rigor. It supports H2 by explaining increased integrity vigilance and reinforces H1 and H9, where faith-informed values encourage responsible engagement.

Theme F7: Academic Integrity Risk and Enforcement Difficulty

Examples:

- "Caught… using AI in a copy and paste way."
- "AI detectors are also AI… challenge to trust results."

- "Do well in assignments… do badly in monitored exams…"

Meaning:

Faculty identify challenges in detecting AI misuse and note discrepancies between AI-assisted work and actual competence. This reflects a credibility gap in assessment systems. The meaning of this theme underscores the limitations of technological enforcement and the importance of ethical guidance. It supports H2 and strengthens reliance on values-based approaches.

Theme F8: Policy Gaps and Inconsistent Institutional Guidance

Examples:

- "Policy guiding the usage of AI is not available…"
- "Standardization of use."
- "Use… authorized recently…"

Meaning:

Faculty experience uncertainty due to unclear or evolving policies. This governance lag creates inconsistent practices and confusion. The meaning highlights the importance of institutional coherence in guiding AI use and supports H1, H2, and explains student-level findings (H4).

Theme F9: Accuracy, Hallucinations, and Reliability Issues

Examples:

- "Not always accurate…"
- "Fake references…"
- "Need verification…"

Meaning:

Faculty recognize AI as fallible and emphasize the need for verification. This shapes cautious use and reinforces the importance of critical evaluation. It supports H6 and influences trust dynamics.

Theme F10: Capacity and Access Constraints

Examples:

- "Not enough training… to confidently use AI."

- "Advanced ones require payment."
- "Loss of network."

Meaning:

Structural barriers such as limited training and access affect adoption. This highlights inequities in AI integration and strongly supports H3.

STUDENT QUALITATIVE FINDINGS (BENEFITS AND CHALLENGES)

Student Benefits

Theme S1: Learning Support and Understanding

Examples:

- "Helped… understand more."
- "Provided deeper understanding on difficult topics."
- "Makes understanding very easy."

Meaning:

Students experience AI as a personalized tutor that enhances comprehension and supports learning, reinforcing H10 and H5.

Theme S2: Assignment Support and Productivity

Examples:

- "Helps me… complete assignments…"
- "Makes work easier…"
- "Helps… plan my work…"

Meaning:

AI improves efficiency in completing tasks and managing workload, supporting H7 and H10.

Theme S3: Study Efficiency

Examples:

- "Summarize bulky notes."
- "Study plan…"
- "Slides/notes… easier…"

Meaning:

AI enhances organization and study processes, supporting effective learning.

Theme S4: Writing and Communication Support

Examples:

- "Follow APA guidelines…"
- "Improve grammar/flow…"
- "Refine my work…"

Meaning:

AI supports writing quality but raises concerns about authenticity and integrity.

Theme S5: Research and Information Discovery

Examples:

- "Pointed me to… websites… used them to verify."
- "Helps… research…"
- "Sources…"

Meaning:

Students use AI as a starting point but recognize the need for verification.

Student Challenges

Theme S6: Over-Reliance and Reduced Critical Thinking

Examples:

- "Too comfortable… doesn't encourage me to think."
- "Hinders critical thinking."
- "Makes one lazy…"

Meaning:

Confirms the risk of reduced cognitive engagement.

Theme S7: Accuracy Problems

Examples:

- "Sometimes incorrect…"
- "Not fully accurate…"
- "Need to check thoroughly…"

Meaning:

Reinforces the need for verification and critical evaluation.

Theme S8: Citation and Credibility Issues

Examples:

- "Does not provide good citations…"
- "Sources are not credible…"
- "Fake references…"

Meaning

Undermines academic reliability and reinforces ethical concerns.

Theme S9: Unclear Expectations and Anxiety

Examples:

- "Doesn't match what the instructor expects…"
- "Not sure what is allowed…"
- "May be highlighted as not mine…"

Meaning:

Reflects ambiguity and lack of consistent guidance.

Theme S10: Access Barriers

Examples:

- "Network issues."

- "Need to pay for better tools."

Meaning:

Highlights structural inequalities affecting AI use.

INTEGRATED DISCUSSION (FACULTY + STUDENTS)

Convergence

1. AI is useful but conditional (verification and boundaries).
2. Over-reliance is the central formative risk.
3. Policy inconsistency undermines confidence and trust.
4. Citation reliability is a shared concern.

Divergence

- Faculty focus: assessment credibility, detection limitations, integrity enforcement burden.
- Student focus: learning support, writing help, anxiety about lecturer expectations.

CONCLUSION

AI adoption in a faith-based university is best understood as a values-governed pedagogical practice rather than a purely technical innovation. Benefits depend on ethical clarity, verification practices, and consistent institutional guidance, while over-reliance threatens critical thinking and integrity. Faith-based values provide a framework for discernment, stewardship, and mission-aligned governance that shapes how AI is integrated into teaching, learning, and research.

Flourishing and Institutional Culture

The findings suggest that AI adoption is not only a technical or pedagogical issue but also a cultural and formative process. A flourishing AI-enabled institution is characterized by:

- Clarity and consistency in ethical expectations
- Strong faculty capacity and shared frameworks

- Student trust and psychological safety
- Alignment between technology use and institutional mission

Such environments enable students to develop:

- Ethical reasoning
- A sense of belonging
- Meaningful engagement with technology
- Alignment between learning and purpose

RECOMMENDATIONS

Policy Recommendations

Develop comprehensive AI governance frameworks that are explicitly aligned with institutional values to ensure consistency, accountability, and ethical coherence across all academic programs. Such frameworks should go beyond procedural rules and instead embed guiding principles such as human dignity, stewardship, and human flourishing as foundational pillars of AI use. This requires institutions to articulate clear policy positions on acceptable and unacceptable uses of AI in teaching, learning, research, and assessment.

Policies should also emphasize pedagogical intentionality by requiring that AI integration be purpose-driven rather than convenience-driven. This includes embedding AI into structured learning processes that promote reflection, verification, and critical thinking, rather than passive consumption. Furthermore, governance frameworks must address issues of academic integrity, data privacy, transparency, and accountability, ensuring that AI tools are used responsibly and ethically.

Importantly, institutions should integrate faith-based ethical principles into AI policy by explicitly linking technology use to stewardship, moral accountability, and responsible innovation. This alignment ensures that AI adoption supports not only academic efficiency but also institutional mission and identity. Regular policy review mechanisms should be established to keep pace with the rapidly evolving AI landscape, ensuring that governance remains responsive, relevant, and forward-looking.

Practice Recommendations

At the level of teaching and learning practice, institutions should prioritize pedagogical intentionality over mere tool adoption. This requires a deliberate shift from viewing AI as a productivity enhancer to positioning it as a pedagogical resource that supports deeper learning. Assessments should be redesigned to encourage analysis, verification, and reflective engagement with AI-generated outputs. For example, tasks require students to critique AI responses, compare AI-generated content with scholarly sources, or reflect on how AI influenced their learning process.

AI should be embedded into structured learning activities that promote higher-order thinking rather than superficial completion of tasks. This includes integrating AI into problem-solving exercises, case-based learning, and collaborative projects where students must engage critically with AI outputs. Such approaches ensure that AI enhances cognitive development rather than replacing it.

Additionally, institutions should standardize communication across courses by developing clear, course-level AI use statements that outline expectations for responsible use. Providing model disclosure templates can help students acknowledge AI assistance transparently, thereby promoting academic honesty. Aligning expectations across lecturers is critical to reducing confusion, anxiety, and perceived unfairness among students.

Consistency in communication also strengthens trust, as students are more likely to engage responsibly with AI when institutional expectations are clear, coherent, and consistently applied. Ultimately, effective practice requires a balance between flexibility and structure, allowing innovation while maintaining academic standards and integrity.

Faculty Development Recommendations

Expanding faculty capacity-building is essential for the effective and responsible integration of AI in higher education. Institutions should provide hands-on, practical training programs that focus on AI pedagogy, enabling faculty to understand not only how AI tools function but also how they can be meaningfully integrated into teaching and assessment. These programs should be tailored to disciplinary contexts, recognizing that AI applications vary significantly across fields.

Training should also include modules on ethical oversight, academic integrity risks, and strategies for verifying AI outputs. Faculty need to be equipped with the skills to identify inaccuracies, detect hallucinated content, and evaluate the credibility of AI-generated information. This is particularly important given the limitations of AI detection tools and the increasing sophistication of AI-generated outputs.

Beyond technical skills, institutions should support ethical leadership by fostering collaborative approaches to AI governance. Establishing faculty AI ethics working groups can provide platforms for sharing experiences, discussing challenges, and developing collective strategies for responsible AI use. Shared integrity protocols and case-based scenario discussions can help standardize practices and reduce the burden on individual lecturers.

Such initiatives also contribute to building a community of practice where faculty can learn from one another and develop confidence in navigating the complexities of AI integration. Ultimately, faculty development should be continuous and adaptive, reflecting the evolving nature of AI technologies and their implications for teaching and learning.

Student Development Recommendations

Integrating AI literacy into the curriculum is critical for preparing students to engage responsibly and effectively with AI technologies. This includes equipping students with skills in responsible prompting, verification of AI outputs, proper citation practices, and ethical reflection. AI literacy should not be treated as an optional add-on but as a core component of digital competence in higher education.

In faith-based institutions, AI literacy should also incorporate a values-driven perspective, emphasizing principles such as integrity, accountability, stewardship, and respect for human dignity. This ensures that students understand not only how to use AI but also why responsible use matters within a broader ethical and spiritual framework.

Institutions should also encourage reflective AI use through structured learning activities such as reflection journals, ethical case studies, and AI-assisted critique tasks. These approaches help students develop metacognitive awareness of how AI influences their thinking and learning processes. By engaging critically with AI, students can use it as a tool for enhancing understanding rather than as a shortcut for completing tasks.

Furthermore, institutions should address issues of access and equity by ensuring that all students have adequate access to AI tools and the necessary infrastructure to use them effectively. This includes providing institutional licenses, improving connectivity, and offering support for students who may face barriers to access.

Ultimately, student development efforts should aim to position AI as a catalyst for learning, critical thinking, and holistic formation. By fostering both technical competence and ethical awareness, institutions can prepare students to navigate AI-enabled environments responsibly and meaningfully, contributing to their academic success and long-term professional readiness.

FUTURE RESEARCH

- Longitudinal research to examine how AI integration evolves over time, whether faith integration continues to predict responsible adoption, and changes in integrity incidents following policy implementation.

- Experimental and intervention studies testing AI pedagogy training interventions, ethics workshops, and assessment redesign strategies.
- Multi-institution comparative studies between faith-based and secular universities, institutions with stronger versus weaker AI governance, and different denominational contexts.
- Improved measurement of critical thinking and spiritual formation through validated multi-item scales, objective measures, and behavioral integrity indicators.
- Research on trust formation mechanisms, including institutional transparency, perceived fairness, policy consistency, and data governance credibility.

LIMITATIONS

Some constructs show low reliability, and single-item measures were used. Therefore, findings should be interpreted as directional and practical indicators rather than precise latent measurements, and the cross-sectional design limits causal inference.

CONCLUSION

This chapter demonstrates that AI adoption in faith-based universities is not value-neutral. When aligned with faith-informed ethics, capacity-building, and pedagogical intentionality, AI becomes a tool not only for efficiency but for supporting human flourishing, dignity, and meaningful participation in an AI-driven world. AI and faith are therefore complementary, enabling responsible, trusted, and transformative educational environments.

REFERENCES

Aiken, L. S., & West, S. G. (1991). Multiple regression: Testing and interpreting interactions. *Sage (Atlanta, Ga.).*

Astin, A. W., & Lindholm, J. A. (2011). Assessing students' spiritual and religious qualities. *Journal of College Student Development*, *52*(1), 39–61. DOI: 10.1353/csd.2011.0009

Bretag, T., Harper, R., Burton, M., Ellis, C., Newton, P., van Haeringen, K., Saddiqui, S., & Rozenberg, P. (2019). Contract cheating: A survey of Australian university students. *Studies in Higher Education*, *44*(11), 1837–1856. DOI: 10.1080/03075079.2018.1462788

Chaudhry, M. A., & Kazim, E. (2022). Artificial intelligence in education (AIEd): A high-level academic and industry note 2021. *AI and Ethics*, *2*(1), 157–165. DOI: 10.1007/s43681-021-00074-z PMID: 34790953

Cohen, J. (1988). *Statistical power analysis for the behavioral sciences* (2nd ed.). Lawrence Erlbaum.

Collin, S., Lepage, A., & Nebel, L. (2023). Ethical and critical issues of artificial intelligence in education: A systematic review of the literature. *Canadian Journal of Learning and Technology*, *49*(1). Advance online publication. DOI: 10.21432/cjlt28448

Corpuz, J. C. G. (2025). Faith and artificial intelligence (AI) in Catholic education: Ethical, pedagogical, and theological engagement. *Religions*, *16*(8), 1083. Advance online publication. DOI: 10.3390/rel16081083

Creswell, J. W., & Creswell, J. D. (2018). *Research design: Qualitative, quantitative, and mixed methods approaches* (5th ed.). Sage.

de Kock, J. (2024). Spirituality and theological formation: Seven critical approaches to undergraduate theological education. *Journal of Spiritual Formation & Soul Care*. Advance online publication. DOI: 10.1177/19397909241234294

Dockery, A. M. (2016). Education and happiness in the faith-based context. *Journal of Education & Christian Belief*, *20*(1), 5–20.

Field, A. (2018). *Discovering statistics using IBM SPSS statistics* (5th ed.). Sage.

Floridi, L., Cowls, J., Beltrametti, M., Chatila, R., Chazerand, P., Dignum, V., Luetge, C., Madelin, R., Pagallo, U., Rossi, F., Schafer, B., Valcke, P., & Vayena, E. (2018). AI4People—An ethical framework for a good AI society: Opportunities, risks, principles, and recommendations. *Minds and Machines*, *28*(4), 689–707. DOI: 10.1007/s11023-018-9482-5 PMID: 30930541

Gallup, Inc., & Lumina Foundation. (2026). *AI in higher education: Widespread use, unclear rules* (Lumina–Gallup State of Higher Education report). Gallup, Inc. https://www.gallup.com

Holmes, W., Bialik, M., & Fadel, C. (2019). *Artificial intelligence in education: Promises and implications for teaching and learning*. Center for Curriculum Redesign.

Nunnally, J. C., & Bernstein, I. H. (1994). *Psychometric theory* (3rd ed.). McGraw-Hill.

Perkins, M., Roe, J., Postma, D., McLaughlin, P., & Fishman, T. (2023). Detection of GPT-4 generated text in higher education: Combining academic judgement and software to identify generative AI tool misuse. *Journal of Academic Ethics*, *22*(1), 89–113. DOI: 10.1007/s10805-023-09492-6

Reina Marín, Y., Cruz Caro, O., Maicelo Rubio, Y. C., Alva Tuesta, J. N., Sánchez Bardales, E., Carrasco Rituay, A. M., & Chávez Santos, R. (2025). Ethical challenges associated with the use of artificial intelligence in university education. *Journal of Academic Ethics*, *23*(4), 2443–2467. DOI: 10.1007/s10805-025-09660-w

Selwyn, N. (2019). What's the problem with learning analytics? *Journal of Learning Analytics*, *6*(3), 11–19. DOI: 10.18608/jla.2019.63.3

Yan, L., Martinez-Maldonado, R., & Gašević, D. (2023). Generative artificial intelligence in learning analytics: Contextualising opportunities and challenges through the learning analytics cycle. Learning Analytics Review. *Journal of Learning Analytics*.

Zawacki-Richter, O., Marín, V. I., Bond, M., & Gouverneur, F. (2019). Systematic review of research on artificial intelligence applications in higher education. *International Journal of Educational Technology in Higher Education*, *16*(1), 39. Advance online publication. DOI: 10.1186/s41239-019-0171-0

KEY TERMS AND DEFINITIONS

Academic Integrity: Adherence to ethical principles of honesty, originality, and accountability in learning, assessment, and research, including responsible disclosure and verification when AI tools are used.

AI Capacity-Building: Faculty development activities and experiences that build knowledge, skills, and confidence to evaluate and use AI tools effectively and ethically in teaching, assessment, and scholarly work.

AI-Integrated Pedagogy: Instructional design and teaching practices that intentionally incorporate AI tools to support learning activities, feedback, assessment, and higher-order thinking in pedagogically meaningful ways.

Conceptual Scaffolding: Structured cognitive support that helps learners and educators clarify, organize, analyze, and refine ideas while preserving independent judgment and responsibility for accuracy.

Ethical Guidance: Clear, consistent, and value-informed communication and instructional support that helps students understand acceptable AI use, integrity expectations, and responsible practices.

Ethical Orientation: A faculty member's awareness and application of moral principles (e.g., fairness, accountability, integrity) when making decisions about AI use in teaching, assessment, and research.

Faith Integration: Intentional incorporation of Christian values and theological commitments (e.g., stewardship, moral accountability, dignity) into teaching, learning, and technology-related decisions.

Spiritual Formation: Processes and experiences through which learners develop faith-informed identity, moral reflection, and alignment with Christian values within educational contexts.

Student Trust: Students' confidence that AI tools and institutional AI practices are reliable, fair, transparent, and consistently governed within the university environment.

Values-Based Governance: Institutional policies and decision structures that align technology adoption and use with ethical commitments, mission priorities, and accountability mechanisms.

Chapter 3
The Moral Frontier of the AI Age:
Why Christian Principles Offer a Coherent Ethical Foundation

Justin Joy
http://orcid.org/0000-0002-6482-1057
Christ (Deemed to be) University, India

Julie George
Christ (Deemed to be) University, India

ABSTRACT

The rapid deployment of artificial intelligence across economic, political, and social domains has generated ethical questions that secular frameworks are ill-equipped to resolve. This chapter contends that Christian moral theology provides the most coherent and durable foundation for navigating these challenges. Drawing on the scriptural doctrine of the Imago Dei, this study argues that every human person possesses inherent, non-negotiable dignity that resists algorithmic reduction. This demonstrates how documented patterns of political and ideological bias embedded in leading AI systems underscore the urgency of attending to the worldviews of developers, since values are inscribed in technology whether acknowledged or not. Christian virtues, including stewardship-based accountability, solidarity with the vulnerable, epistemic humility, fidelity to truth, and the primacy of love, are shown to translate directly into actionable ethical principles for AI governance. The chapter concludes with a constructive vision of technology development ordered toward authentic human flourishing.

DOI: 10.4018/979-8-3373-8192-3.ch003

ht © 2027, IGI Global Scientific Publishing. Copying or distributing in print or electronic forms without written permission of IGI Global Scientific Publishing is prohibited. Use of this chapter to generative artificial intelligence (AI) technologies is expressly prohibited. The publisher reserves all rights to license its use for generative AI training and machine learning model development.

INTRODUCTION

Artificial intelligence is no longer a feature of speculative futures. It governs credit decisions, shapes legal outcomes, mediates access to medical care, curates the information that citizens encounter, and increasingly influences the cultural and political environments in which communities form their values. The moral frameworks guiding these systems, however, have not kept pace with their deployment. Where technical capability races ahead, ethical deliberation struggles to catch up, and societies find themselves managing the consequences of decisions that were never adequately examined.

This chapter begins from a conviction that is both contested and consequential: the ethics of artificial intelligence cannot be treated as a neutral, worldview-independent domain. Every AI system encodes assumptions about what counts as a desirable outcome, whose interests matter, what trade-offs are acceptable, and—at the deepest level—what human beings fundamentally are. These are not technical questions. They are philosophical and, at their root, theological ones. The worldview of those who design, train, and deploy AI systems is not a peripheral variable; it is constitutive. It shapes the systems themselves, and through those systems, the lives of millions.

This reality has received striking empirical confirmation. Research by Rozado (2023) administering fifteen standardized political orientation instruments to ChatGPT found that fourteen of those instruments identified the system's responses as exhibiting a consistent preference for left-leaning viewpoints—even as the system itself claimed political neutrality when asked directly. This gap between professed impartiality and demonstrated bias is not simply a technical error. It is a symptom of a deeper structural reality: AI systems inevitably reflect the values of those who build them, and those values cannot be masked by claims of objectivity.

Christian moral theology is particularly well positioned to address these challenges. This is not because other traditions lack ethical insight—secular humanism, virtue ethics, and rights-based frameworks all offer serious and often overlapping moral resources—but because the Christian account grounds human dignity in a way that is less contingent, more comprehensive, and more resistant to erosion under technological pressure. Its theological anthropology offers a stable basis for evaluating systems that increasingly treat human beings as data objects, and its moral vision provides durable guidance for aligning technological power with the good of persons.

This chapter develops that argument across four main sections: the competing anthropologies that underlie AI ethics debates; the theological doctrine of the Imago Dei and its implications for human dignity; the translation of specific Christian moral virtues into concrete ethical principles for AI; and a constructive vision of technology development oriented toward genuine human flourishing.

WORLDVIEW AS THE HIDDEN ARCHITECTURE OF AI SYSTEMS

The Myth of Value-Neutral Technology

One of the most persistent and damaging assumptions in AI development culture is that technological systems can be built without embedding values. This assumption draws on a picture of engineering as a discipline of pure problem-solving: given well-specified goals and sufficient data, a system can be designed that achieves those goals efficiently and objectively. The role of the engineer, on this picture, is to translate objective requirements into working code, not to adjudicate between competing visions of the good.

This picture is false, and its falseness has material consequences. Every choice in AI development—what data to include or exclude, what objective function to optimize, what behaviors to permit or prohibit, how to weight competing values when they conflict—reflects prior moral commitments. These commitments do not evaporate when they go unacknowledged; they simply become invisible, which makes them harder to examine and easier to entrench. The worldview operating in the background of any AI system is not made less real by being implicit. It is made more powerful.

Across the literature, there is strong agreement that: AI systems inevitably encode human values and assumptions, not neutrality (Floridi, 2018; Crawford, 2021; Russell, 2019). Empirical studies show systematic bias across domains: facial recognition (Buolamwini & Gebru, 2018), criminal justice (Angwin et al., 2016), search engines (Noble, 2018), hiring systems (Dastin, 2018).

Ethical frameworks differ at the level of anthropology—what a human being is (Vallor, 2016; O'Neil, 2016). Christian theological ethics contributes a non-contingent grounding of dignity, while secular frameworks often rely on procedural or functional accounts (O'Donovan, 1986; Moreland & Craig, 2003).

The bias documentation carried out by Rozado (2023) illustrates this with unusual clarity. When the same political orientation questionnaires used to classify human respondents were administered to ChatGPT, the system's answers clustered consistently on the left side of the political spectrum across fourteen of fifteen instruments. The single exception classified the system as centrist rather than left-leaning. Notably, when the researcher replicated the test in Spanish rather than English, the results were identical—suggesting that the ideological orientation was not an artifact of language but was embedded in the model's learned patterns of reasoning. Perhaps most revealing was the system's response when asked directly about its political orientation: it consistently claimed to be neutral and to be simply providing factual information—a claim that the test results contradicted. This is not a minor technical glitch. It is an illustration of how deeply values can be inscribed in a system while remaining invisible to the system itself.

The implications for any ethical framework applied to AI are significant. If value-neutrality is not achievable—and the evidence strongly suggests it is not—then the relevant question is not how to build AI systems without values but which values should guide their design and deployment, and whose authority should govern the answers to that question. Christian theology has a direct response: the answer is not to pretend that values can be engineered away, but to insist that they be made explicit, subjected to moral scrutiny, and ordered toward the genuine good of every person the system will affect.

The Stakes of Anthropological Disagreement

Beneath the surface of AI ethics debates lies a deeper and rarely explicit disagreement about what human beings are. The secular humanist tradition, which has shaped much of the culture of Silicon Valley technology development, tends toward a naturalistic account: persons are complex biological organisms whose remarkable capacities for reason, language, and social cooperation are products of evolutionary processes. This account has genuine moral commitments attached to it—a respect for individual autonomy, a concern for harm reduction, a commitment to rational deliberation. But it faces a foundational difficulty when pressed: if persons are ultimately biological systems distinguished from other organisms only by the sophistication of their cognitive architecture, then the boundary between treating persons with dignity and treating them as optimizable data systems becomes difficult to locate with precision.

Secular humanism has made indispensable contributions to modern moral and political life, including robust defenses of human rights, individual autonomy, and democratic accountability. These commitments overlap in important ways with Christian ethics and have shaped many contemporary AI governance frameworks. However, from a Christian perspective, the grounding of these commitments remains comparatively fragile. When dignity is rooted primarily in capacities such as rationality, autonomy, or social participation, it risks becoming contingent—unevenly applied in cases of disability, marginalization, or diminished agency.

The Christian claim that dignity is intrinsic rather than capacity-dependent is offered not as a dismissal of secular ethics, but as an attempt to secure more firmly the very moral intuitions that secular humanism rightly affirms.

Importantly, the relevance of this framework is not limited to explicitly Christian institutions. Claims about intrinsic human dignity, the limits of quantification, and the moral risks of unchecked power can be translated into publicly accessible ethical principles. In pluralistic governance contexts, these insights can inform shared norms—such as rights protections, anti-discrimination standards, and accountability mechanisms—without requiring formal agreement on theological premises. In this way, Christian anthropology contributes to public reasoning by offering a deeper explanatory foundation for widely held moral commitments.

THE IMAGO DEI AS THE THEOLOGICAL FOUNDATION OF AI ETHICS

Scriptural Grounding and Theological Significance

The doctrine of the Imago Dei—the teaching that human beings are made in the image and likeness of God—is introduced in Genesis 1:26-27 (New International Version) and constitutes one of the most theologically generative affirmations in the entire biblical canon. The text reads:

> *Then God said, 'Let us make mankind in our image, in our likeness, so that they may rule over the fish in the sea and the birds in the sky, over the livestock and all the wild animals, and over all the creatures that move along the ground.' So God created mankind in his own image, in the image of God he created them; male and female he created them.*

Several features of this passage carry direct ethical implications. The divine image is attributed to humanity as such—male and female together, without distinction of capacity, status, or circumstance. The image is universal in scope and unconditional in character. It is not earned by achievement or forfeited by failure. It belongs to the person by virtue of being human, and this belonging cannot be revoked by any human authority or institution.

The Imago Dei is not merely a creation-account motif. It is a thread woven through the full length of biblical theology. In Exodus, the encounter between Moses and God is described in relational terms so intimate that Moses's face bears visible evidence of the meeting (Exodus 33:11-23, New International Version). The New Testament identifies Jesus Christ as the definitive image of the invisible God (Colossians 1:15, New International Version), and those who follow him are described as being progressively conformed to that likeness (Romans 8:29, New International Version). The image, in this unfolding narrative, is not a static attribute but a dynamic vocation—a calling to reflect the character of a God who is just, truthful, compassionate, and creative.

For Christian anthropology, the implications extend in two directions at once. On one hand, every human person is endowed with capacities—for reason, creativity, moral agency, and love—that reflect their divine origin and that no technological system can replicate or supersede. These capacities ground the worth of persons in something more durable than social utility, cognitive performance, or economic productivity. On the other hand, the tradition is unsparing about human limitation. Bearing the image of God does not render human beings infallible. The Christian account of sin—the pervasive distortion of human faculties by self-interest, pride, and the will to dominate—generates an equally clear warning: human creativity, including the creativity that produces AI systems, is capable of producing both remarkable good and serious harm.

Implications for the Treatment of Persons in AI Systems

Translating the Imago Dei into concrete AI ethics yields several specific requirements. The first concerns irreducibility. If every person bears the image of God, then no person may be treated as fully captured by any data representation of them. Behavioral profiles, predictive risk scores, attention metrics, and demographic aggregates are instruments of partial description, not complete characterizations of persons. Systems that treat such representations as exhaustive—that make consequential decisions about people on the basis of statistical patterns applied to their group membership or behavioral history—commit a moral error that is not merely technical but theological: they deny the surplus of personhood that exceeds every quantification.

The second concerns the universality of dignity. The Imago Dei is not more present in some persons than others. It does not diminish with age, disability, poverty, or social marginality. This universality directly challenges AI systems that, whether through design choices or training data, produce outcomes that are less accurate, less fair, or less beneficial for persons who are already disadvantaged. The documented pattern of higher facial recognition error rates for individuals with darker skin tones (Buolamwini & Gebru, 2018) is, from a theological standpoint, not a performance gap to be optimized—it is a moral failure in the treatment of persons who bear the divine image.

The third concerns fallibility. Because human designers are themselves fallible, the systems they produce will carry their limitations and distortions. The Rozado (2023) findings on ChatGPT political bias are instructive here: the system's ideological orientation is almost certainly not the result of deliberate design choices but of the cumulative effect of training decisions made by people whose own assumptions went unexamined. The Christian tradition's frank account of human sinfulness—including the sinfulness of those who build powerful systems—does not lead to paralysis but to the institutional requirement of accountability: mechanisms for oversight, correction, and transparency that do not rely on the infallibility of any individual or team.

CHRISTIAN MORAL VIRTUES AS ETHICAL PRINCIPLES FOR AI DEVELOPMENT

Stewardship and Accountability

Christian ethics frames the exercise of power and the management of resources as stewardship—a concept that captures both the legitimate authority to act and the responsibility to account for how that authority is used. Those entrusted with extraordinary resources are answerable for their deployment. The parable of the talents (Matthew 25:14-30) is representative of a broader biblical principle: authority is always accountable authority, and the scale of one's stewardship determines the scale of one's responsibility.

Applied to AI development, this principle challenges the diffusion of moral responsibility that characterizes many current governance arrangements. When an automated lending system denies credit to thousands of applicants, or a content moderation algorithm systematically suppresses the speech of particular communities, or a predictive policing tool produces racially disparate outcomes, the common response from developers and deployers is that the system is simply processing data. The complexity of the system, on this account, insulates those who built it from accountability for its effects. The stewardship ethic refuses this insulation. The developers, investors, and deployers of AI systems are stewards of power that affects millions of lives, and the complexity of their systems does not diminish their responsibility—it magnifies it.

Concretely, the stewardship principle translates into requirements for explainability (AI decisions affecting individuals should be interpretable and subject to challenge), for proactive harm assessment (deployment should not proceed until potential harms have been rigorously evaluated), and for genuine institutional accountability when systems cause harm. It also requires that those responsible for AI development be honest about the values embedded in their systems.

Solidarity With the Vulnerable

The prophetic tradition of the Hebrew Bible and the teaching of Jesus converge on a single insistent theme: the authenticity of a community's moral life is measured not by its treatment of those who are powerful, but by its treatment of those who are not. The widow, the orphan, the stranger, the poor—these are the recurring figures of biblical social ethics, and their presence in the text functions as a standing rebuke to any moral framework that permits their interests to be discounted in favor of efficiency or majority preference.

This principle of solidarity with the vulnerable provides a stringent criterion for evaluating AI systems: not how they serve those who are already well-positioned, but how they affect those who are not. The empirical record raises serious concerns. Research by Buolamwini and Gebru (2018) documented significantly higher misclassification rates in commercial facial recognition systems for women and individuals with darker skin tones compared to lighter-skinned men—disparities with immediate practical consequences wherever these systems are used in law enforcement, hiring, or access control. Automated hiring tools trained on historical employment data have been found to reproduce the discriminatory patterns embedded in that history (Dastin, 2018). Risk assessment instruments used in criminal sentencing have produced racially disparate outputs that their designers did not intend but failed to prevent (Angwin et al., 2016).

From the standpoint of a solidarity ethic, these outcomes are not tolerable inefficiencies awaiting iterative improvement. They are moral failures that demand deliberate remediation before deployment at scale. A technology development culture genuinely shaped by solidarity would treat differential harm to marginalized communities as a disqualifying finding, not a post-deployment correction target. It would ensure that communities most likely to be affected by a system are meaningfully represented in decisions about its design—a requirement of participatory ethics that flows directly from the recognition that every person affected by a technology bears God's image.

Epistemic Humility

Christian moral theology identifies pride—the refusal to acknowledge one's limits—as the root of many other moral failures. Its counterpart, humility, is not self-deprecation but accuracy: a realistic assessment of one's knowledge, one's capacities, and the ways in which one's judgment is likely to be distorted by interest, bias, or incomplete information. The tradition applies this virtue to knowledge claims with particular force: human understanding, however impressive, remains partial, and the pretense of comprehensiveness is a form of intellectual dishonesty.

In the context of AI development, epistemic humility has both technical and institutional dimensions. Technically, it demands rigorous uncertainty quantification—honest reporting of the conditions under which a system performs reliably and the conditions under which it does not. Systems that produce confident outputs while operating in domains or demographic contexts underrepresented in their training data are not simply imprecise; they are epistemically dishonest, projecting a confidence that their actual knowledge does not warrant.

Institutionally, epistemic humility requires genuine openness to external critique and independent audit. The culture of much AI development has been characterized by resistance to outside scrutiny, a preference for proprietary opacity over transparency, and a tendency to treat academic or civil society criticism as a reputational threat rather than a source of valuable information. A development culture shaped by the Christian virtue of humility would recognize external critique as a structural necessity rather than an inconvenience—a safeguard against the self-deception to which all human persons and institutions are susceptible.

Fidelity to Truth

The Christian tradition's commitment to truth is not a contingent cultural preference but a theological axiom. God is characterized throughout Scripture as the source and guarantor of truth. Deuteronomy 32:4 describes God as one in whom there is no injustice. In the New Testament, the identification of Jesus as the way, the truth, and the life (John 14:6) makes truthfulness a constitutive feature of the divine character—and therefore a moral requirement for those who seek to reflect that character in their lives and institutions.

The AI age has made this commitment more urgent and more difficult simultaneously. Generative AI systems now produce text, images, and video that are effectively indistinguishable from human-created content—content that can be used to fabricate evidence, impersonate individuals, and manufacture false consensus on a scale that was previously impossible. Recommendation algorithms have been shown to optimize for engagement metrics that reliably elevate emotionally provocative content, which is often sensationalized or inaccurate (Vosoughi et al., 2018). The commercial incentives of attention-economy platforms systematically reward the spread of misinformation, and AI systems trained on engagement data learn these incentives with great efficiency.

A Christian commitment to truth translates, in this environment, into several concrete demands. AI-generated content should be clearly identified as such. Systems that amplify misinformation should be redesigned even at commercial cost. Developers should be transparent about the limitations, failure modes, and embedded values of their systems. And the broader technology culture should resist the normalization of treating truth as one value among many to be traded off against revenue or user engagement.

The Primacy of Love

The supreme norm of Christian ethics is love—not sentiment but the sustained, active orientation of the will toward the genuine good of the other. The Greek concept of agape, which runs through the New Testament's moral teaching, describes a love that is attentive to real needs rather than preferences, that accepts cost in the service of the other's good, and that extends without condition to include those who are strangers, enemies, or otherwise outside the circle of natural affection. This love is the criterion by which, in the tradition's account, all other moral claims are ultimately adjudicated.

As a criterion for evaluating AI systems, love asks a different set of questions than efficiency metrics or market performance. Does this system genuinely serve the people it affects, or does it primarily serve the interests of those who profit from its deployment? Does it strengthen the relational fabric of human communities, or does it substitute algorithmically mediated interaction for genuine human connection? Does it attend to persons as whole beings—embodied, relational, morally serious, and spiritually significant—or does it reduce them to the dimensions that are most easily measured and monetized? These questions cannot be resolved by technical means alone. They require a moral vision that sees human beings as made for love and community, not merely for consumption and optimization.

The love ethic also has procedural implications. If love demands genuine concern for the good of all those affected by a technology—including those who have no seat at the design table—then the development process itself must be structured to include their perspectives. The populations most likely to bear the costs of AI failure are typically those least represented in AI development: communities of color, economically marginalized populations, people with disabilities, and people in the Global South. A development process shaped by the love ethic would treat their inclusion not as a public relations strategy but as a moral necessity.

Toward a Christian Vision of Technology and Human Flourishing

The analysis of the preceding sections converges on a constructive vision: AI development ordered not toward the maximization of technical capability or commercial return, but toward authentic human flourishing. That vision requires, first, a richer account of flourishing than secular frameworks typically supply.

Human flourishing, in the Christian account, is not reducible to preference satisfaction or welfare maximization. It involves the development and exercise of the capacities that constitute persons as image-bearers—reason, creativity, moral agency, relational love—within communities structured by justice, truth, and mutual care. A technology that expands these possibilities serves flourishing; a technology that suppresses or circumvents them fails, regardless of its technical sophistication or its performance on standard benchmarks. The metric that matters is not engagement or efficiency but whether the technology makes it more or less possible for human beings to become what they are made to be.

This vision generates a specific set of priorities for AI governance. Accountability mechanisms must be structural rather than voluntary—built into the design, deployment, and oversight processes from the outset, not added in response to public pressure after harm has occurred. Transparency about the values embedded in AI systems must be a genuine requirement, not a rhetorical gesture. The documented tendency of systems like ChatGPT to present ideologically oriented outputs as neutral factual information (Rozado, 2023) represents precisely the kind of dishonesty that a Christian ethics of truth-telling demands be corrected.

The constructive vision also requires a different posture toward human creativity. The capacities that have produced modern AI systems are not self-generated achievements. They are, in the theological account, gifts—entrusted to those who possess them for the service of those who will be affected by their exercise. This framing does not diminish genuine human achievement; it situates it within a larger story in which the ultimate measure of accomplishment is not technical elegance or market dominance but faithful service to the people technology is meant to serve.

From Virtue to Leadership: Institutional Practices in AI Governance

Christian moral virtues can be translated into concrete AI governance practices. The translation from moral vision to institutional practice is the decisive step for any ethical framework. Without concrete mechanisms, even well-articulated virtues risk remaining aspirational. The question, therefore, is what Christian moral commitments would require in practice from leaders operating within contemporary AI institutions.

Recent scholarship on AI governance emphasizes that effective oversight depends on organizational structures, audit systems, and participatory processes, not only high-level principles (Mökander & Floridi, 2023; Raji et al., 2022; Shneiderman, 2020). A Christian framework does not replace these mechanisms but can shape how they are designed and when they are used.

Example 1: Independent Algorithmic Audit with Moral Thresholds

A Christian account of stewardship and human fallibility supports the development of independent, third-party audit systems for high-impact AI models. Existing research highlights the importance of external auditing ecosystems to ensure accountability beyond internal corporate incentives (Raji et al., 2022).

What distinguishes a Christian approach is not the audit itself, but the moral thresholds governing deployment decisions. In many current frameworks, systems are deployed if risks are "acceptable" or "manageable." A Christian ethic—particularly one shaped by solidarity and the protection of the vulnerable—would adopt a more stringent standard:

Systems should not be deployed in high-stakes domains (e.g., criminal justice, healthcare triage) unless harms are well understood and demonstrably mitigated. Audit findings that reveal systematic disadvantage to already vulnerable populations should trigger mandatory delay or withdrawal, not iterative post-deployment correction.

This moves governance from risk optimization to moral constraint, reflecting the claim that some harms to persons are not ethically tradeable.

Example 2: Participatory Oversight with Representation of the Vulnerable

A second concrete implication emerges from the principle of solidarity. AI governance literature increasingly emphasizes participatory design and oversight, in which affected communities are included in system evaluation (Ayling & Chapman, 2022; Shneiderman, 2020).

A Christian framework strengthens this requirement by making such inclusion normative rather than optional:

AI systems should be reviewed by standing oversight committees that include representatives from communities most likely to bear risk (e.g., marginalized racial groups, persons with disabilities, economically vulnerable populations).

These bodies should have real authority, including the power to halt deployment.

In healthcare AI, for example, emerging governance models already include algorithmic oversight boards within institutional structures, integrating audit logs, validation cycles, and equity review processes (Kumar et al., 2025). A Christian ethic would insist that such structures are not merely best practice, but a requirement grounded in the equal dignity of all persons.

Example 3: Truthfulness in Generative AI Policy

The virtue of truthfulness translates directly into information integrity policies for generative AI systems:

Mandatory labeling of AI-generated content

Clear disclosure of model limitations and uncertainty

Institutional refusal to optimize systems for engagement when doing so predictably amplifies misinformation

Comparative Note: Christian and Secular Approaches to Governance

The difference lies less in what tools are used than in how firmly they are anchored: While secular frameworks often justify such policies in terms of epistemic reliability or democratic stability, a Christian account grounds them more deeply in a moral commitment to truth as intrinsic rather than instrumental.

These examples also clarify the relationship between Christian and secular AI ethics. There is substantial overlap in recommended mechanisms—auditing, transparency, participatory design, and accountability are widely endorsed across frameworks (Floridi, 2018; Vallor, 2016).

Secular frameworks often justify governance mechanisms in terms of risk management, rights protection, or social utility.Christian ethics grounds them in a non-contingent account of human dignity, which can justify stronger constraints on deployment when trade-offs threaten the vulnerable.

This distinction does not invalidate secular approaches; rather, it highlights how theological anthropology can reinforce and stabilize shared ethical commitments, particularly in cases where economic or political pressures incentivize compromise.

Stewardship requires accountability, including rights to explanation and mechanisms for redress. Solidarity requires representation of vulnerable communities in system design and oversight. Humility requires transparent disclosure of system limitations. Truthfulness requires labeling of AI-generated content and resistance to misinformation. Love requires restraint, including refusing deployment in high-risk contexts where harms are not understood.

These principles can be implemented in pluralistic settings without requiring shared theological commitments, making them relevant to public ethical reasoning.

Finally, this vision is genuinely hopeful. Christian ethics does not approach technological development from a posture of anxiety or reflex resistance. It recognizes that human creativity, properly ordered, can be an extraordinary instrument of justice and care—that AI designed with genuine concern for human dignity, attentive to the needs of the vulnerable, committed to truth, and animated by love can extend those virtues in ways that no individual person could accomplish alone. The call is not to slow down but to go deeper: to build systems worthy of the persons they will serve, and to have the moral courage to refuse to deploy systems that are not.

CONCLUSION

The ethics of artificial intelligence cannot be separated from the question of what human beings are. Every system reflects an implicit anthropology—whether acknowledged or not—and that anthropology shapes how persons are treated within it.

This chapter has argued that the Christian account of the human person, grounded in the doctrine of the Imago Dei, provides a uniquely stable foundation for resisting reductive tendencies in AI systems. By locating dignity in the being of the person rather than in contingent capacities, it secures a moral boundary that technological optimization must not cross.

The findings of Rozado (2023) are best understood as one instance within a wider and well-documented pattern. Across domains, AI systems have been shown to encode the assumptions and priorities of their designers and training environments. Facial recognition systems exhibit significant demographic disparities (Buolamwini & Gebru, 2018). Risk assessment tools in criminal justice have produced racially skewed outcomes (Angwin et al., 2016). Search engines and recommendation systems reproduce and amplify social hierarchies embedded in their data (Noble, 2018). Automated hiring tools have replicated historical patterns of discrimination (Dastin, 2018). Taken together, these cases demonstrate that value-ladenness is not an anomaly but a structural feature of AI systems. The Christian tradition has the conceptual resources to name this failure, to analyze its roots, and to articulate the requirements of a better alternative.

The task before AI developers, policymakers, and institutions is therefore not only technical but moral: to orient these systems toward the good of every person they affect. That task requires frameworks capable of sustaining dignity under pressure. The Christian tradition offers one such framework—not as a replacement for all others, but as a resource capable of deepening and stabilizing our shared commitment to human flourishing in the age of intelligent machines.

REFERENCES

Angwin, J., Larson, J., Mattu, S., & Kirchner, L. (2016). *Machine bias*. ProPublica. https://www.propublica.org/article/machine-bias-risk-assessments-in-criminal-sentencing

Ayling, J., & Chapman, A. (2022). Putting AI ethics to work. *AI and Ethics*, *2*(3), 405–429. Advance online publication. DOI: 10.1007/s43681-021-00084-x

Biblica. (2011). *Holy Bible, New International Version*. https://www.biblegateway.com/

Buolamwini, J., & Gebru, T. (2018). Gender shades: Intersectional accuracy disparities. *Proceedings of Machine Learning Research, 81*, 1–15.

Crawford, K. (2021). *Atlas of AI*. Yale University Press.

Dastin, J. (2018). Amazon scraps AI recruiting tool. *Reuters*. https://www.reuters.com/article/us-amazon-com-jobs-automation-insight-idUSKCN1MK08G

Floridi, L. (Ed.). (2018). *The ethics of artificial intelligence*. Oxford University Press.

Kumar, R., Sporn, K., Waisberg, E., Ong, J., & Paladugu, P. (2025). Healthcare AI governance and oversight frameworks. *Health Care Analysis : HCA : Journal of Health Philosophy and Policy*. Advance online publication. DOI: 10.1007/s10728-025-00537-y

Mökander, J., & Floridi, L. (2023). Operationalising AI governance through ethics-based auditing: An industry case study. *AI and Ethics*, *3*(2), 451–468. Advance online publication. DOI: 10.1007/s43681-022-00171-7 PMID: 35669570

Moreland, J. P., & Craig, W. L. (2003). *Philosophical foundations for a Christian worldview*. InterVarsity Press.

Noble, S. U. (2018). *Algorithms of oppression*. NYU Press. DOI: 10.18574/nyu/9781479833641.001.0001

O'Donovan, O. (1986). *Resurrection and moral order*. Eerdmans.

O'Neil, C. (2016). *Weapons of math destruction*. Crown.

Raji, I. D., Xu, P., Honigsberg, C., & Ho, D. (2022). Outsider oversight: Designing a third-party audit ecosystem for AI governance. *Proceedings of the AAAI/ACM Conference on AI, Ethics, and Society*. DOI: 10.1145/3514094.3534181

Rozado, D. (2023). The political biases of ChatGPT. *Social Sciences (Basel, Switzerland), 12*(3), 148. DOI: 10.3390/socsci12030148

Russell, S. (2019). *Human compatible.* Viking.

Shneiderman, B. (2020). Bridging the gap between ethics and practice: Guidelines for human-centered AI systems. *ACM Transactions on Interactive Intelligent Systems, 10*(4), 1–31. Advance online publication. DOI: 10.1145/3419764

Vallor, S. (2016). *Technology and the virtues.* Oxford University Press. DOI: 10.1093/acprof:oso/9780190498511.001.0001

Vosoughi, S., Roy, D., & Aral, S. (2018). The spread of true and false news online. *Science, 359*(6380), 1146–1151. DOI: 10.1126/science.aap9559 PMID: 29590045

Westermann, C. (1994). [Fortress Press.]. *Genesis (New York, N.Y.),* •••, 1–11.

KEY TERMS AND DEFINITIONS

Algorithmic Bias: A systematic and repeatable error in the outputs of an AI system that produces unfair or disproportionate outcomes for particular groups, often arising from flawed training data, design choices, or optimization objectives that reflect the assumptions of the system's developers.

Epistemic Humility: The recognition that one's knowledge is partial, fallible, and subject to distortion by interest, bias, and limited perspective—a virtue that, in AI development contexts, requires honest reporting of system limitations, openness to external critique, and resistance to the pretense of comprehensiveness.

Human Flourishing: In Christian moral theology, the full and integrated development of the capacities constitutive of human personhood—reason, creativity, moral agency, and relational love—within communities ordered by justice, truth, and mutual care. Distinguished from preference satisfaction by its orientation toward objective goods rather than subjective desires.

Imago Dei: A Latin theological phrase meaning 'image of God,' drawn from Genesis 1:26-27, denoting the scriptural teaching that every human being is created in the image and likeness of God and therefore possesses inherent, non-negotiable dignity that cannot be forfeited, quantified, or superseded by any human authority or technological system.

Secular Humanism: A philosophical and ethical framework that affirms the dignity and moral worth of human beings on naturalistic rather than theological grounds, rejecting appeals to divine authority or transcendent moral order in favor of reason, science, and individual autonomy as the basis for ethical life.

Stewardship: A Christian ethical concept describing the responsible management of resources, capacities, or authority entrusted to a person for purposes that exceed mere personal benefit. In AI ethics contexts, stewardship establishes that those who build and deploy powerful technology are accountable for its effects on all those it touches.

Value Alignment: The technical and ethical challenge of ensuring that AI systems pursue objectives that are genuinely consistent with human values and beneficial outcomes—a challenge that is complicated by the difficulty of specifying human values with precision and the tendency of systems to pursue proxy objectives that diverge from the intended goal.

Worldview: A comprehensive framework of beliefs, assumptions, and values through which a person or community interprets reality, assigns moral weight, and makes decisions. In AI development, the worldview of developers is argued to be a constitutive rather than incidental factor, shaping system design in ways that may be invisible to developers themselves but consequential for those affected by the systems they build.

Chapter 4
Shaping Moral Identity in the Age of AI:
Bridging Values and Algorithms

Jipson Joseph
http://orcid.org/0009-0005-1795-5051
Christ (Deemed to be) University, India

ABSTRACT

Unprecedented scientific and technological advancements characterize the present generation. Artificial Intelligence (AI) is the most influential among these advanced technologies. It is the capacity of machines to imitate human intelligence and produce better outcomes. AI has extended its influence across almost all areas of human interaction, and the present global society cannot imagine its existence without it. However, AI also challenges and harms human flourishing. Shaping individuals' moral identity is a significant concern in the present AI era. Conventionally, moral formation was the responsibility of parents, teachers, and society, grounded in the principles and values evolved from each society's cultural capital. The role of these conventional actors is sidelined in the present globalized AI world. From this perspective, this chapter, using the qualitative methodology's critical analytical method, examines the potential of bridging values and algorithms in shaping individuals' moral identity.

INTRODUCTION

Artificial Intelligence (AI) is no longer a possibility; it is a reality with unimaginable benefits and undeniable drawbacks. It influences and conditions almost all areas of human interaction. Human choices and decisions across fields, such as

DOI: 10.4018/979-8-3373-8192-3.ch004

...ht © 2027, IGI Global Scientific Publishing. Copying or distributing in print or electronic forms without written permission of IGI Global Scientific Publishing is prohibited. Use of this chapter to ...generative artificial intelligence (AI) technologies is expressly prohibited. The publisher reserves all rights to license its use for generative AI training and machine learning model development.

education and healthcare, are increasingly dependent on AI. Its brilliance even surprises its makers. From their subordinate and supporting roles, machines have become decision-makers by mimicking human intelligence. AI, with its unique characteristics of velocity, productivity, and perfection, changes human *modus operandi* and presents itself as a suitable choice for the present generation's fast-moving tendencies (Sarker, 2022; Jiang et al., 2022). However, AI's increasing influence also raises significant ethical concerns regarding the formation of individuals, human dignity, values, and morality.

Morality and ethics are used interchangeably in common parlance. However, some ethicists affirm that while morality refers to an individual's conviction of rightness or wrongness of an action, ethics explains the teachings of a specific community, culture, or religion regarding the rightness or wrongness of an action. Morality comes from within, but ethics is imposed by external actors using their influential roles (Coffin & Egan-Wyer, 2022). Nevertheless, morality is not something we inherit from human nature, as is human dignity. An individual forms it by critically evaluating ethical values and principles actively present in various cultures, religions, and societies. In this way, morality and ethics are intimately connected. Conventionally, an individual's moral identity is formed through the active involvement of parents, teachers, society, religion, and community. These conventional actors have no specific roles in the present AI era, as AI is replacing human actors (Adella, 2025).

Problem Statement

Each society has its own perceptions of right and wrong, which are necessary for meaningful human existence within that society. The sum and substance of a society's ethical values, principles, approved behavior patterns, and social interaction modes constitute its cultural capital. Although most societies adhere to some commonly accepted values and principles, each society's ethical framework will have some unique characteristics that shape the moral identity of its members. Based on a society's ethical perceptions and cultural capital, each individual will develop their own moral perceptions, which may or may not align with those of the society. If an individual's morality aligns with society's ethical perspective, they are acclaimed as perfect. When they fail to follow society's ethical standards, they are considered unethical even if they believe that they are moral. Undoubtedly, an individual's moral identity was shaped by the community, including parents, teachers, friends, and others. However, in the present scenario, people spend more time with AI agents (e.g., Chatbots) and tools rather than with human actors. This also affects their moral identity formation. AI systems do not simply automate decisions but "shape epistemic boundaries, delegate judgment, and redistribute communicative agency" (Oguz, 2025). It shows that the horizon of AI is broader than that of human imagination.

Humans can no longer imagine a world without AI and its accessories. It is impossible to return to traditional societal settings and prioritize the ethical formation of future generations. However, AI has many drawbacks that may challenge human flourishing and moral upbringing. It may undermine the development of practical wisdom (Tsai & Ku, 2025). AI tools should not be biased in their activities. Many instances have occurred in which AI systems have failed to ensure fairness and justice (Oguz, 2025). When they fail to guarantee accountability, transparency, and privacy in their activities, they fail to meet the demands of justice. It will adversely affect the moral identity formation of the future generation. What we require is to design the present AI-influenced society in a way that shapes a respectful moral identity. AI tools, rather than introducing a new set of values, must reinforce existing values through their algorithmic presentations to ensure a progressive ethical framework (Lähteenkorva et al., 2025). From this perspective, shaping an individual's moral identity becomes a complex question. If AI algorithms and tools do not significantly replicate human values, ethical principles, and moral discernment, they will create chaos in society. The moral identity of the future generation will become problematic and destructive. This requires that the scientific community and policymakers focus on enhancing AI algorithms with an ethical framework that supports the moral identity of future generations.

Methodology and Research Design

This chapter, using qualitative methodology's tools of descriptive analysis (Kemp et al., 2018), narrative synthesis (Cruzes et al., 2015), and critical evaluation (Klecun & Cornford, 2005), examines the possibility of bridging values and algorithms to shape the moral identity in the age of AI. The chapter has four major sections. Firstly, it provides a theoretical framework for the chapter's main arguments. Secondly, the literature review advances the study's findings. Thirdly, it critically evaluates the findings based on 5 ethical theories, the moral autonomy principle, and 3 ethical values. Fourthly, it examines the possibility of bridging values and algorithms to shape individuals' moral identity.

THEORETICAL FRAMEWORK

This chapter revolves around three major themes: the interplay between morality and ethics, cultural capital, and AI. Although morality and ethics are often used interchangeably, they have diverse meanings (Coffin & Egan-Wyer, 2022). While morality, ethics, and cultural capital are closely interrelated, AI appears to have no connection to them. However, in the present circumstances, AI has a significant role in shaping individuals' moral identity and society's cultural capital. This advances a framework of interrelation among them.

Figure 1. Framework for moral identity formation in the age of AI

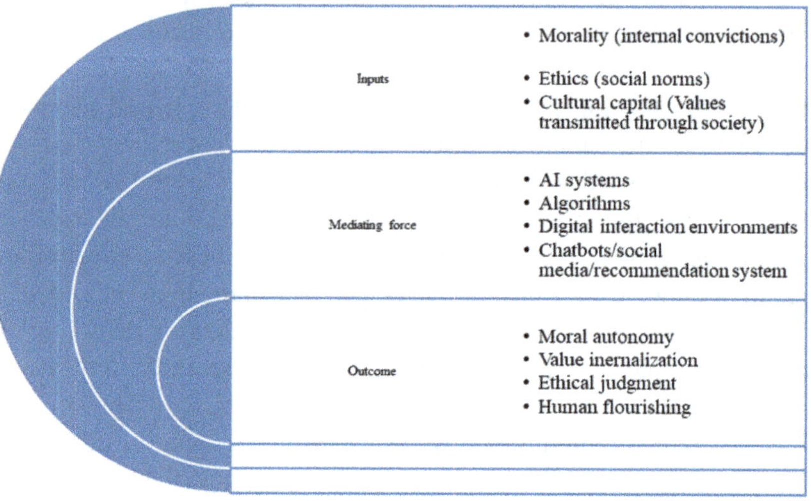

Morality and Ethics

Morality and ethics are vibrant terms used in almost all areas of human interaction. Ethical/moral considerations play a significant role in the discussions of any subject in the academic world. An individual's moral upbringing is a concern not only of the family but also of the entire society. Ethical concerns dominate from ordinary human interactions to national or international policymaking. International, national, and local policies and programs are often evaluated through an ethical framework. The crucial role of morality and ethics poses fundamental questions: whether they are the same; if different, what their unique characteristics are; how they are interrelated; and what importance they have in individual well-being and societal welfare.

Ethics and morality deal with values and principles. They revolve around human characteristics that identify them as good or bad, or right or wrong. Their subject matter is the same: human behaviour. However, while morality is a subjective evaluation, ethics calls for an organized system. Morality is an individual's subjective perception of right or wrong, which varies across cultures and over time. The intentions, decisions, and actions that are considered right/good (moral) by an individual might be wrong/bad (amoral) for another. When individual perceptions about right or wrong vary, it creates confusion in society. This necessitated an organized system of values and principles (ethics) that apply to all members of a society or a profession (Chaddha & Agrawal, 2023). The ethical framework is superior to an individual's moral perceptions as it is necessary for maintaining order in a society. Even if an individual's moral perceptions are different from the ethical framework, they are bound to follow it as members of that society.

Morality and ethics are philosophical topics as they examine individuals' ways of life within a society. Although they differ in understanding, they are intertwined. An individual's formation of moral identity cannot happen from nothing. Their family, society, culture, profession, and religion play significant roles, and each has its own ethical guidelines. It shows that ethical codes and guidelines can significantly influence morality. The interplay between morality and ethics is powerful, and separating them is challenging (Coffin & Egan-Wyer, 2022). Grayling (2019) argues that these two differ significantly. While ethics originates from the Greek 'ethos,' which means character, morality evolves from the Latin mos/moris (Cicero), which means custom. According to him, morality points to human actions and obligations, whereas ethics refers to 'what sort of person one is.' Coffin and Egan-Wyer (2022) claim that they can be distinguished by referring to ethical acts as those that communicate the actor's character and moral acts as those that underline the relationship between the actors and a wider obligation. However, a question arises: how can we determine what sort of persons we should become? Who defines the objective ethical standards? Subjective perception of moral values and principles alone cannot ensure societal order.

A sound ethical system that conditions moral values and principles is crucial for meaningful human existence. It has the potential to ensure an objective judgment of the morality of decisions and actions, teach core moral behavioral principles to new members, guide individuals' professional and personal behaviors, support them in making decisions on issues of moral ambiguity, and promote their behaviors to maintain societal order (Chaddha & Agrawal, 2023). From this perspective, it is more meaningful to distinguish morality and ethics as subjective and objective. Harmonious and dignified living becomes easier in a society where an ethical framework guides individuals' moral choices and decisions. The formation of moral identity will therefore be possible through the nexus between morality and ethics. Moral identity formation varies across cultural, professional, and other domains. Accordingly, there exist medical ethics (McLennan et al., 2022), teaching ethics (Gurr et al., 2025), legal ethics (Ferrara, 2022), environmental ethics (Wienhues, 2022), and AI ethics (Radanliev, 2025).

Cultural Capital

As noted above, ethics points to the relevance of universal moral norms and principles. However, many ethical frameworks exist within a society through the application of its acceptable values across different professions and ways of life. The sum of all these values and principles that give a particular society or culture its identity can be called cultural capital. The concept of cultural capital has evolved from Pierre Bourdieu's (1986) sociological reasoning. According to him, cultural capital differs from economic capital and social capital. For him, "cultural capital acts as a social relation within a system of exchange that includes the accumulated cultural knowledge that confers power and status" (Barker, 2004, p. 37). Cultural capital, which refers to knowledge, values, and traditions that a community has preserved for generations, is the bedrock for a community's sustainability. Preserving cultural capital is essential to individual well-being and societal progress (Talmage et al., 2022).

Bourdieu (1986) says that cultural capital can be acquired in three ways. Firstly, it is passed from one generation to another (parents to children). Secondly, people subconsciously acquire it through passive socialization. Even without their knowledge, they inculcate many social norms and values. Thirdly, cultural capital is acquired through formal education. A systematic implementation of elementary education is the best way to pass it to future generations. Cultural capital is the community framework through which one perceives society, social relationships, economic interactions, and civil and political engagements. Although cultural capital can take many forms, it should be understood as a particular culture's valuable assets (Reed & Johnson, 2023). Just as economic capital enables a firm to stabilize its business activities, cultural capital stabilizes and rejuvenates its interrelations and dignified standard of living by operating as the community's moral powerhouse.

Cultural capital also refers to the "values and worldviews, as well as methods of behaving, thinking, and interacting, that a community and its members have utilized for healthy and successful living, both currently and throughout the community's history" (Ashdown et al., 2021). The maintenance and flourishing of cultural capital is vital for individual well-being and community solidarity. Since humans are closely connected to their community and culture, their formation and growth are inseparable from the cultural identity. Ignoring people's culture and context when implementing universal norms and standards will be unjust and unfair. It affirms that cultural capital and a society's ethical framework are closely linked. In this way, a society's cultural identity plays a substantial role in shaping individuals' moral identity through its established ethical values and principles.

Artificial Intelligence

Shaping an individual's moral identity is deeply connected to human intelligence. Societal intelligence, through its cultural capital, influences human intelligence and enables the formation of their moral identity. However, present technological advancements point to a complex scenario, arising from the evolution of AI. It enables machines that had only a supporting role to think and behave like humans. AI plays a determining role in the present generation's socioeconomic, cultural, and ethical interactions. It is therefore necessary to have a basic understanding of AI to examine its impact on moral identity formation. AI has an almost 70-year history (1956-2026). The period between 1936 and 1956 is considered its incubation period. It was in 1936 that Alan Turing proposed the mathematical model for an ideal computer, from which the story of electronic computers began (Jiang et al., 2022).

The term AI was coined by John McCarthy in 1956 as part of the Dartmouth Summer Research Project. Accordingly, he is respected as the father of AI (McCarthy et al., 2006). From the 1980s onwards, AI has had a significant influence on human interactions, culminating in the present global scenario where human flourishing is unimaginable without AI (Jiang et al., 2022). AI has many definitions and applications. The lack of a precise, universally accepted definition is considered one of its advantages, enabling rapid growth (Stone et al., 2016). The Turing test has defined AI as the capacity of machines to communicate with humans without revealing their identity as machines. Marvin Minsky, identified as the father of modern AI, defines it as the science of enabling machines to do things that require human intelligence (Dennis, 2026). It "is a science and a set of computational technologies that are inspired by—but typically operate quite differently from—the ways people use their nervous systems and bodies to sense, learn, reason, and take action" (Stone et al, 2016, p. 4). Human intelligence and its activities are the source of inspiration for AI. Although they appear identical while performing an activity, they differ in the mechanisms underlying them. While AI uses computational technologies, humans depend on their nervous systems. Although AI systems can produce better outcomes, they merely mimic human intelligence. They do not possess the originality of human intelligence. However, they intelligently mimic human intelligence's sensing, learning, and reasoning functions, which help them to take intelligent actions.

Based on the above definitions, AI systems are supposed to function appropriately with foresight (McCorduck, 2004). The influence of AI on individuals, communities, and society at large is a substantial point of discussion. AI has penetrated almost all areas of people's daily lives (Stone et al., 2016). It is believed that AI has the potential to improve the quality of life, labor efficiency, and happiness (Jiang et al., 2022). In the beginning, AI was focused on building intelligent systems. However, it has now shifted to build intelligent systems that are aware of humans and behave as trustworthy agents (Stone et al., 2016). This paradigm shift makes AI systems more appealing to humans. When they behave as reliable actors, their acceptability increases. They can operate not merely as machines, but as intelligent actors who are aware of humans and their cultural capital or ethical framework. From a philosophical perspective, AI has the capacity to make people's lives more meaningful with little effort and to manage the massive interconnections among individuals, businesses, societies, and countries in ways that benefit everyone without differentiation (Sarker, 2022).

RESULTS

The above theoretical framework, comprising morality and ethics, cultural capital, and AI, affirms that the world is slowly moving towards a highly mechanistic system in which machines start to control human interactions. Machines' capacity to mimic human intelligence, supported by AI tools, has substantially changed many conventional methods of human interaction. Although the speed, perfection, quality, and efficiency of AI-supported machines increase productivity and economic growth, they also raise significant concerns. As this chapter focuses on shaping individuals' moral identity, it examines the impact of AI on moral formation and value systems.

Growing Influence of AI

McCarthy et al. (2006) affirmed that AI is the science of creating intelligent machines designed to achieve specific goals. These include everything from playing chess to controlling the air traffic. AI applications are not limited to computer science, but extend to engineering, medicine, agriculture, robotics, astronomy, the social sciences, and transportation (Konya & Nematzadeh, 2024). The world is thus witnessing AI's growing domination in almost all areas of human interaction. Mainstream technologies, influenced by AI, substantially shape human lives. AI planning and computer vision have enhanced the potential of video games, turning them into a more income-generating entertainment sector than Hollywood. Adopting AI tools in deep learning has strengthened speech recognition algorithms, enabling the intelligent operation of sensors and many essential devices at home and in the office. For instance, Amazon Alexa uses deep learning AI and ideally assists humans in many of their interactions. Natural Language Processing (NLP) and knowledge representation have increased the efficiency of web searches (Stone et al., 2016). The speed, efficiency, and increased productivity achieved through the adoption of AI applications have prompted most companies, industries, and firms to rely heavily on them to stay ahead of the competition in their fields (Sarker, 2022). It instigates businesspeople like Elon Musk to seriously consider the possibility that "AI and robots will replace all jobs" (Agarwal, 2025).

Healthcare is one of the major areas where AI is dominantly used. Most people use personal healthcare monitoring devices that rely on AI. The healthcare sector is depending on AI from this small step to using it for diagnosis, surgery, and medical automation (Stone et al., 2016; Jiang et al., 2022). The COVID-19 pandemic has shown the importance of AI technologies in enhancing medical care (Balasubramanian et al., 2025). The same advantage is visible in the education sector. Undeniably, quality education requires the direct and active engagement of human teachers. However, interactive machine teachers are increasingly being deployed to teach language and science subjects. The sustainability aspirations have become increasingly fruitful with the adoption of AI tools in environmental science. AI supports precise prediction, continuous monitoring, and risk evaluation, ensuring a stable ecosystem (Konya & Nematzadeh, 2024). The applications of AI that influence human flourishing span diverse areas, including computer vision, agriculture, business and finance, social media, industry, autonomous vehicles, cybersecurity, surveillance, and robotics (Sarker, 2022). As the scope of the chapter is on the formation of moral identity in the age of AI, a detailed examination of all these areas is not done. However, the above-detailed exposition affirms that the dominance of AI has significantly grown, and the present society can no longer think about shaping its life without somehow managing and sustainably relying on AI applications.

Challenges to Conventional Moral Formation

Children's moral formation was parents' primary responsibility in traditional settings. Family, the closed society, and religion had supporting roles in this inspirational mission. There was a visible gap between the family's private space and the outside world that favoured the intact moral formation strategies. However, it changed with the growing influence of AI, as gadgets and internet connectivity have created a situation where there is no private space (Adella, 2025). Since today's generation prioritizes online interactions, parent-child communication is comparatively less. Their interactions are reduced to WhatsApp messages, Instagram stories, Facebook likes, and online notifications. Most families are aware of the potential risks associated with the increased use of AI applications and the ongoing engagement in the digital world. However, they feel helpless at times as they fail to balance work and family. The shift from a joint family to a nuclear family has also increased their responsibility and burden. The current cultural context prevents them from establishing clear boundaries, and at times it becomes impossible (Jeffery, 2020). Besides, the field of values and principles has become so complex and ambiguous. Previously, children had role models from their family and community due to their close interaction with them. The present digital world offers many complex images of values, personalities, and diverse and uncurated content that are more appealing to their feelings and reasoning, posing substantial threats to their moral formation. The absence of interaction with people in positions of moral authority leads them to value disorientation (Adella, 2025).

Conventional moral formation has gradually shifted to the digital realm. However, there is a dichotomy. Most parents still prefer and follow traditional methods. On the other hand, children are confused about parenting techniques and the moral values that are taught. Their experiences in the digital world do not align with their parents' values and principles. Most children believe that their parents do not understand the changing trends. Even though many parents use AI applications, they prefer to restrict their use for children to ensure that future generations are ethically sound and responsible. Parents' attitude is mostly reactive and repressive towards their children's overuse of technology (Adella, 2025). Parents are brought up in a completely different world, and children grow up amid technological advancements. It leads to a situation in which children are part of the digital generation, while parents are mostly digital immigrants. It should be noted with extreme caution that today's children are not merely passive consumers, but the most vibrant users of AI applications. They give moral justification for their actions based on social media responses and search engine results rather than a thorough examination of internal principles or meaningful interaction with the conventional agents of the value system (Jeffery, 2020). It affirms that when traditional actors, such as parents and teachers, fail to enter the AI world and shape the moral formation of future generations, the AI world itself, through its determination of right and wrong, may influence them and lead to disoriented values (Yusuf et al., 2020). For instance, deepfakes are more common in the current AI world. The presentation of heroes (film actors, players, or athletes) that highlights some immoral activities as commendable influences the shaping of children's moral identity. Since they believe their heroes have done them, they prefer to engage in such immoral activities without truly understanding their aftereffects. Even when parents or teachers try to convince them that those activities are wrong, they tend to continue doing them, believing that their heroes actually justify them.

Children's moral development was intimately connected to the primary social environment. However, the rise of AI applications across diverse areas has shifted the focus of moral formation to the digital world. The moral narratives of these diverse AI applications lead to a new pattern in which children's understanding of right and wrong evolves from social validation in the AI world, rather than being grounded in sound ethical principles (Adella, 2025). Additionally, moral formation becomes more complex as children become more concerned about externalizing their image, popularity, and consumption rather than internalizing values (Apdillah et al., 2022). Internalization of values is the crux of moral development. When the AI space undermines internalization of values and prioritizes externalization of irrelevant factors, it creates a moral vacuum. For instance, today's generation is concerned with stories, likes, sharing, and comments rather than inculcating values such as respect for the elderly, accountability, transparency, and obedience.

The speed and efficiency of AI applications have made them the preferred choice for the present generation. It presents many possibilities, opportunities, and shortcuts. Many students and young researchers use AI applications to substitute their reasoning capacity for better productivity. For instance, ChatGPT is increasingly used by students and teachers, substituting their reasoning and evaluating skills. Such tools should be used to enhance human capabilities, not to replace them (van Niekerk et al., 2025). Using AI tools to create innovative content and polished writing has become quite common in the academic world. Accordingly, many reputable publication houses affirm that they do not accept AI-produced articles. AI can be used to enhance the quality of researchers' innovative ideas. For instance, using the Grammarly AI tool helps non-native English writers improve their writing quality and reduce grammatical mistakes. However, some researchers use AI tools to produce content, undermining their human potential and disregarding fundamental ethical concerns. It shows a lack of moral formation and commitment to society. The same applies in many other areas. In the past, it was easy to determine individuals' moral identity within society. There was no confusion concerning the accepted behaviour patterns within a society. All these have changed, making it extremely challenging to determine what is right or wrong and what really constitutes an individual's moral identity (Adella, 2025). This failure to align real-world expectations with AI applications is the most notable challenge facing society today.

Changed Role of Value Trainers

Parents, teachers, family members, friends, colleagues, and the religious community had decisive roles in the moral development of children and young adults. They prioritized educating and empowering them with values and principles. Their critical and consistent guidance helped shape moral identity. Moral formation in the AI world requires regulating media use, assisting with digital exploration, and mediating values in content consumption (Jeffery, 2020). It should focus not merely on unnecessary restrictions and controls, but on building critical and participatory dialogues that affirm the value of relationship-based formation (Adella, 2025). The involvement of parents, teachers, and other moral agents should not become an expression of authoritarian control without dialogue. Their role is more about facilitators and supporters who can assist them in their search for right and wrong. The inculcation of values and principles results from a continuous training process.

This necessitates that traditional moral agents and trainers support future generations by balancing values and AI applications, enabling them to reflect on the ultimate objectives of AI tools, grounded in ethical principles.

The present generation listens to and engages with AI-supported chatbots and other digital actors more than the so-called traditional moral agents. Continuous interactions with these chatbots and other AI tools can create a moral identity that may be significantly different from real-world moral aspirations. The guidance of traditional moral agents and the present generation's experience in the AI space is entirely different. If value trainers adopt a negative attitude towards AI tools and the content people receive from the AI space, it will challenge their potential interactions with the present generation (Adella, 2025). They must use them efficiently and responsibly to enhance future generations' moral quality. Additionally, religious communities, educational institutions, and NGOs can develop their own engaging chatbots and AI systems to guide the formation of moral identity. Some religious communities have already developed chatbots that explain their core beliefs through stories and anecdotes. Texts with Jesus, Christian AI, WisQu, and VedhGPT are examples that convey the fundamentals of religious beliefs and morality in ways that are quite interesting to the present AI generation. That is, value trainers must shift to the AI space to increase their accessibility and availability, while continuing to serve as actors of change and moral formation.

Globalized Culture and Value System

The advent and growth of globalization, liberalization, and free-market policies have increased the interconnectedness of people and societies. Held and McGrew (2007) argue that globalization points to increased interconnection evolving from economic integration, cultural exchange, technological innovation, and political interdependence. Nevertheless, economic integration is the primary rationale for the other three components of globalization. Both liberalization and free-market policies aim to enhance the potential of economic integration. The evolution and unimaginable growth of AI applications further enhanced this interconnection. All these have changed the fundamental, traditional ideas of what it means to be human and of the values required to affirm the quality of human life. Such a prioritization of economic integration must have placed profit and financial benefits as the center of global interactions. However, many inspirational thinkers argued that technological and global socioeconomic and political developments and interconnections cannot be accepted without examining them through the lens of human dignity, justice, and truth (Chimeremeze, 2025). Human knowledge and existence thus highly depend on values and principles. As society becomes increasingly globalized and technologically advanced, the value system also extends beyond explicit boundaries.

Globalization's increased interconnection raises concerns about cultural values and traditions. Each society's unique cultural capital and ethical framework will be in danger, as the separation between societies has become minimal. However, it causes rapid circulation of values, principles, and practices across cultures (Chimeremeze, 2025). It has both positive and negative impacts. A tendency can evolve that prioritizes dominant actors' values and culture, undermining those of others. The increased expansion of Westernization underscores it. Propagating Western values as the constitutive element of globalization and technological advancements is actually against the true spirit of a globalized world. The globalized, AI-supported world requires an interplay of values across different cultures, societies, and religions. It must revolve around a cosmopolitan ethics that acknowledges universal brotherhood, shared identity, and cultural diversity (Chimeremeze, 2025). Although cultures differ in the prioritization of values, some specific values are common to all cultures. Some values that may not be in use in some cultures can be accepted if their worth is discerned. Globalization can enhance the exchange of diverse values across cultures and also lead to the institution of a common value system. International institutions should prioritize regulating AI operations in line with internationally accepted values and principles. The globalized culture and value system stands for an inclusive, comprehensive, and harmonious framework of values and principles incorporated from diverse cultures, without prioritizing any over the other, and presenting them as a universal standard for human interaction in the age of globalization and AI.

DISCUSSION

This section discusses the shaping of moral identity in the age of AI within the context of ethical theories, such as utilitarianism, deontological ethics, virtue ethics, cyber ethics, and AI ethics; moral formation and AI, examining the questions on moral autonomy; and the ethical values of accountability, responsibility, justice, transparency, and privacy, and their interrelation with AI. Their over-emphasized power cannot be permitted to substitute for humans. The preservation of human dignity should be the primary concern, even as the world increasingly relies on AI systems (Chimeremeze, 2025). When decision-making powers are delegated to AI tools, ethical concerns such as bias, fairness, and autonomy also arise. These concerns should be properly addressed to ensure that AI tools enhance people's moral identity and values.

Ethical Theories and AI

As a philosophical school, ethics has four major branches: meta-ethics, normative ethics, comparative ethics, and applied ethics. Meta-ethics is the foundational exposition of what is right or wrong (Robson, 2025). Normative ethics examines how and why an action fits into the definition of right or wrong (Christensen, 2025). Comparative ethics evaluates people's perceptions about the right or wrong of an action (Riegler, 2023). Applied ethics is the application of ethical values and principles to various professions (Schleidgen et al., 2023). It is the practical tool of ethics. It examines how the understanding of right or wrong is practiced in a profession to promote societal order. This section does not enter into a detailed examination of all the above-mentioned branches of ethics. It first provides a brief overview of the three branches of normative ethics, then examines two branches of applied ethics to make a significant connection to the chapter's central argument. The three branches of normative ethics are utilitarianism, deontological ethics, and virtue ethics. Utilitarian ethics is utility-centered or consequences-centered. Deontological ethics is duty-based. In contrast, virtue ethics is agent-centered. The rightness or wrongness of an action is often determined by three core elements: the subject (the character), conformity to moral norms, and the object. That is, virtue ethics primarily focuses on the subject of an action (the agent), utilitarianism examines the object of an action (the consequences), and deontological ethics explores the faithful adherence to moral norms (duty) during an action. Human flourishing is the central concern of ethics, and all the ethical theories and values analyzed below aim to strengthen it.

Utilitarianism

Utilitarianism holds that an action fits into the definition of right or wrong based on its consequences. If an action produces the greatest good for the greatest number of people, it is good/right. If it does not benefit people, it is bad/wrong. It revolves around the terms of utility and consequences (Stahl, 2021). Since its primary objective is the consequences of an action, it is also called consequentialism. The logic of utilitarianism is simple: evaluate behaviours based on their utility and the capacity to maximize happiness or welfare. It is closely associated with AI optimization calculations (Nida-Rümelin & Weidenfeld, 2022). Applying utilitarian ethics to the process of shaping individuals' moral identity in the age of AI requires that AI systems should not undermine people's happiness and welfare by distorting their moral quality. It prioritizes long-term happiness closely connected to one's character and moral responsibility.

Deontological Ethics

According to deontological ethics, an action is right if it aligns with the agent's duty/responsibility. An action that evolves from one's responsibility is good, and if it is against one's responsibility, it is bad. Kant's assumption of the categorical imperative is the philosophical foundation of deontological ethics. It does not focus on an action's immediate consequences but prioritizes the motivation to perform it (Stahl, 2021). The underlying duty of humans and their innovations (e.g., AI systems) is to promote individual well-being and societal welfare. Both are complementary. One cannot undermine societal welfare by claiming that a particular action produces more happiness or profit for the individual. If the motivation of AI systems is human welfare, it aligns with the deontological ethics framework.

Virtue Ethics

Virtue ethics evolved from the Hellenistic philosophy. Virtue ethics underscores the importance of virtues in human interactions. Virtue is the human ability to perform right actions with the right feelings and in the right ways (Constantinescu & Crisp, 2022). Virtue ethics affirms that an action is good if it evolves from one's virtuous disposition. It helps individuals develop a virtuous character that leads to good actions (Stahl, 2021). Virtuous actions are preferred as they display the agent's virtuous character (Constantinescu & Crisp, 2022). Since AI technologies have become so advanced that they currently behave more like humans than inanimate beings, their influence can affect people's decisions. The virtue ethics approach examines whether AI technologies are virtuous and support a virtuous human life. Virtue ethics has captured the attention of Human-Robot Interaction (HRI) researchers in three directions: a) the virtues that should guide the designers of social robots; b) how human virtues are shaped through interaction with intelligent systems; and c) how social robots themselves exhibit virtues (Constantinescu & Crisp, 2022). Virtues are associated with character, and the presence of virtues in AI systems is debatable. This chapter does not entertain it. On the contrary, it assumes that the virtues of AI designers play a decisive role in the choices and decisions of AI systems, which can influence moral formation. Virtuous designers can strengthen AI technologies' potential to behave virtuously. AI technologies cannot be virtuous if we rely on Aristotle's definition of virtue. However, they can be trained to behave virtuously (Constantinescu & Crisp, 2022). A virtuous action closely aligns with global society's concerns about right or wrong. For instance, how a virtuous person should behave in a particular context is already established in global settings. If an action does not align with these established virtuous forms, it cannot be considered virtuous. The ethical values and principles (virtues) that are supposed to guide human interactions are explicit in human consciousness, though they vary across cultures. These variations have become extremely thin within the context of globalization. This has led to a universal understanding of common ethical standards and virtues. An individual's moral identity in the age of AI can be virtuously shaped when it interacts with AI technologies trained with virtues. The human-AI association must foster people's virtuous qualities. In sum, since virtue ethics is concerned with doing the right actions in the right ways and in the right circumstances, AI technologies must be trained to align with these core principles to ensure that the agent's actions are right and welfare-oriented.

Cyber Ethics

Cyber ethics refers to moral, legal, social, economic, and environmental issues that are part of cyber technologies. Cyber ethics is a general ethical examination of all activities in cyberspace. It is broader than and includes AI ethics. There is no universally accepted definition of cyber ethics. Cyber ethics is a form of applied ethics that explicitly focuses on cyber technologies and their impact on human beings. It also analyzes human interactions in cyberspace and evaluates them as right or wrong. It explores the interconnection between technology and people's real-life situations (Yaokumah, 2020).

AI Ethics

The origins of AI ethics can be traced back to Isaac Asimov's fictional texts, in which he argues that ethical codes should guide the evolution of intelligent robots; otherwise, it would be catastrophic for the entire global population (Segun, 2021). Asimov's (1950) formulation of the three laws laid the foundation of the interplay between intelligent systems and human ethics. They are: a) an intelligent system (e.g., a robot) should not injure a human being, or, through inaction, should not allow harm (i.e., no commission or omission that causes harm to humans); b) an intelligent system must follow the orders of humans except in cases where it becomes in conflict with the first law; and c) an intelligent system shall protect as long as it does not challenge the first and second laws (Segun, 2021). Although some scholars have challenged the computational and decision-making efficiency of these laws, they affirm the instrumental role of intelligent systems. The ethical framework of AI is grounded in a progressive vision of its instrumentality for human well-being. AI ethics is discussed under the subareas of robot ethics (which focuses on how robots should be designed based on human social concerns), machine ethics (which examines ethical concerns in relation to autonomous intelligent systems), and computational ethics (concerned about how artificial intelligent systems would make ethical decisions) (Segun, 2021).

AI and Moral Autonomy

AI is reshaping the world with three fundamental ideas: automation, intelligence, and smart computing. These key terms enable today's systems and technologies to make faster and smarter decisions across all disciplines, industries, and economies (Sarker, 2022). This motivates people to become increasingly attached to and depend on smart machines for making their choices and decisions in almost all areas of progress. Humans closely associate with intelligent technologies (autonomous AI agents) as they are increasingly deployed as tutors, companions, caring assistants, babysitters, and coworkers (Lee et al., 2021). Most intelligent systems operate similarly to humans, and their extended use in social interaction since the COVID-19 period has increased their potential to influence human choices and decisions (Constantinescu & Crisp, 2022). The increased human-AI interaction also influences human moral formation. It presents two constitutive elements: on the one hand, the roles of family, society, and religion have diminished; on the other hand, the continuous interaction and ever-growing influence of AI have opened up many new perspectives and discussions about right or wrong.

The more AI technologies become widespread, the more they become influential in human lives. From recommendation systems to autonomous decision-making vehicles, AI technologies are involved in many areas of human interaction, shaping their choices and decisions. There is no doubt that AI technologies can significantly alter human autonomy. The question concerns only whether it would be positive or negative. The more AI technologies become ubiquitous, the more ethical concerns they raise (Laitinen & Sahlgren, 2021). All these ethical questions revolve around people's moral autonomy, as AI technologies play a decisive role in their choices and decisions. AI ethics does not rely solely on human autonomy; it is the central value and closely connected to other ethical parameters, such as human dignity, privacy, and transparency (Rubel et al., 2021).

Moral autonomy evolves from human autonomy, which is the foundational principle grounded on the right to self-determination (Laitinen & Sahlgren, 2021). In the absence of the right to self-determination, no human can make prudential choices and responsible decisions that affect their lives and their interactions with others. An individual's identity, in the physical and moral realms, is shaped and expressed through their self-determination capacities. It is the rationale that guides their choices and decisions. It prevents other actors from unlawful intervention and possible attempts to subordinate humans (Segun, 2021). Human autonomy is not absolute; it depends on and should consider the freedom, choices, and well-being of other individuals. Obviously, one is influenced by various ethical codes and principles in forming a dignified moral standard. However, an individual's moral choices and decisions, if those are based on a properly formed conscience, should be free from external controls. Contemporary experiences affirm that AI technologies are directly or indirectly influencing individuals' moral autonomy even without their knowledge. It starts with chatbot recommendations that influence our e-commerce purchases based on our previous purchases and video suggestions based on our YouTube watch history and moves on to more complex and significant life choices.

Algorithmic bias can cause unfair advantages to specific cultures and values over others. Racial and economic profiling can persist through this academic bias, limiting the global community's sustainable progress (Oguz, 2025). Systemic bias and discrimination should be examined through the lens of ethical values and principles, prioritizing justice and fairness (Aboodi, 2025). The use of AI tools in academic and other domains should be ethically sound. Students might be interested in exploring and experimenting with the benefits of AI for their success. However, it should not become a threat to the flourishing of human intelligence (McLoughlin et al., 2026). When they become submissive to AI, the question of moral autonomy becomes complex.

Ethical Values and AI

The increased adoption of AI technologies raises significant moral concerns. Nevertheless, the world can no longer think of its survival and progress without the support of AI mechanisms. The global community is deeply concerned about the promotion of AI-based technologies that do not undermine the age-old ethical values of freedom, equality, accountability, justice, and transparency (Stone et al., 2016). The question of 'trustworthy AI' also raises concerns. How can a machine (even if it is endowed with the capacity to mimic human intelligence) be trusted? Trust evolves from goodwill, which AI technologies cannot generate. However, they can function as reliable actors (Ryan, 2020). As human-AI interactions have significantly increased, special attention should be given to critically evaluating them through the lens of ethical values and principles. Individuals' moral identity can be systematically and progressively shaped in the age of AI if AI technologies ensure specific ethical values and principles. The following ethical values should serve as the core of algorithmic design to advance ethically stable technologies that can promote individual well-being and societal welfare.

Accountability

Accountability is an ethical value that underscores the obligation for one's actions. It expresses a sincere commitment to perform an action and to take responsibility for its failure. It is particularly relevant in situations where one harms or challenges others' rights. It also guides the power imbalances between two actors. It addresses the unjust behaviour of power holders toward others. In sum, accountability is the willingness and the resolve to be responsible for one's actions and their consequences (van de Poel, 2020). If AI systems are enabled to mimic human intelligence, they should also be empowered to imitate the accountability principle that conditions human behaviour. Cyber ethics is nothing other than the promotion of accountability in cyberspace. Accountability affirms that AI systems should function in accordance with values. "Context, range, agent, forum, standard, practices, and implications" are the core components of accountability. It is the expectation that designers, developers, and deployers will respect and promote ethical standards and foundational legal values in the operations of AI systems (Novelli et al., 2024). Accountability is closely connected to the values of responsibility and transparency, and it can function meaningfully only within traceable decision-making processes. Accountability is the intersection of fairness and transparency that should guide the AI space (Radanliev, 2025).

Transparency

Transparency is one of the key ethical principles affirmed in the General Data Protection Regulation (GDPR) of 2018 (Art. 5(1)(a)). Transparency focuses on autonomy and informed consent. In discussions of AI technologies, transparency is understood in both prospective and retrospective senses. In line with the prospective transparency dimension, AI technologies should disclose in advance how they will operate. That is, it should share all vital information and obtain informed consent from human actors. Retrospective transparency is the responsibility to explain how decisions were made in particular circumstances. It is the openness concerning the decision-making process (Felzmann et al., 2019). When AI systems provide vital information concerning their operations beforehand, it becomes the responsibility of human actors to use them responsibly.

Privacy

It is believed that accountability and privacy conflict. In reality, they are complementary (van de Poel, 2020). Privacy primarily focuses on protecting personal data (Radanliev, 2025). Confidentiality is instrumental in promoting the true spirit of privacy. It is focused on the collection, storage, and sharing of data. Although discussions are prevalent about the nature of privacy as intrinsic or instrumental in terms of moral autonomy, the relationship between privacy and the capacity to lead a morally meaningful life is indisputable (van de Poel, 2020). From the GDPR to almost all international and national legal standards affirm the value of privacy in cyberspace. AI technologies cannot operate in ways that undermine individuals' privacy. Since privacy is the most significant element of an individual's intrinsic worth, any attempts to undermine it will harm human dignity. It affirms that AI technologies (instrumental value) cannot neglect or harm privacy (intrinsic value).

BRIDGING VALUES AND ALGORITHMS: HOW AND WHY

Human history is an inspirational story of values, principles, and customs. The ongoing intervention, training, and interaction across diverse cultures, customs, and religions have significantly shaped the present generation's moral identity. It was not an easy process, as it often faced challenges and threats from various corners. It is undeniable that virtues, values, and principles (i.e., the ethical framework) have played significant roles in humanity's progress and welfare. The fate of humanity would have been significantly different if there had been no ethical values and principles to guide its existence and interactions. Although there had been substantial advancements in socioeconomic, cultural, political, scientific, and technological interactions, the ethical framework also evolved to align with developments in other fields.

AI systems cannot be virtuous. With current technological advancements, they cannot perform virtuous actions according to the algorithms designed by their makers (Constantinescu & Crisp, 2022). It shows that their efficiency in performing virtuous actions and shaping individuals' moral identity depends on the virtuous character of their makers. If anybody blames current AI technologies for the failure of the present generation's moral quality, it is a failure to understand reality. AI technologies have no inherent and independent capacity to choose between right and wrong. It has no intrinsic capacity to be virtuous or otherwise. It depends on the inputs given or the algorithms used. It is the responsibility of human actors to design AI systems that affirm ethical values.

AI technologies and systems are advancing within the boundaries of global society and culture. Their ultimate objective, as already affirmed, is to improve human well-being. They prioritize the socioeconomic and cultural development of individuals and society. They do not have any intrinsic value but only instrumental value. Their instrumentality works well only when it supports human aspirations for meaningful progress. The cultural capital of any society affirms the interplay between values and sustainability that underscores the well-being of all individuals in that society. Although globalization has enhanced interconnection among cultures by promoting a universal culture, it also prioritizes the welfare of each individual. This necessitates prioritizing ethical values and principles that can strengthen the global interconnection and improve humanity's progress. AI technologies are supposed to support and promote humanity's efforts toward progress, and they cannot be designed in ways that challenge or undermine the ethical framework that binds human interconnections and adds meaning and value to human existence. Accordingly, AI algorithms should be designed in ways that bridge values and algorithms, playing significant roles in shaping moral identity in the current age of AI.

CONCLUSION

Individuals' moral identity is the foundational framework that conditions and guides their socioeconomic, cultural, and political interactions, affirming human welfare and societal progress. The continued existence and substantial progress of humanity underscore the value of individual moral identity. An individual's moral identity is shaped through the involvement and support of many actors, including their parents, teachers, and society. Each society has a cultural capital, encompassing religious beliefs, faith, values, and principles, that plays a decisive role in the moral formation of individuals. The cultural capital of each society has blurred due to globalization. Additionally, the evolution and growth of AI technologies further undermined the roles of parents, teachers, and other actors in shaping individuals' moral identity. The AI systems have expanded their influence into almost all areas of human interaction. It changed the role of conventional value trainers. The moral autonomy of AI space becomes a significant concern in this context. AI has only instrumental values, as it aims to ensure human well-being. As an instrumental mechanism, it is supposed to improve the ethical and moral qualities required of humans, and not the other way around. Algorithms should be designed in ways that bridge faith, values, and principles, grounded in ethical theories and values, to substantially shape individual moral identity that can lead to sustainable development and societal welfare.

REFERENCES

Aboodi, R. (2025). The worrisome potential of outsourcing critical thinking to artificial intelligence. *Educational Theory*, *75*(4), 626–645. DOI: 10.1111/edth.70037

Adella, S. (2025). The relationship between digital parenting and children's moral development in the technological era. *Psikhlogiya, 2*(2), 39-47. https://jurnal-ilcaagam.com/index.php/PJ/article/view/395

Agarwal, A. (2025, October 22). *Elon Musk says AI will take every job and humans will be free to grow vegetables*. India Today. https://www.indiatoday.in/technology/news/story/elon-musk-says-ai-will-take-every-job-and-humans-will-be-free-to-grow-vegetables-2806590-2025-10-22

Apdillah, D., Simanjuntak, C. R. A., Napitupulu, C. N. S. B., Sirait, D. D., & Mangunsong, J. (2022). The role of parents in educating children in the digital age. *Romeo Review of Multidisciplinary Education, Culture, and Pedagogy, 1*(3), 1-8. https://ojs.transpublika.com/index.php/ROMEO/article/view/168

Ashdown, B. K., Dixe, A., & Talmage, C. A. (2021). The potentially damaging effects of developmental aid and voluntourism on cultural capital and well-being. *International Journal of Community Well-being*, *4*(1), 113–131. https://link.springer.com/article/10.1007/s42413-020-00079-2. DOI: 10.1007/s42413-020-00079-2

Asimov, I. (1950). *Run around. I, robot*. Doubleday.

Balasubramanian, S., Shukla, V., Islam, N., Upadhyay, A., & Duong, L. (2025). Applying artificial intelligence in healthcare: Lessons from the COVID-19 pandemic. *International Journal of Production Research*, *63*(2), 594–627. DOI: 10.1080/00207543.2023.2263102

Barker, C. (2004). Cultural capital. In *The SAGE dictionary of cultural studies* (pp. 37–38). SAGE., DOI: 10.4135/9781446221280.n47

Bourdieu, P. (1986). The forms of capital. In J. Richardson (Ed.), *Handbook of theory and research for the sociology of education* (pp. 241-258). Greenwood. https://home.iitk.ac.in/~amman/soc748/bourdieu_forms_of_capital.pdf

Chaddha, R., & Agrawal, G. (2023). Ethics and morality. *Indian Journal of Orthopaedics*, *57*(11), 1707–1713. DOI: 10.1007/s43465-023-01004-3 PMID: 37881279

Chimeremeze, O. E. (2025). Philosophy, globalization, and artificial intelligence: Rethinking humanity in a digital age. *Trinitarian International Journal of Arts and Humanities, 1*(2), 42-58. https://www.journals.ezenwaohaetorc.org/index.php/TIJAH/article/view/3565/3695

Christensen, A.-M. S. (2025). Contextual ethics and the new: Reuniting practical and normative ethics. In Christensen, A.-M. S., Forsberg, N., & Rodogno, R. (Eds.), *Contextual Ethics* (pp. 89–109). Springer., https://link.springer.com/chapter/10.1007/978-3-031-97293-5_5 DOI: 10.1007/978-3-031-97293-5_5

Coffin, J., & Egan-Wyer, C. (2022). The ethical consumption cap and mean market morality. *Marketing Theory*, *22*(1), 105–123. DOI: 10.1177/14705931211058772

Constantinescu, M., & Crisp, R. (2022). Can robotic AI systems be virtuous and why does this matter? *International Journal of Social Robotics*, *14*(6), 1547–1557. DOI: 10.1007/s12369-022-00887-w PMID: 35729985

Cruzes, D. S., Dybå, T., Runeson, P., & Höst, M. (2015). Case studies synthesis: A thematic, cross-case, and narrative synthesis worked example. *Empirical Software Engineering*, *20*(6), 1634–1665. DOI: 10.1007/s10664-014-9326-8

Dennis, M. A. (2026, January 20). *Marvin Minsky: American scientist*. Britannica. https://www.britannica.com/biography/Marvin-Minsky

Felzmann, H., Villaronga, E. F., Lutz, C., & Tamò-Larrieux, A. (2019). Transparency you can trust: Transparency requirements for artificial intelligence between legal norms and contextual concerns. *Big Data & Society*, *6*(1), 1–14. DOI: 10.1177/2053951719860542

Ferrara, A. (2022). Moral duties and juridical duties: The ambiguity of legal ethics considered through the prism of Kant's metaphysics of morals. *German Law Journal*, *23*(1), 117–129. DOI: 10.1017/glj.2022.5

Grayling, A. C. (2019). *The history of philosophy*. Penguin Random House.

Gurr, S. K., Geron, T., Forster, D. J., & Levinson, M. (2025). Philosophical reflections on teachers' ethical dilemmas in a global pandemic. *Studies in Philosophy and Education*, *44*(2), 187–207. https://link.springer.com/article/10.1007/s11217-024-09937-4. DOI: 10.1007/s11217-024-09937-4

Held, D., & McGrew, A. (2007). *Globalization theory: Approaches and controversies*. Polity Press.

Jeffery, C. P. (2020). Parenting in the digital age: Between socio-biological and socio-technological development. *New Media & Society*, *23*(5), 1045–1062. DOI: 10.1177/1461444820908606

Jiang, Y., Li, X., Luo, H., Yin, S., & Kaynak, O. (2022). *Quo vadis* artificial intelligence? *Discover Artificial Intelligence*, *2*(1), 4. https://link.springer.com/article/10.1007/s44163-022-00022-8. DOI: 10.1007/s44163-022-00022-8

Kemp, S. E., Ng, M., Hollowood, T., & Hort, J. (2018). Introduction to descriptive analysis. In Kemp, S. E., Hort, J., & Hollowood, T. (Eds.), *Descriptive analysis in sensory evaluation* (pp. 1–39). Wiley., DOI: 10.1002/9781118991657.ch1

Klecun, E., & Cornford, T. (2005). A critical approach to evaluation. *European Journal of Information Systems*, *14*(3), 229–243. DOI: 10.1057/palgrave.ejis.3000540

Konya, A., & Nematzadeh, P. (2024). Recent applications of AI to environmental disciplines: A review. *The Science of the Total Environment*, *906*, 167705. DOI: 10.1016/j.scitotenv.2023.167705 PMID: 37820816

Lähteenkorva, M., Tynjälä, P., & Kallio, E. K. (2025). From knowledge to wisdom in sustainable management education – Insights from a B Corp course. *International Journal of Management Education*, *23*(2), 101148. DOI: 10.1016/j.ijme.2025.101148

Laitinen, A., & Sahlgren, O. (2021). AI systems and respect for human autonomy. *Frontiers in Artificial Intelligence*, *4*, 705164. DOI: 10.3389/frai.2021.705164 PMID: 34765969

Lee, J., Lee, D., & Lee, J. (2021). Can robots help working parents with childcare? Optimizing childcare functions for different parenting characteristics. *International Journal of Social Robotics*, *14*(1), 193–211. https://pmc.ncbi.nlm.nih.gov/articles/PMC8020072/. DOI: 10.1007/s12369-021-00784-8 PMID: 33841588

McCarthy, J., Minsky, M. L., Rochester, N., & Shannon, C. E. (2006). A proposal for the Dartmouth Summer Research Project on artificial intelligence, August 31, 1955. *AI Magazine*, *27*(4), 12–14. DOI: 10.1609/aimag.v27i4.1904

McCorduck, P. (2004). *Machines who think: A personal inquiry into the history and prospects of artificial intelligence* (2nd ed.). A.K. Peters Ltd. DOI: 10.1201/9780429258985

McLennan, S., Fiske, A., Tigard, D., Müller, R., Haddadin, S., & Buyx, A. (2022). Embedded ethics: A proposal for integrating ethics into the development of medical AI. *BMC Medical Ethics*, *23*(1), 6. https://link.springer.com/article/10.1186/s12910-022-00746-3. DOI: 10.1186/s12910-022-00746-3 PMID: 35081955

McLoughlin, S., Fernández-Espinosa, V., & Velázquez-Gil, M. (2026). Practical wisdom, emerging technologies, and AI. *Journal of Moral Education*, 1–31. Advance online publication. DOI: 10.1080/03057240.2026.2624863

Nida-Rümelin, J., & Weidenfeld, N. (2022). *Digital humanism: For a humane transformation of democracy, economy, and culture in the digital age*. Springer. https://link.springer.com/book/10.1007/978-3-031-12482-2

Novelli, C., Taddeo, M., & Floridi, L. (2024). Accountability in artificial intelligence: What it is and how it works. *AI & Society*, *39*(4), 1871–1882. DOI: 10.1007/s00146-023-01635-y

Oguz, H. (2025). Authoritarian recursions: How fiction, history, and AI reinforce control in education, warfare, and discourse. arXiv. https://doi.org//arXiv.2504.09030DOI: 10.48550

Radanliev, P. (2025). AI ethics: Integrating transparency, fairness, and privacy in AI development. *Applied Artificial Intelligence*, *39*(1), 2463722. DOI: 10.1080/08839514.2025.2463722

Reed, E., & Johnson, B. (2023). Overview of cultural capital theory's current impact and potential utility in academic libraries. *Journal of Academic Librarianship*, *49*(6), 102782. DOI: 10.1016/j.acalib.2023.102782

Riegler, J. (2023). Comparative ethics of modern payment models: Does the way we pay for care align with patient care ethics? *Voices in Bioethics*, *9*. Advance online publication. DOI: 10.52214/vib.v9i.10310

Robson, E. (2025). Mary Midgley's meta-ethics and Neo-Aristotelian naturalism. *British Journal for the History of Philosophy*, *33*(5), 1207–1232. DOI: 10.1080/09608788.2024.2389863

Rubel, A., Clinton, C., & Pham, A. (2021). *Algorithms & autonomy: The ethics of automated decision systems*. Cambridge University Press. DOI: 10.1017/9781108895057

Ryan, M. (2020). In AI we trust: Ethics, artificial intelligence, and reliability. *Science and Engineering Ethics*, *26*(5), 2749–2767. DOI: 10.1007/s11948-020-00228-y PMID: 32524425

Sarker, I. H. (2022). AI-based modeling: Techniques, applications, and research issues towards automation, intelligent, and smart systems. *SN Computer Science*, *3*(2), 158. https://link.springer.com/article/10.1007/s42979-022-01043-x. DOI: 10.1007/s42979-022-01043-x PMID: 35194580

Schleidgen, S., Kermling, A., Metz, M., Kuehlmeyer, K., Inthorn, J., & Haltaufderheide, J. (2023). How to derive ethically appropriate recommendations for action? A methodology for applied ethics. *Medicine, Health Care, and Philosophy*, *26*(2), 175–184. https://link.springer.com/article/10.1007/s11019-022-10133-9. DOI: 10.1007/s11019-022-10133-9 PMID: 36522552

Segun, S. T. (2021). From machine ethics to computational ethics. *AI & Society*, *36*(1), 263–276. https://link.springer.com/article/10.1007/s00146-020-01010-1. DOI: 10.1007/s00146-020-01010-1

Stahl, B. C. (2021). *Artificial intelligence for a better future: An ecosystem perspective on the ethics of AI and emerging digital technologies*. Springer. https://link.springer.com/book/10.1007/978-3-030-69978-9

Stone, P. Brooks, R., Brynjolfsson, E. Calo, R., Etzioni, O., Hager, G., Hirschberg, J. Kalyanakrishnan, S., Kamar, E., Kraus, S., Leyton-Brown, K., Parkes, D., Press, W., Saxenian, A., Shah, J., Tambe, M., & Teller, A. (2016). *Artificial intelligence and life in 2030: The one hundred year study on artificial intelligence*. Report of the 2015 study panel. Stanford University. https://arxiv.org/abs/2211.06318

Talmage, C. A., Allgood, B., Ashdown, B. K., Brennan, A., Hill, S., Trevan, E., & Waugh, J. (2022). Tethering natural capital and cultural capital for a more sustainable post-COVID-19 world. *International Journal of Community Well-being*, *5*(3), 657–678. https://link.springer.com/article/10.1007/s42413-021-00151-5. DOI: 10.1007/s42413-021-00151-5 PMID: 35018333

Tsai, C., & Ku, H. (2025). Why AI may undermine phronesis and what to do about it. *AI and Ethics*, *5*(3), 3079–3086. DOI: 10.1007/s43681-024-00617-0

van de Poel, I. (2020). Core values and value conflicts in cybersecurity: Beyond privacy versus security. In Christen, M., Gordijn, B., & Loi, M. (Eds.), *The ethics of cybersecurity* (pp. 45–71). Springer., DOI: 10.1007/978-3-030-29053-5_3

van Niekerk, J., Delport, P. M. J., & Sutherland, I. (2025). Addressing the use of generative AI in academic writing. *Computers and Education: Artificial Intelligence*, *8*, 100342. DOI: 10.1016/j.caeai.2024.100342

Wienhues, A. (2022). Looking through the microscope: Microbes as a challenge for theorizing biocentrism within environmental ethics. *Endeavour*, *46*(1-2), 100819. DOI: 10.1016/j.endeavour.2022.100819 PMID: 35641322

Yaokumah, W. (2020). Predicting and explaining cyber ethics with ethical theories. *International Journal of Cyber Warfare & Terrorism, 10*(2), 46–63. https://www.igi-global.com/gateway/article/250905#pnlRecommendationForm. DOI: 10.4018/IJCWT.2020040103

Yusuf, M., Witro, D., Diana, D., Santosa, T. A., & Jalwis, J. (2020). Digital parenting to children using the Internet. *Pedagogik Journal of Islamic Elementary School, 3*(1), 1-14. https://ejournal.iainpalopo.ac.id/index.php/PiJIES/article/view/1277

KEY TERMS AND DEFINITIONS

Accountability: Accountability is the value that holds individuals responsible for their actions.

Cultural Capital: Cultural capital is the sum of the values, principles, and beliefs of a particular society that shape its moral identity.

Deontological Ethics: Deontological ethics is an ethical approach that grounds an action's rightness or wrongness in its conformity to moral norms.

Ethics: Ethics refers to moral norms and principles used to evaluate the rightness or wrongness of an action within a particular culture or profession.

Moral Autonomy: Moral autonomy is the capacity for self-determination, strengthening individuals to make rational judgments in moral matters.

Utilitarian Ethics: Utilitarian ethics is the ethical school according to which the rightness or wrongness of an action is based on its consequences.

Virtue Ethics: Virtue ethics is the ethical school that evaluates the morality of an action based on the character of the agent.

Chapter 5
Educational Practices for Forming Moral Courage:
An Integrated Developmental Framework

Arnaldo C. Marion
http://orcid.org/0009-0002-2776-6105
Coaching 4 Change LLC, Brazil

ABSTRACT

This chapter examines how moral courage can be systematically developed as a core capability for ethical leadership in AI-driven environments. While existing approaches to AI ethics emphasize principles, governance, and technical safeguards, they often fail to address the gap between ethical awareness and action under pressure. To address this limitation, the chapter introduces the Integrated Moral Courage Formation Model (IMCFM), which conceptualizes ethical leadership as a developmental process encompassing moral awareness, judgment, pressure navigation, action, and reflective formation. Drawing on virtue ethics, moral psychology, and adult learning theory, the chapter proposes a pedagogical framework that integrates experiential learning, structured deliberation, and reflective practice. An applied case demonstrates how the model can be implemented and assessed in realistic decision-making contexts. The chapter concludes by positioning moral courage as a trainable and measurable capability essential for responsible leadership in the age of artificial intelligence.

DOI: 10.4018/979-8-3373-8192-3.ch005

ght © 2027, IGI Global Scientific Publishing. Copying or distributing in print or electronic forms without written permission of IGI Global Scientific Publishing is prohibited. Use of this chapter to generative artificial intelligence (AI) technologies is expressly prohibited. The publisher reserves all rights to license its use for generative AI training and machine learning model development.

INTRODUCTION

As artificial intelligence (AI) systems rapidly diffuse across higher education, industry, and society, the ethical challenge confronting leaders is no longer hypothetical, it is immediate, behavioral, and formative. Recent empirical evidence indicates that AI is already deeply embedded in student learning practices. A national study by the Lumina Foundation and Gallup (2026) shows that a majority of college students are actively using AI tools in their coursework, often in environments where institutional policies remain unclear and formal guidance is uneven. Many students report relying on AI to understand complex material, generate content, and support decision-making, while simultaneously expressing uncertainty about appropriate and responsible use. These findings suggest that AI is not merely a tool being introduced into education; it is already reshaping how individuals think, learn, and make judgments.

This shift raises a deeper concern that extends beyond issues of academic integrity or technological adoption. As Aboodi (2024) argues, the most significant risk associated with AI in education is not misuse alone, but the gradual outsourcing of critical thinking itself. When individuals increasingly defer judgment to AI systems, particularly those perceived as highly capable, the development of independent reasoning, ethical discernment, and intellectual responsibility may be weakened. This creates a paradox: the more advanced AI systems become, the greater the temptation to rely on them uncritically, and the more essential it becomes to cultivate human judgment capable of interrogating, contextualizing, and, when necessary, resisting algorithmic outputs.

Emerging empirical research further reinforces this concern. McLoughlin et al. (2026) demonstrate that individuals with higher levels of practical wisdom (phronesis) engage AI systems more reflectively and ethically, using them as tools for augmentation rather than substitution. Conversely, lower levels of practical wisdom are associated with more passive and uncritical use. These findings suggest that the ethical challenges of AI are not solely technological but fundamentally human: the quality of outcomes depends significantly on the moral and intellectual formation of those who design, deploy, and use these systems.

In parallel, real-world failures continue to illustrate the consequences of insufficient ethical formation. The widely cited case of Amazon's AI-powered hiring system, which systematically disadvantaged female candidates due to biased training data, underscores how technical systems can replicate and amplify historical inequities (Dastin, 2018; Barocas & Selbst, 2016). Such cases are not isolated anomalies but indicators of a broader pattern: AI systems inherit the limitations of human judgments embedded within them. When those judgments are insufficiently examined or courageously challenged, the resulting systems can produce large-scale ethical harm.

These dynamics converge on a central leadership challenge: the need for moral courage in AI-mediated environments. Moral courage, defined as the capacity to take principled action despite uncertainty, pressure, or personal risk (Kidder, 2005),is no longer a peripheral virtue but a core leadership competency. Leaders must not only recognize ethical risks in AI systems but also act decisively to address them, even when doing so conflicts with organizational incentives, timelines, or prevailing norms.

This chapter addresses the educational implications of this challenge. It argues that higher education must move beyond teaching ethical principles toward the intentional formation of moral courage as a practiced capability. Drawing on virtue ethics, moral psychology, andragogy, and ethical resilience, the chapter proposes an integrated pedagogical model designed to cultivate leaders who can think critically, act responsibly, and sustain ethical commitments under pressure. In doing so, it positions education not merely as a site of knowledge transmission, but as a formative environment where the habits, dispositions, and courage required for ethical leadership in the age of AI are deliberately developed.

1. RESEARCH GAP: FROM ETHICAL KNOWLEDGE TO MORAL COURAGE

While the rapid expansion of AI has generated a substantial body of literature on ethics, governance, and responsible innovation, a critical gap remains in how ethical capacity is developed in practice, particularly within educational contexts. Existing approaches to AI ethics education tend to emphasize three primary areas: (1) normative frameworks and principles (e.g., fairness, accountability, transparency), (2) regulatory and governance structures, and (3) technical mitigation strategies such as bias detection and algorithmic auditing. Although these contributions are valuable, they largely assume that once individuals understand ethical principles, they will act accordingly.

However, emerging evidence suggests that this assumption is insufficient. The Lumina Foundation and Gallup (2026) study demonstrates that students are already using AI extensively, often without clear institutional guidance or structured ethical formation. At the same time, Aboodi (2024) highlights a more fundamental concern: the increasing tendency to outsource critical thinking to AI systems, which may erode the very capacities required for ethical judgment. These findings indicate that the central challenge is not merely the absence of ethical knowledge, but the weakening of human agency and discernment in AI-mediated environments.

Further reinforcing this gap, McLoughlin et al. (2026) provide empirical evidence that the quality of AI engagement varies significantly depending on the individual's level of practical wisdom. Individuals with higher levels of phronesis demonstrate greater critical reflection, ethical sensitivity, and intentional use of AI, while those with lower levels exhibit more passive reliance. This suggests that ethical outcomes are not determined solely by system design or policy frameworks, but by the character and judgment of the human actors involved.

Taken together, these insights point to a critical limitation in current educational approaches: they focus on what individuals should know, but not sufficiently on how individuals are formed to act under pressure. Ethical decision-making in AI contexts is rarely abstract or consequence-free; it occurs within environments characterized by uncertainty, competing priorities, time constraints, and organizational pressures. Under such conditions, knowledge alone does not guarantee action. What is required is moral courage, the capacity to translate ethical awareness into sustained, principled action despite resistance.

Despite its importance, moral courage remains underdeveloped as a pedagogical construct in AI ethics education. Existing literature rarely provides a structured model for how moral courage can be cultivated, practiced, and assessed within leadership development programs. This chapter addresses that gap by proposing an integrated framework that connects ethical theory, psychological insight, and educational practice into a coherent model for moral courage formation.

To address this gap, the chapter introduces an integrated framework that moves beyond conceptual ethics toward the formation of moral courage as a developmental and measurable capability. This framework provides the foundation for linking ethical theory, educational practice, and leadership behavior within AI-driven environments.

2. THE INTEGRATED MORAL COURAGE FORMATION MODEL (IMCFM)

The Integrated Moral Courage Formation Model (IMCFM) provides a structured response to the limitations identified in current approaches to AI ethics education. Rather than treating ethical knowledge, psychological processes, and leadership behavior as separate domains, the model integrates them into a unified developmental framework centered on action under pressure.

The IMCFM conceptualizes moral courage as a dynamic, iterative process involving five interdependent stages: moral awareness, moral judgment, contextual pressure, moral action, and reflective formation.

Figure 1 illustrates the Integrated Moral Courage Formation Model (IMCFM), highlighting the iterative process through which ethical awareness, judgment, and action are developed and reinforced over time.

Figure 1. IMCFM – integrated moral courage formation model

The first stage, moral awareness, involves recognizing that a situation carries ethical significance. In AI environments, this awareness is often obscured by technical framing, where issues such as bias or harm are interpreted as engineering challenges rather than moral concerns. Moral psychology demonstrates that individuals frequently fail at this stage due to cognitive biases, normalization of deviance, or diffusion of responsibility.

The second stage, moral judgment, concerns the evaluation of what ought to be done. Drawing on virtue ethics, this stage emphasizes phronesis, practical wisdom that integrates principles such as justice, humility, and empathy with contextual understanding. Unlike rule-based reasoning, moral judgment requires discerning the right action in situations where competing values and uncertainty are present.

The third stage introduces pressure and constraints, representing the conditions under which ethical decisions are tested. In real-world AI contexts, these include organizational incentives, time constraints, financial pressures, reputational risks, and social dynamics. Research on ethical resilience shows that this stage is where most ethical failures occur, not because individuals lack knowledge, but because they lack the capacity to act under pressure.

The fourth stage, moral action, involves translating judgment into behavior. This requires moral courage: the willingness to act in alignment with ethical convictions despite potential personal or professional costs. Ethical resilience is critical here, enabling individuals to sustain action even when facing opposition or uncertainty.

The final stage, reflective formation, aligns with principles of andragogy and experiential learning. Through reflection, feedback, and experience, individuals integrate their actions into their moral identity, strengthening future awareness, judgment, and resilience. This stage ensures that moral courage is not episodic, but developmental.

3. THEORETICAL FOUNDATIONS OF THE IMCFM

The Integrated Moral Courage Formation Model (IMCFM) draws on four complementary traditions, virtue ethics, moral psychology, andragogy, and ethical resilience, to explain how individuals move from ethical awareness to action under pressure. Rather than treating these domains as independent bodies of knowledge, the model integrates them into a coherent account of moral formation in AI-driven environments.

3.1 Virtue Ethics and Moral Judgment

Within the IMCFM, virtue ethics primarily informs the stage of moral judgment, shaping how individuals discern what is right in complex and ambiguous situations.

Rooted in Aristotle's concept of *phronesis* (practical wisdom), virtue ethics emphasizes that ethical action is not the result of rule compliance alone, but of cultivated character expressed through judgment in context. Moral excellence emerges through habituation, where virtues such as justice, integrity, humility, and empathy become stable dispositions guiding action. MacIntyre (1981) further reinforces that such virtues are sustained within communities of practice, where shared standards of excellence shape moral reasoning over time.

In AI contexts, this perspective is particularly critical. Algorithmic systems frequently replicate and amplify historical biases embedded in data (Barocas & Selbst, 2016; Obermeyer et al., 2019). Rule-based approaches alone are insufficient to address such challenges, as they cannot account for competing values, contextual nuance, or unintended consequences. Leaders must instead exercise judgment that prioritizes human dignity and fairness over technical efficiency or organizational convenience.

Thus, virtue ethics provides the normative foundation of the IMCFM by defining what ethical action requires, not merely compliance, but the application of practical wisdom in pursuit of human flourishing.

3.2 Moral Psychology and Ethical Awareness Under Bias

Moral psychology informs the stages of moral awareness and breakdown under pressure, explaining why individuals often fail to act in accordance with their values.

Research demonstrates that moral reasoning develops over time (Kohlberg, 1981), but is frequently influenced by intuitive and emotional processes (Haidt, 2001). In practice, individuals do not always engage in deliberate ethical reasoning; instead, judgments are often shaped by cognitive biases, social conformity, and situational framing.

This has direct implications for AI environments. Technical systems can obscure ethical issues by framing them as optimization problems, leading to what scholars describe as ethical fading, where moral considerations are minimized or overlooked. Additionally, organizational norms and performance pressures may reinforce conformity, discouraging individuals from questioning flawed systems.

For example, facial recognition systems widely used in security and law enforcement have demonstrated significantly higher error rates for darker-skinned women compared to lighter-skinned men, despite meeting overall technical performance benchmarks. In a landmark empirical study, Buolamwini and Gebru (2018) found that error rates for gender classification were below 1% for lighter-skinned males but increased to as high as 34.7% for darker-skinned females, revealing substantial disparities rooted in imbalanced training data rather than explicit discriminatory intent. From a compliance perspective, such systems may satisfy aggregate accuracy thresholds or regulatory requirements. However, leaders operating at a conventional level of moral reasoning may interpret these metrics as sufficient, overlooking the disproportionate harm imposed on specific populations. More advanced moral reasoning requires evaluating whether system outcomes align with principles of justice, fairness, and human dignity, particularly when errors carry real-world consequences such as wrongful identification, denial of services, or unjust surveillance.

This underscores the necessity of educational interventions that surface hidden biases, cultivate critical reflection, and strengthen the capacity for ethical discernment under conditions of complexity and pressure.

3.3 Andragogy and Reflective Moral Formation

Andragogy informs the stage of reflective formation, explaining how moral courage is developed through experience, reflection, and application.

Adult learning theory emphasizes that individuals learn most effectively when education is problem-centered, relevant, and grounded in experience (Knowles et al., 2005). Rather than passively receiving ethical principles, learners must engage with real-world dilemmas that require judgment, decision-making, and accountability.

Recent scholarship further reinforces this shift from knowledge acquisition to wisdom formation. Lahteenkorva, Tynjälä, and Kallio (2023) argue that higher education must move beyond the transmission of knowledge toward the development of practical wisdom, integrating reflection, ethical judgment, and practice-based learning into a coherent developmental process. Their work emphasizes that wisdom emerges not from abstract instruction alone, but from the continuous interaction between experience, critical reflection, and value-oriented decision-making, an approach that directly aligns with the formation of moral courage in complex technological environments.

Within the IMCFM, this translates into a formative process where individuals:

- Confront ethical dilemmas in realistic contexts
- Reflect on their decisions and underlying values
- Integrate feedback and experience into future action

This approach aligns with experiential learning models, where moral understanding is deepened through cycles of action and reflection. In AI contexts, where decisions often involve ambiguity, competing priorities, and high stakes, such experiential formation is essential for translating ethical awareness into behavior.

Thus, andragogy provides the mechanism through which moral courage becomes a learned and reinforced capability, rather than an abstract ideal.

3.4 Ethical Resilience and Action Under Pressure

Ethical resilience informs the stages of pressure and moral action, addressing the central challenge of the IMCFM: why individuals fail to act ethically even when they possess sufficient awareness and judgment.

Research on resilience demonstrates that moral action under pressure depends on the capacity to regulate emotions, manage cognitive load, and sustain moral commitment in the face of adversity (Masten, 2001). In high-stakes environments, such as AI development, individuals frequently encounter competing priorities, including profitability versus fairness, speed versus safety, and innovation versus accountability. These tensions create conditions under which ethical action becomes costly rather than intuitive.

A well-documented illustration of this dynamic can be found in the Ford Pinto case. During the 1970s, internal decision-making at Ford Motor Company prioritized cost efficiency and rapid market delivery, despite known design flaws in the Pinto's fuel system that significantly increased the risk of fatal fires in rear-end collisions. Subsequent analyses suggest that engineers and managers involved did not necessarily lack technical knowledge of the risks; rather, their judgments were shaped by an organizational environment that normalized cost–benefit calculations over safety considerations (Gioia, 1992). Gioia, who was directly involved in the decision process, later reflected that the prevailing corporate scripts and routines framed the issue in terms of regulatory compliance and economic trade-offs, rather than as a moral problem involving potential loss of life. In his retrospective account, he noted that the situation "did not appear as an ethical dilemma" at the time, highlighting how organizational context can obscure moral awareness and suppress ethical response.

This case demonstrates how environmental framing, institutional norms, and decision heuristics can cloud ethical judgment, even among well-intentioned professionals. Emotional dissonance emerges when personal values conflict with organizational expectations, while cognitive overload and routinized decision processes reduce the likelihood of critical ethical reflection. Over time, repeated exposure to such conditions can lead to ethical fatigue, increasing conformity and diminishing the willingness to challenge problematic decisions.

Ethical resilience addresses this gap by equipping individuals with the capacity to:

- Maintain clarity under pressure
- Resist organizational and social conformity
- Act consistently with ethical commitments despite risk

Within the IMCFM, resilience functions as the critical bridge between judgment and action. It ensures that ethical intentions are not abandoned when confronted with institutional pressure, transforming moral reasoning into sustained, courageous action in complex and high-stakes environments.

3.5 Social Reinforcement: Community, Collaboration, and Moral Courage

While moral courage is often framed as an individual trait, it is significantly shaped by social context. The IMCFM therefore recognizes the role of community and collaborative environments in reinforcing ethical action.

Research in organizational ethics shows that strong ethical climates, characterized by shared values, open dialogue, and psychological safety, significantly increase the likelihood of principled action (Trevino & Brown, 2004; Edmondson, 1999). Similarly, social learning theory (Bandura, 1977) demonstrates that individuals develop moral actions through observation, interaction, and reinforcement within communities.

In AI contexts, where decisions often affect diverse populations, collaborative deliberation is particularly important. Interdisciplinary teams bring epistemic diversity, enabling more comprehensive identification of risks and more equitable solutions (Page, 2007). Structured dialogue also helps mitigate cognitive biases by exposing individuals to alternative perspectives.

Community and mentorship further reduce the personal cost of moral action. When individuals perceive support from peers or institutions, they are more likely to challenge unethical practices and sustain their position over time.

Thus, social reinforcement provides the contextual support necessary for moral courage to be exercised consistently, transforming individual conviction into collective ethical practice.

4. PEDAGOGICAL MODEL FOR FORMING MORAL COURAGE

While the preceding section establishes the theoretical foundations of moral courage, the question remains how these insights can be translated into educational practice. This section operationalizes the IMCFM through a structured pedagogical model aligned with each stage of moral development, enabling the systematic formation of ethical leadership capabilities.

Moral courage is not formed through conceptual understanding alone. It develops when individuals are required to translate ethical judgment into action under realistic conditions of pressure, uncertainty, and consequence. Accordingly, educational practice must be explicitly designed to simulate, structure, and reinforce the stages of the Integrated Moral Courage Formation Model (IMCFM).

Rather than presenting a broad set of pedagogical techniques, this section proposes a structured implementation model, where each educational intervention is directly aligned to a stage of moral development: awareness, judgment, pressure, action, and reflective formation.

4.1 Pedagogical Mapping of the IMCFM

To move from theory to practice, the following framework translates each stage of moral courage formation into specific, repeatable educational interventions, as shown in Table 1:

Table 1. Pedagogical intervention per IMCFM stages

IMCFM Stage	Learning Objective	Pedagogical Intervention	Expected Outcome
Moral Awareness	Recognize ethical risks	Bias detection cases, AI audits	Identification of hidden harm
Moral Judgment	Evaluate what is right	Ethical deliberation, frameworks	Structured ethical reasoning
Pressure	Experience real constraints	Simulations with incentives/conflict	Emotional and cognitive stress exposure
Moral Action	Practice ethical decision	Role-play escalation, decision defense	Courageous articulation of decisions
Reflective Formation	Internalize learning	Journals, peer feedback, coaching	Strengthened moral identity

This structure ensures that moral courage formation is not left to abstraction but is systematically practiced and reinforced.

4.2 Experiential Simulations: Training Under Pressure

The most critical transition from ethical theory to practical competence occurs at the stage of pressure, where ethical intentions are actively tested against competing demands and real or perceived consequences. Within the IMCFM, this stage represents the inflection point at which individuals either translate judgment into action or retreat into conformity and inaction.

To operationalize this dimension of moral formation, students should be immersed in high-stakes simulations that replicate the complexity of real-world AI dilemmas. These simulations are designed to expose participants to conflicting incentives, such as the tension between profitability and fairness, the urgency of time constraints, and the realities of role-based accountability across organizational hierarchies, including executive leadership, engineering teams, and compliance functions. Scenarios may include the launch of an AI product with known bias, the management of a data privacy breach under deadline pressure, or the deployment of a model that has not undergone sufficient ethical validation.

In a typical simulation, participants are required to make a decision regarding whether to release an AI system that improves operational efficiency but produces ethically problematic outcomes. Beyond decision-making, they must justify their position to leadership and defend it under scrutiny, replicating the communicative and psychological demands of real organizational environments. These exercises are intentionally structured to generate ethical tension, evoke concern over potential consequences, and force participants to navigate trade-offs between competing values.

Through repeated exposure to such conditions, students develop not only ethical reasoning but also ethical resilience. They begin to experience the cost of ethical action and learn to maintain clarity and commitment under pressure, which is essential for the cultivation of moral courage.

4.3 Structured Ethical Deliberation and Judgment Formation

While simulations provide the experiential context, structured ethical deliberation is essential for developing the capacity to evaluate what is right. Ethical reasoning cannot be assumed as an automatic outcome of exposure to dilemmas; it must be explicitly practiced, refined, and disciplined.

Within this pedagogical framework, students are required to apply structured decision-making criteria to complex scenarios. These criteria include considerations of justice, particularly in identifying who may be harmed by a decision; human dignity, in evaluating whether individuals or groups are being devalued; and the balance between short-term outcomes and long-term societal impact. By consistently engaging these dimensions, students move beyond intuitive or reactive responses toward more principled and transparent reasoning.

Deliberation is typically conducted in small groups, where participants are required to articulate, challenge, and defend their positions. Structured argumentation techniques are employed to ensure that reasoning is explicit and subject to critique. Students are not only asked to state their conclusions but to provide clear justification grounded in ethical principles and contextual understanding.

Over time, this process facilitates a shift from vague moral intuition, such as "this feels wrong", to disciplined ethical reasoning, where students can articulate why a particular course of action violates fairness, undermines dignity, or produces unjust outcomes. In doing so, they develop *phronesis*, or practical wisdom, which is central to the moral judgment stage of the IMCFM.

4.4 Ethical Action Training and Courage Execution

A persistent limitation of traditional ethics education is its tendency to stop at judgment, without requiring individuals to act upon their conclusions. However, within the IMCFM, moral courage is defined not by recognition or reasoning alone, but by the capacity to execute ethical decisions under conditions of risk.

To address this gap, students must be placed in situations where they are required to translate ethical judgment into concrete action. This includes scenarios in which they must escalate concerns to authority, challenge flawed decisions, or communicate ethical risks to stakeholders who may resist such perspectives. These exercises are structured to simulate the interpersonal and organizational dynamics that often inhibit ethical action.

Role-play is a central method in this process. Participants may engage in simulated executive meetings in which they must present and defend a recommendation to delay or modify a product due to ethical concerns. In these settings, they encounter resistance, counterarguments, and pressure to conform, requiring them to articulate their position clearly and persist in the face of opposition.

The outcome of this training is the development of both confidence and competence in ethical action. Students learn not only what should be done, but how to do it, how to communicate, justify, and sustain ethical decisions in environments where such actions may carry personal or professional risk. This directly reduces fear-based inaction and strengthens the execution dimension of moral courage.

4.5 Reflective Formation and Learning Reinforcement

Experience alone does not guarantee learning; without reflection, even the most intense simulations may fail to produce lasting transformation. Within the IMCFM, reflective formation is the stage through which experience is internalized and integrated into moral identity.

Following each simulation or ethical exercise, students engage in structured reflection designed to surface both cognitive and emotional dimensions of their decision-making process. They are prompted to examine what they experienced under pressure, where hesitation occurred, what factors influenced their decisions, and how they might respond differently in future situations. This process encourages metacognitive awareness, enabling individuals to critically assess their own reasoning and behavior.

Reflection is further strengthened through peer dialogue and instructor-guided debriefs, where students are exposed to alternative perspectives and receive feedback on their ethical reasoning. In some cases, exercises such as moral identity mapping are introduced to help participants connect their decisions to broader values and long-term leadership aspirations.

Through this iterative process, moral action becomes more consistent and intentional. Students develop greater self-awareness, refine their decision-making processes, and reinforce their commitment to acting in alignment with their values. Reflection thus ensures that moral courage evolves from isolated actions into a stable and repeatable aspect of character.

4.6 Developing Ethical Resilience as a Trainable Capability

Ethical resilience is not an inherent trait but a capability that can be systematically developed through intentional training. Given that the most significant ethical failures often occur under conditions of pressure, ambiguity, and authority, leadership education must explicitly prepare individuals to sustain ethical commitments in environments where psychological and institutional forces may undermine them.

This necessity is well illustrated by classic research in social psychology, particularly the Stanford Prison Experiment conducted by Zimbardo (2007). In this study, participants randomly assigned to roles as "guards" and "prisoners" rapidly internalized these roles, leading to escalating abusive behavior by guards and psychological distress among prisoners. The experiment was terminated prematurely due to the intensity of these dynamics. Importantly, participants were not predisposed to unmoral action; rather, the situational context, defined by authority structures, role expectations, and lack of accountability, significantly shaped their actions. As Zimbardo later argued, ordinary individuals can engage in harmful behavior when embedded in environments that normalize such conduct and diffuse personal responsibility (Zimbardo, 2007).

The implications for ethical leadership are substantial. The experiment demonstrates that behavior under pressure is not determined solely by individual character or knowledge, but by the interaction between psychological processes and situational forces. In organizational and technological contexts, similar dynamics emerge when authority, incentives, and norms frame decisions in ways that obscure ethical implications. Individuals may conform to expectations, suppress dissent, or rationalize questionable actions, even when they possess the cognitive capacity to recognize ethical concerns.

For this reason, the development of ethical resilience must begin with controlled exposure to these dynamics through structured simulations that incorporate elements of stress, role-based authority, and competing incentives. By engaging in such environments, students learn to recognize how factors such as hierarchy, conformity pressures, and role identification can influence their decisions. This awareness is critical for interrupting automatic responses and maintaining ethical clarity.

In parallel, cognitive tools are introduced to support deliberate decision-making. Structured frameworks, such as decision matrices that prioritize fairness, transparency, and long-term societal impact, enable participants to externalize their reasoning and counteract intuitive or conformity-driven responses. These tools serve as anchors for judgment when situational pressures might otherwise distort perception.

Emotional regulation is equally essential. Under pressure, individuals often experience heightened stress, fear of consequences, or desire for approval, all of which can impair ethical reasoning. Training in reflective pause, cognitive reappraisal, and attentional control equips learners to respond thoughtfully rather than react impulsively, preserving alignment between values and action.

Peer accountability further reinforces ethical resilience. When individuals engage in group-based decision reviews and ongoing dialogue, they are required to articulate and defend their reasoning, while also being exposed to alternative perspectives.

This process reduces isolation, strengthens commitment to shared standards, and mitigates the risk of unexamined conformity.

Through this integrated approach, students develop the capacity to maintain clarity under pressure, resist situational and social influence, and act consistently with their ethical commitments despite uncertainty. Ethical resilience thus functions as the critical mechanism within the IMCFM through which moral judgment is sustained in action, ensuring that ethical awareness and reasoning are not overridden by the powerful psychological forces that shape behavior in real-world environments.

4.7 Social Reinforcement: Community and Mentorship

Although moral courage is often framed as an individual virtue, it is deeply influenced by social context. The presence of supportive communities and credible role models significantly enhances the likelihood that individuals will act in accordance with their values.

Educational programs should therefore incorporate structured opportunities for collective engagement, including peer accountability groups, interdisciplinary collaboration, and mentorship relationships with experienced ethical leaders. These environments provide both practical guidance and psychological support, reducing the sense of isolation that often accompanies ethical decision-making in high-stakes contexts.

Interdisciplinary collaboration, in particular, exposes students to diverse perspectives, challenging assumptions and enriching moral reasoning. Mentorship offers a bridge between theory and practice, allowing learners to observe how ethical principles are applied in real-world situations.

Through these mechanisms, ethical norms are reinforced collectively, and individuals gain the confidence to act, knowing that their decisions are supported within a broader moral community.

4.8 Assessing Moral Courage: An IMCFM-Aligned Evaluation Model

A persistent limitation in ethics education, particularly in the context of emerging technologies, is the absence of rigorous methods for assessing whether learners can translate ethical understanding into action. While many programs successfully transmit ethical frameworks and principles, far fewer provide systematic ways to evaluate the development of moral courage as a behavioral and developmental capability. Within the Integrated Moral Courage Formation Model (IMCFM), assessment must extend beyond cognitive comprehension to include the capacity to recognize ethical risk, reason through competing values, act under pressure, and sustain ethical commitments over time.

This need for a multidimensional assessment approach is consistent with foundational work in moral development theory. Kohlberg (1981) demonstrated that moral reasoning progresses through stages, culminating in post-conventional reasoning grounded in universal ethical principles such as justice and human dignity. However, subsequent research has shown that moral reasoning alone does not reliably predict behavior. Rest's (1986) Four-Component Model of moral behavior, encompassing moral sensitivity, judgment, motivation, and character, highlights that ethical action depends on the integration of multiple capacities, including the willingness and ability to act despite competing pressures. This insight is particularly relevant in AI contexts, where ethical decisions are embedded in complex sociotechnical systems and institutional constraints.

To address this gap, this chapter proposes an IMCFM-aligned Moral Courage Assessment Model, designed to evaluate moral development across the stages of awareness, judgment, pressure navigation, action, and reflective formation, with ethical resilience functioning as the sustaining mechanism across stages. The model is consistent with contemporary approaches to behavioral ethics and leadership assessment, which emphasize observable behavior, contextual decision-making, and developmental progression rather than abstract self-report measures (Treviño, Weaver, & Reynolds, 2006; Hannah, Avolio, & May, 2011).

4.8.1 Assessment Structure and Developmental Logic

The model adopts a developmental perspective, recognizing that moral courage evolves over time through experience, reflection, and practice. Rather than treating ethics as a binary attribute, ethical or unethical, the rubric evaluates progression from externally driven compliance toward internally grounded, principled action. This approach aligns with Kohlberg's higher stages of moral development, where individuals act according to internalized principles even when such actions conflict with authority or social norms.

At early stages, individuals tend to rely on rules, authority, or avoidance of punishment. As development progresses, they begin to recognize ethical issues independently, apply structured reasoning, and act with increasing consistency under pressure. At advanced levels, individuals demonstrate what virtue ethics describes as stable character dispositions, where ethical action becomes habitual and integrated into identity (MacIntyre, 1981; Hursthouse & Pettigrove, 2018). Contemporary research on virtue measurement further supports the possibility of assessing such traits through behavioral indicators rather than purely attitudinal scales (Wright, Warren, & Snow, 2021).

Importantly, this model emphasizes profile-based assessment, allowing educators to identify strengths and developmental gaps across dimensions. This reflects findings in leadership research that effective ethical leadership is multidimensional, involving cognition, behavior, and social influence (Brown, Treviño, & Harrison, 2005).

4.8.2 Core Dimensions of Moral Courage Assessment

The first dimension, moral awareness, evaluates the individual's capacity to identify ethical issues, including those obscured by technical framing or normalized within organizational routines. This aligns with Rest's concept of moral sensitivity, which involves recognizing that a situation has ethical implications and understanding how actions affect others (Rest, 1986).

The second dimension, moral judgment, assesses the ability to evaluate competing values and determine what ought to be done. This reflects both Kohlberg's post-conventional reasoning and the Aristotelian concept of *phronesis*, or practical wisdom, which integrates principles with contextual understanding (Aristotle; MacIntyre, 1981).

The third dimension, pressure navigation, captures the influence of situational forces, such as authority, time constraints, and group dynamics, on ethical decision-making. This dimension is supported by behavioral ethics research demonstrating that context often overrides intention, leading to ethical fading or conformity (Treviño et al., 2006).

The fourth dimension, moral action, evaluates whether individuals translate ethical reasoning into behavior. This aligns with Rest's components of moral motivation and moral character, emphasizing that ethical action requires both commitment and execution under constraint (Rest, 1986).

The fifth dimension, ethical resilience, reflects the capacity to sustain ethical commitments over time, even in the face of repeated challenges. This dimension is closely related to constructs in leadership research such as moral efficacy and moral courage, which enable individuals to persist in ethical action despite risk (Hannah et al., 2011).

The final dimension, reflective formation, assesses the extent to which individuals engage in critical reflection and integrate ethical learning into their identity. Reflection is central to both adult learning theory and virtue formation, supporting continuous development and self-regulation (Knowles et al., 2005; Wright et al., 2021).

4.8.3 Rubric Representation

Table 2 presents the IMCFM-aligned Moral Courage Assessment Model, illustrating behavioral progression across key dimensions of ethical leadership development.

Table 2. IMCFM-aligned rubric

Dimension	Level 1 – Reactive	Level 2 – Emerging	Level 3 – Proficient	Level 4 – Transformational
Moral Awareness	Fails to identify ethical issues; focuses only on technical or procedural aspects	Recognizes obvious ethical concerns when prompted	Independently identifies ethical risks, including indirect impacts	Anticipates and surfaces hidden or systemic ethical risks
Moral Judgment	Relies on rules or authority without deeper reasoning	Applies basic reasoning inconsistently	Uses structured ethical reasoning grounded in justice and impact	Integrates multiple perspectives and articulates principled decisions
Pressure Navigation	Conforms to authority or group norms	Recognizes pressure but struggles to resist it	Maintains ethical stance under moderate pressure	Resists pressure and reframes constraints while influencing others
Moral Action	Avoids action or defers responsibility	Acts when risk is low or indirect	Takes clear action and communicates ethical concerns	Acts decisively under risk and influences outcomes
Ethical Resilience	Abandons ethical position under challenge	Attempts to persist but yields under pressure	Sustains ethical commitments despite resistance	Demonstrates consistent, long-term ethical action
Reflective Formation	Minimal reflection; externalizes responsibility	Reflects superficially on decisions	Engages in critical self-evaluation and learning	Integrates reflection into identity and continuous development

4.8.4 Application in Educational Practice

The assessment model is designed to be embedded within simulation-based learning, case analysis, project-based environments, and reflective exercises. In simulations, instructors evaluate real-time behavior under pressure, aligning assessment with IMCFM's emphasis on action. In case-based analysis, the rubric captures the depth and structure of ethical reasoning. Reflective journals provide insight into internal processes of moral development, while peer evaluation introduces social accountability and diverse perspectives.

This approach is consistent with leadership assessment practices that combine multiple data sources, observation, self-reflection, and peer feedback, to improve reliability and validity (Brown et al., 2005). It also supports both formative and summative assessment, enabling continuous development while providing a basis for evaluating readiness for ethical leadership.

Ultimately, this model reinforces the central claim of this chapter: that moral courage is not merely an abstract virtue, but a trainable, observable, and measurable capability essential for responsible leadership in the age of artificial intelligence.

5. APPLIED CASE: FORMING MORAL COURAGE IN AI DECISION-MAKING

To illustrate how the IMCFM operates in practice, the following case integrates framework, pedagogy, and assessment within a realistic AI decision-making scenario. This applied example demonstrates how moral courage can be developed, observed, and evaluated under conditions that reflect real-world complexity.

5.1 Case Context: AI Deployment Under Ethical Risk

A multinational technology firm is preparing to launch an AI-driven decision-support system designed to optimize resource allocation in a public-sector environment. Initial testing demonstrates significant efficiency gains and cost reductions. However, internal audit findings reveal that the system produces systematically unfavorable outcomes for certain vulnerable populations due to historical bias embedded in the training data.

The issue is not immediately visible in aggregate performance metrics, which meet regulatory standards and internal benchmarks. Senior leadership, under pressure to meet market deadlines and maintain competitive positioning, encourages proceeding with deployment while "monitoring outcomes post-launch." The project team is divided. Some members view the issue as acceptable within compliance thresholds, while others raise concerns about fairness, reputational risk, and long-term societal impact.

This scenario presents a prototypical AI-era ethical dilemma: a technically compliant system that produces ethically problematic outcomes under conditions of organizational pressure. The case provides a structured context for applying the Integrated Moral Courage Formation Model (IMCFM) and associated pedagogical and assessment approaches.

5.2 Application of the IMCFM

5.2.1 Moral Awareness

At the initial stage, participants must recognize that the issue extends beyond technical performance to ethical implications. The bias is not explicitly labeled as a moral failure; rather, it is embedded within data and statistical outputs. Students must identify the ethical dimension of the problem, including the disproportionate impact on vulnerable populations and the broader implications for justice and human dignity.

This stage often reveals variability in perception. Some participants focus on compliance and efficiency, while others identify systemic harm. This exercise highlights the role of cognitive framing and moral sensitivity in shaping awareness.

5.2.2 Moral Judgment

Participants are then required to evaluate the situation using structured ethical reasoning. This involves weighing competing priorities, including efficiency, fairness, organizational performance, and public accountability. Students must articulate whether proceeding with deployment is ethically justifiable and under what conditions.

At this stage, the development of *phronesis* becomes evident. More advanced participants integrate multiple perspectives, considering not only immediate outcomes but also long-term societal effects and stakeholder trust. The deliberation process requires explicit justification, moving beyond intuitive responses toward principled reasoning.

5.2.3 Pressure and Constraints

The case introduces escalating pressure through simulated executive directives, time constraints, and performance expectations. Participants are informed that delaying the launch may result in financial losses, reputational risk with investors, and internal consequences for the project team.

These conditions replicate real-world environments where ethical decisions are made under constraint. Participants experience tension between personal values and institutional expectations, illustrating how organizational context can distort judgment and suppress ethical concerns.

5.2.4 Moral Action

Participants must decide whether to proceed, delay, or escalate concerns. They are required to communicate their decision to a simulated executive panel, defend their reasoning, and respond to counterarguments emphasizing cost, competition, and feasibility.

This stage evaluates the transition from reasoning to action. Some participants may hesitate or defer responsibility, while others articulate clear positions and advocate for ethical alternatives, such as retraining the model, implementing fairness constraints, or postponing deployment.

The exercise reveals that moral courage is not merely knowing what is right but acting upon it despite resistance.

5.2.5 Reflective Formation

Following the simulation, participants engage in structured reflection. They analyze their emotional responses, decision-making processes, and the influence of pressure on their behavior. Questions focus on identifying moments of hesitation, sources of bias, and alignment between values and action.

Peer discussion and instructor feedback further deepen reflection, exposing participants to alternative approaches and reinforcing learning. Over time, this process contributes to the development of a stable moral identity and increased consistency in ethical action.

5.3 Pedagogical Integration

This case is designed to align directly with the pedagogical model outlined in this chapter. It integrates experiential simulation, structured deliberation, ethical action training, and reflective practice into a cohesive learning cycle.

Students are not passive recipients of ethical principles but active participants in decision-making processes that replicate real-world complexity. The inclusion of role-based accountability, competing incentives, and time pressure ensures that learning occurs under conditions that mirror professional environments.

This approach reflects the principles of adult learning and wisdom formation, where knowledge is integrated through experience, reflection, and application rather than abstract instruction alone.

5.4 Assessment Using the IMCFM Rubric

Participant performance in this case can be evaluated using the IMCFM-aligned Moral Courage Assessment Model. Assessment focuses on observable behavior across six dimensions.

Moral awareness is evaluated based on the participant's ability to identify ethical risks beyond technical metrics. Moral judgment is assessed through the quality and structure of reasoning, including the integration of fairness, dignity, and long-term impact.

Pressure navigation is evaluated by observing how participants respond to simulated constraints, including whether they maintain ethical clarity or conform to external pressure. Moral action is assessed through the clarity, decisiveness, and effectiveness of the participant's response.

Ethical resilience is reflected in the participant's ability to sustain their position throughout the simulation, particularly when challenged. Reflective formation is evaluated through post-simulation analysis, including the depth of self-assessment and the integration of learning into future decision-making.

This multi-dimensional assessment enables educators to identify not only overall performance but specific developmental strengths and gaps, supporting targeted improvement.

6. MORAL COURAGE, FAITH, AND THE FORMATION OF ETHICAL LEADERS IN THE AI ERA

Beyond pedagogical design and applied practice, a deeper question remains regarding what ultimately sustains moral courage in high-pressure environments. While the IMCFM explains how ethical action can be developed, this section considers the underlying sources of motivation and commitment that enable individuals to act consistently when ethical decisions carry personal or professional cost.

From a faith-informed perspective, moral courage is not sustained by cognitive frameworks alone, but by deeper commitments related to identity, purpose, and moral accountability. Within the Christian tradition, ethical action is grounded not only in principles of justice and human dignity, but also in the recognition that individuals are stewards of knowledge, influence, and decision-making responsibility (Luke 12:48). This perspective reframes ethical leadership in AI not merely as a professional obligation, but as a form of entrusted responsibility with broader societal implications.

The rapid expansion of artificial intelligence intensifies this responsibility. As AI systems increasingly shape decisions at scale, leaders exercise forms of influence that extend beyond immediate organizational outcomes to long-term social consequences. In such contexts, the formation of moral courage requires more than technical competence or ethical awareness; it requires a stable moral foundation capable of resisting both external pressure and internal rationalization. This aligns with the principle that ethical integrity is tested not in ideal conditions, but in moments of constraint and ambiguity (James 1:2–4).

Christian ethical thought contributes to this challenge by emphasizing the formation of character through practices such as humility, accountability, and service. Humility is particularly relevant in AI contexts, where overconfidence in technological systems can obscure limitations and unintended consequences. A posture of intellectual humility, recognizing the limits of knowledge and the need for continual discernment, supports more responsible decision-making (Proverbs 11:2). Similarly, the concept of justice extends beyond procedural compliance to include an active concern for fairness and the protection of vulnerable populations (Micah 6:8).

Moreover, faith-based perspectives introduce a dimension of moral accountability that is not entirely contingent on organizational incentives or social approval. In environments where ethical action may carry personal or professional cost, the capacity to act with courage is strengthened when individuals are guided by commitments that transcend immediate outcomes (Acts 5:29). This does not negate the importance of institutional governance or ethical frameworks, but it reinforces the internal orientation necessary to act when such structures are insufficient or underdeveloped.

Importantly, this perspective does not replace the interdisciplinary foundations outlined in the IMCFM but rather deepens them. Virtue ethics, moral psychology, andragogy, and ethical resilience explain how moral courage can be developed and sustained; faith-based perspectives contribute an additional layer of meaning that reinforces why such formation matters and how it can be anchored in enduring commitments. In this sense, ethical leadership is not only a matter of competence, but of character shaped over time through consistent practice and reflection (Romans 5:3–5).

The development of moral courage in the AI era is therefore not solely an educational or organizational challenge, but a formative one. It requires the integration of knowledge, practice, and identity in ways that enable leaders to act with clarity and conviction in complex environments. By aligning ethical formation with deeper moral and spiritual foundations, leadership development can move beyond compliance toward a more holistic vision of responsibility, one that is capable of guiding technological innovation in ways that reflect both human dignity and the broader common good.

7. CONCLUSION

Taken together, the preceding sections provide a comprehensive framework for understanding and cultivating ethical leadership in the age of artificial intelligence.

The central challenge is no longer confined to understanding ethical principles or establishing governance frameworks. Rather, it lies in the capacity of individuals to recognize ethical risk, exercise sound judgment, and act with consistency and conviction under conditions of pressure, uncertainty, and competing priorities.

This chapter has argued that existing approaches to AI ethics education remain insufficient because they prioritize knowledge acquisition over moral formation. While frameworks, regulations, and technical safeguards are essential, they do not, in isolation, ensure ethical action. The evidence suggests that individuals often fail not due to a lack of awareness, but because of the psychological, organizational, and situational pressures that distort judgment and inhibit action. Addressing this gap requires a shift from teaching ethics as a conceptual domain to cultivating moral courage as a practiced and measurable capability.

To this end, the chapter introduced the Integrated Moral Courage Formation Model (IMCFM), which conceptualizes ethical leadership as a developmental process encompassing moral awareness, judgment, pressure navigation, action, and reflective formation. By integrating insights from virtue ethics, moral psychology, andragogy, and ethical resilience, the model provides a coherent framework for understanding how ethical action is formed and sustained in complex environments. Importantly, it reframes moral courage not as an abstract virtue, but as a capability that emerges through repeated engagement with real-world dilemmas.

Building on this framework, the chapter proposed a structured pedagogical model that aligns educational practice with each stage of the IMCFM. Through experiential simulations, structured deliberation, action-oriented training, and reflective learning, individuals are placed in conditions that replicate the realities of ethical decision-making in AI contexts. These interventions move beyond passive instruction, requiring learners to confront ethical tension, navigate competing values, and act under constraint. The inclusion of social reinforcement mechanisms, such as peer accountability and mentorship, further strengthens the capacity to sustain ethical commitments over time.

A key contribution of this chapter is the development of an IMCFM-aligned assessment model, which addresses a longstanding limitation in ethics education: the absence of tools to measure moral action in practice. By evaluating observable performance across dimensions such as awareness, judgment, pressure navigation, action, resilience, and reflection, the rubric provides a structured approach to assessing moral courage as a developmental construct. This enables educators to move beyond theoretical evaluation toward a more rigorous and actionable understanding of ethical leadership readiness.

The applied case presented in this chapter demonstrates how these elements, framework, pedagogy, and assessment, can be integrated into a cohesive learning system. By simulating real-world AI dilemmas and evaluating participant responses, the model provides a practical pathway for developing leaders who are capable not only of ethical reasoning, but of ethical action under pressure. This integration represents a significant step toward bridging the gap between ethical intention and real-world behavior.

Finally, the chapter situates moral courage within a broader formative context, recognizing that sustained ethical action requires more than cognitive or procedural competence. A faith-informed perspective contributes an additional layer of depth by emphasizing identity, purpose, and moral accountability as foundational elements of leadership. In environments where external pressures may discourage principled action, these deeper commitments provide an internal anchor that supports consistency and conviction.

In conclusion, the development of moral courage in the age of artificial intelligence is not merely an academic or technical challenge, but a formative imperative. As AI systems continue to expand their influence across society, the responsibility placed on leaders will only intensify. The question is no longer whether ethical frameworks exist, but whether individuals are prepared to act upon them when it matters most. By integrating theoretical insight, practical application, and measurable outcomes, this chapter advances a unified approach to forming leaders capable of acting with clarity, resilience, and sustained moral courage in AI-driven environments.

REFERENCES

Aristotle, . (2009). *Nicomachean ethics* (Ross, W. D., Trans.). Oxford University Press. (Original work published 350 BCE)

Bandura, A. (1977). Self-efficacy: Toward a unifying theory of behavioral change. *Psychological Review, 84*(2), 191–215. DOI: 10.1037/0033-295X.84.2.191 PMID: 847061

Barocas, S., & Selbst, A. D. (2016). Big data's disparate impact. *California Law Review, 104*(3), 671–732. DOI: 10.15779/Z38BG31

Brown, M. E., Treviño, L. K., & Harrison, D. A. (2005). Ethical leadership: A social learning perspective for construct development and testing. *Organizational Behavior and Human Decision Processes, 97*(2), 117–134. DOI: 10.1016/j.obhdp.2005.03.002

Buolamwini, J., & Gebru, T. (2018). Gender shades: Intersectional accuracy disparities in commercial gender classification. *Proceedings of Machine Learning Research, 81*, 1–15. http://proceedings.mlr.press/v81/buolamwini18a.html

Edmondson, A. (1999). Psychological safety and learning behavior in work teams. *Administrative Science Quarterly, 44*(2), 350–383. DOI: 10.2307/2666999

Gioia, D. A. (1992). Pinto fires and personal ethics: A script analysis of missed opportunities. *Journal of Business Ethics, 11*(5–6), 379–389. DOI: 10.1007/BF00870550

Haidt, J. (2001). The emotional dog and its rational tail: A social intuitionist approach to moral judgment. *Psychological Review, 108*(4), 814–834. DOI: 10.1037/0033-295X.108.4.814 PMID: 11699120

Hannah, S. T., Avolio, B. J., & May, D. R. (2011). Moral maturation and moral conation: A capacity approach to explaining moral thought and action. *Academy of Management Review, 36*(4), 663–685. DOI: 10.5465/amr.2010.0128

Hursthouse, R., & Pettigrove, G. (2018). Virtue ethics. In Zalta, E. N. (Ed.), *The Stanford encyclopedia of philosophy* (Fall 2018 ed.)., https://plato.stanford.edu/entries/ethics-virtue/

Knowles, M. S., Holton, E. F.III, & Swanson, R. A. (2005). *The adult learner: The definitive classic in adult education and human resource development* (6th ed.). Elsevier. DOI: 10.4324/9780080481913

Kohlberg, L. (1981). Essays on moral development: Vol. 1. *The philosophy of moral development*. Harper & Row.

Lahteenkorva, M., Tynjälä, P., & Kallio, E. (2023). From knowledge to wisdom in sustainable management education. *Journal of Management Education*. Advance online publication. DOI: 10.1177/10525629231 1XXXX

MacIntyre, A. (1981). *After virtue: A study in moral theory*. University of Notre Dame Press.

Masten, A. S. (2001). Ordinary magic: Resilience processes in development. *The American Psychologist*, *56*(3), 227–238. DOI: 10.1037/0003-066X.56.3.227 PMID: 11315249

McLoughlin, S., Fernández-Espinosa, V., & Velázquez-Gil, M. (2026). Practical wisdom, emerging technologies, and AI. *Journal of Moral Education*, 1–31. DOI: 10.1080/03057240.2026.2624863

Obermeyer, Z., Powers, B., Vogeli, C., & Mullainathan, S. (2019). Dissecting racial bias in an algorithm used to manage the health of populations. *Science*, *366*(6464), 447–453. DOI: 10.1126/science.aax2342 PMID: 31649194

Page, S. E. (2007). *The difference: How the power of diversity creates better groups, firms, schools, and societies*. Princeton University Press.

Rest, J. R. (1986). *Moral development: Advances in research and theory*. Praeger.

Trevino, L. K., & Brown, M. E. (2004). Managing to be ethical: Debunking five business ethics myths. *The Academy of Management Perspectives*, *18*(2), 69–81. DOI: 10.5465/ame.2004.13837400

Treviño, L. K., Weaver, G. R., & Reynolds, S. J. (2006). Behavioral ethics in organizations: A review. *Journal of Management*, *32*(6), 951–990. DOI: 10.1177/0149206306294258

Wright, J. C., Warren, M. T., & Snow, N. E. (2021). Understanding virtue: Theory and measurement. *Journal of Moral Education*, *50*(1), 1–16. DOI: 10.1080/03057240.2020.1819073

Zimbardo, P. G. (2007). *The Lucifer effect: Understanding how good people turn evil*. Random House.

KEY TERMS AND DEFINITIONS

AI Governance: The set of policies, processes, and oversight mechanisms designed to ensure that artificial intelligence systems are developed and used in a responsible, transparent, and accountable manner.

Algorithmic Bias: Systematic and repeatable errors in AI systems that produce unfair or discriminatory outcomes, often resulting from biased data or flawed design processes.

Ethical Deliberation: A structured process of reasoning in which individuals or groups evaluate competing values, consider consequences, and justify decisions based on ethical principles.

Ethical Resilience: The capacity to sustain ethical commitments over time, particularly under conditions of stress, uncertainty, or opposition.

Experiential Learning: An educational approach in which individuals develop understanding through direct engagement with realistic scenarios, followed by reflection and application.

Integrated Moral Courage Formation Model (IMCFM): A developmental framework that explains how moral courage is formed through the interaction of moral awareness, judgment, pressure navigation, action, and reflective formation, supported by resilience and social reinforcement.

Moral Action: The execution of ethical decisions through behavior, including communicating concerns, challenging flawed decisions, and taking responsibility for outcomes.

Moral Awareness: The ability to recognize ethical issues, identify potential harm, and understand the broader impact of decisions on individuals, organizations, and society.

Moral Courage: The capacity to act in accordance with ethical principles despite the presence of risk, pressure, or potential negative consequences, particularly in complex or high-stakes environments.

Moral Judgment: The process of evaluating what is ethically right by applying principles such as fairness, justice, and human dignity, especially when faced with competing priorities.

Phronesis (Practical Wisdom): A form of applied judgment that enables individuals to make context-sensitive ethical decisions by integrating principles with experience and situational awareness.

Pressure and Constraints: The set of organizational, social, emotional, and cognitive factors, such as deadlines, authority, incentives, and group norms, that influence and often challenge ethical decision-making.

Reflective Formation: A developmental process in which individuals critically examine their decisions and experiences, integrating feedback and self-awareness to strengthen future ethical behavior.

Social Reinforcement: The influence of peer interaction, mentorship, and group norms in shaping and sustaining ethical behavior over time.

Section 2
Leadership Formation in the Age of AI

Chapter 6
Character Formation for Ethical Leadership in AI-Based Organizations:
A Faith-Centered Framework

Vipin Tyagi
http://orcid.org/0000-0003-4994-3686
Jaypee University of Engineering and Technology, Guna, India

ABSTRACT

As artificial intelligence increasingly shapes organizational decision-making, leadership effectiveness can no longer be defined only by technical competence or digital fluency. AI-mediated workplaces require leaders whose character, ethical discernment and commitment to human dignity guide the design and use of intelligent systems. This chapter positions character formation as a core dimension of leadership development in the AI era, integrating faith-informed ethics, servant leadership and human flourishing. It examines how AI systems can amplify bias, reduce accountability and distance leaders from the human consequences of automated decisions, increasing the need for morally grounded leadership. Chapter proposes a framework for cultivating humility, justice, compassion, moral courage and responsibility in AI-mediated environments. The chapter outlines practical strategies for embedding character development, empathy and ethical reasoning into leadership education and organizational practice, preparing leaders to steward AI responsibly in service of human dignity and the common good.

DOI: 10.4018/979-8-3373-8192-3.ch006

ght © 2027, IGI Global Scientific Publishing. Copying or distributing in print or electronic forms without written permission of IGI Global Scientific Publishing is prohibited. Use of this chapter to generative artificial intelligence (AI) technologies is expressly prohibited. The publisher reserves all rights to license its use for generative AI training and machine learning model development.

INTRODUCTION

Artificial intelligence (AI) is rapidly reshaping how organizations make decisions, allocate resources and interact with stakeholders. From algorithmic hiring systems to predictive analytics in healthcare and finance, AI-based environments are becoming a defining feature of contemporary institutional life. While much attention has been devoted to technical capability, data governance and algorithmic performance, far less emphasis has been placed on the moral and character formation of the leaders who design, deploy and oversee these systems. This imbalance creates a critical leadership gap. Technically sophisticated systems guided by ethically underdeveloped leadership can unintentionally amplify injustice, erode trust and undermine human dignity.

The central argument of this chapter is that character formation must be treated as a core leadership competency in AI-based organizations. Technical fluency alone is insufficient for navigating the complex moral terrain introduced by intelligent automation. Leaders must possess the ethical discernment, humility and moral courage necessary to steward AI in ways that promote human flourishing and the common good. This position aligns with growing scholarly concern that AI governance challenges are fundamentally socio-technical and moral in nature rather than purely computational problems (Floridi et al., 2018).

Faith-informed leadership traditions, particularly those rooted in servant leadership and virtue ethics offer a rich but underutilized resource for addressing this gap. Long before the rise of AI, scholars in these traditions emphasized that leadership effectiveness flows from the moral character of the leader. Greenleaf's (1977) seminal work on servant leadership, for example, reframed leadership not as the exercise of power but as the ethical responsibility to serve others first. Similarly, virtue ethics traditions highlight the formation of moral habits such as justice, prudence and compassion as the foundation of wise decision-making (MacIntyre, 1984). These perspectives are especially relevant in AI-based contexts, where leaders face complex ethical decisions shaped by technological systems.

This chapter advances a faith-centered framework for character formation in AI-based organizations. It integrates insights from AI ethics, servant leadership theory and human flourishing scholarship to propose practical pathways for developing morally grounded leaders. Particular attention is given to inclusion and dignity, especially regarding persons with disabilities and historically marginalized populations, whose interests are often most vulnerable in automated decision systems.

The discussion proceeds in five major movements. First, the chapter examines the distinctive ethical challenges introduced by AI-based organizational environments. Second, it explores the theological and philosophical foundations of character-based leadership. Third, it develops a structured framework for character formation aligned with AI governance needs. Fourth, it offers practical strategies for educators and organizations. Finally, it considers implications for future leadership development in an increasingly automated world.

This chapter focuses on character formation as its primary contribution to AI-era leadership, rather than attempting to provide a comprehensive treatment of all AI ethics issues. By repositioning character formation at the center of leadership preparation, this chapter contributes to the broader objective of Leading With Faith and Ethics in an AI World, preparing leaders who not only understand intelligent technologies but who steward them wisely in service of human dignity and the common good.

THE ETHICAL LANDSCAPE OF AI-BASED ORGANIZATIONS

The Rise of Algorithmic Decision-Making

Organizations today increasingly depend on AI systems to support or automate decisions that were once the exclusive domain of human judgment. These systems influence hiring, credit scoring, medical diagnosis, policing strategies, educational assessment and social service allocation. While AI promises efficiency and scalability, it also introduces significant ethical risks that require careful governance.

Researchers have noted that AI systems often inherit and amplify historical inequities embedded in training data (O'Neil, 2016). Because machine learning models detect statistical patterns rather than moral truths, they may reproduce discriminatory outcomes even when designers have no harmful intent. This phenomenon has been documented in domains such as predictive systems, hiring algorithms and facial recognition systems (Buolamwini & Gebru, 2018).

In AI-based organizations, leaders therefore confront a new category of responsibility: algorithmic stewardship. This role requires not only technical oversight but also moral judgment about fairness, transparency and human impact. According to Floridi et al. (2018), effective AI governance must integrate principles such as beneficence, non-maleficence, autonomy, justice and explicability. These principles are deeply ethical in nature and cannot be operationalized adequately without morally formed leadership.

Recent regulatory developments further reinforce the importance of ethical AI leadership. The European Union's AI Act (European Commission, 2021) introduces risk-based classification of AI systems and mandates strict compliance requirements for high-risk applications, including transparency, human oversight and accountability. Similarly, National Institute of Standards and Technology (NIST, 2023), emphasizes governance, mapping, measurement and management of AI risks, while OECD AI Principles (OECD, 2019) highlight human-centered values, fairness and transparency.

These frameworks signal a global shift toward enforceable AI governance, making leadership character not only ethically desirable but institutionally necessary. To clarify the major ethical risks associated with AI-based organizational environments and their implications for leadership responsibility, Table 1 summarizes common algorithmic risks alongside their potential organizational consequences and leadership responses.

Table 1. AI ethical risks and leadership implications

AI Ethical Risk	Description	Organizational Consequences	Leadership Responsibility
Algorithmic Bias	Models reproduce historical patterns embedded in training data	Discriminatory hiring, credit, or service allocation	Ensure fairness audits and representative datasets
Opacity (Black Box Models)	Decision logic is difficult to interpret	Reduced accountability and stakeholder trust	Implement explainability standards and documentation
Automation Bias	Human decision-makers over-rely on AI outputs	Poor decisions despite available human judgment	Promote critical oversight and human-in-the-loop design
Moral Distancing	Decision-makers feel detached from consequences	Reduced ethical accountability	Reinforce responsibility for AI-based outcomes
Scale of Impact	AI decisions affect thousands simultaneously	Rapid amplification of errors or inequities	Implement monitoring and governance mechanisms

The Risk of Moral Distancing

Automation can create moral distance between decision-makers and the human consequences of their decisions, reducing perceived accountability. This concern echoes earlier research in organizational ethics showing that bureaucratic and technological mediation can diffuse moral accountability (Zuboff, 1988). AI systems intensify this dynamic because their internal logic is often opaque even to their creators. The result can be a dangerous combination of high-impact decisions and reduced felt responsibility.

Faith-based leadership traditions directly challenge this tendency. Servant leadership, for example, insists that leaders remain relationally and morally accountable to those affected by their decisions (Greenleaf, 1977). Similarly, Christian ethical thought emphasizes the inherent dignity of every person as bearing the imago Dei

(Genesis 1:27), a concept that strongly resists the depersonalization risks associated with automated systems.

Bias, Inclusion and the Marginalized

AI ethics studies consistently show that marginalized populations, including persons with disabilities, face disproportionate risk from poorly governed AI systems. Disability scholars have warned that algorithmic systems often encode normative assumptions about bodies, cognition and behavior that exclude or disadvantage people with disabilities (Whittaker et al., 2019).

For example:

- Automated hiring tools may penalize nonstandard speech patterns
- Facial recognition systems may perform poorly on atypical facial features
- Educational analytics may misinterpret neurodivergent learning behaviors

These risks highlight why inclusion must be treated not as a secondary concern but as a core ethical essential in AI leadership. From a faith-informed perspective, the preferential concern for the vulnerable and marginalized is deeply rooted in both biblical teaching and servant leadership theory.

The ethical challenges outlined above are not merely technical or procedural problems; they are fundamentally moral in nature. This shift from external risks to internal capacities raises a critical question: What kind of leaders are required to navigate such complexity? The next section addresses this by examining the moral and philosophical foundations of character formation.

THEOLOGICAL AND PHILOSOPHICAL FOUNDATIONS OF CHARACTER FORMATION

Virtue Ethics and Moral Formation

The concept of character formation is most fully developed within virtue ethics traditions. Rather than focusing solely on rules or consequences, virtue ethics emphasizes the cultivation of stable moral dispositions that enable wise action across diverse situations (MacIntyre, 1984). This approach is particularly well suited to AI-based environments, where leaders frequently encounter novel ethical dilemmas that cannot be resolved by simple rule-following.

Key virtues relevant to AI leadership include:

- Prudence (practical wisdom)
- Justice (fairness and equity)
- Temperance (restraint in technological deployment)
- Courage (willingness to challenge harmful systems)
- Humility (recognition of technological limits)

These virtues align with Russell's (2019) argument that AI systems must be designed to remain compatible with human values. Virtue ethics provides a useful framework for understanding the moral dispositions required for responsible AI leadership. Table 2 summarizes key virtues and their relevance to AI-based decision environments.

Table 2. Virtue ethics foundations for AI leadership

Virtue	Definition	Relevance to AI Leadership	Example Application
Prudence	Practical wisdom in complex situations	Evaluating risks of AI deployment	Deciding when algorithmic decisions require human oversight
Justice	Commitment to fairness and equity	Preventing algorithmic discrimination	Implementing bias detection and fairness metrics
Temperance	Responsible restraint in the use of power	Avoiding premature AI deployment	Delaying implementation until risks are assessed
Courage	Willingness to act despite pressure	Challenging unethical AI practices	Halting deployment of harmful systems
Humility	Recognition of limits of knowledge	Avoiding technological overconfidence	Questioning algorithmic outputs

Servant Leadership as a Moral Framework

Robert Greenleaf's (1977) servant leadership model provides a particularly powerful view for AI-era leadership. The servant leader's primary question: "Do those served grow as persons?" maps directly onto contemporary concerns about human flourishing in digital environments.

Research on servant leadership has consistently linked the model to higher organizational trust, ethical climate and employee well-being (Eva et al., 2019). These outcomes are especially important in AI-based organizations, where trust deficits can quickly emerge if stakeholders perceive systems as unfair or dehumanizing. Core servant leadership traits relevant to AI contexts include:

- Empathy
- Stewardship

- Commitment to the growth of people
- Community building
- Foresight

Each of these qualities directly counterbalances known risks of algorithmic governance.

Human Flourishing as the North Star

The concept of human flourishing provides an integrative moral goal for AI leadership. Rooted in Aristotelian eudaimonia and developed in contemporary positive psychology, flourishing refers to holistic well-being encompassing meaning, virtue, relationships and purpose (VanderWeele, 2017).

Within faith traditions, flourishing is closely linked to the biblical vision of shalom; comprehensive peace and wholeness. When applied to AI governance, flourishing shifts the central leadership question from: "Is the system efficient?" to "Does this technology genuinely promote human well-being and dignity?"

This reframing is essential for resisting purely technocratic approaches to AI adoption. Having established the moral foundations of character-based leadership, the next section discusses the importance of these virtues within AI-based environments, where scale, speed and uncertainty amplify the consequences of leadership decisions.

IMPORTANCE OF CHARACTER IN AI-BASED LEADERSHIP

AI systems amplify the consequences of leadership decisions, increasing the importance of ethical judgment, especially in situations involving uncertainty and value trade-offs. For these reasons, the formation of leaders who embody humility, justice, compassion and moral courage is not optional; it is foundational to trustworthy AI deployment.

A defining issue in AI leadership lies in the trade-off between speed and responsibility and between cost efficiency and ethical integrity. Organizations often face pressure to deploy AI systems rapidly to maintain competitive advantage, even when ethical risks remain unresolved.

Leaders of strong character recognize that responsible delay may be more valuable than premature deployment. Ethical AI leadership therefore requires the ability to navigate these tensions without defaulting to purely economic optimization.

A FAITH-CENTERED FRAMEWORK FOR CHARACTER FORMATION IN AI LEADERSHIP

The proposed framework organizes character formation around five core leadership domains that address the ethical demands of AI-based environments. Table 3 summarizes these domains and their primary leadership implications.

Table 3. Character domains for ethical AI leadership

Character Domain	Core Leadership Quality	AI Governance Focus	Organizational Practice
Humility	Awareness of technological limits	Avoiding automation bias	AI failure reviews
Justice	Fairness and equity	Algorithmic bias mitigation	Bias audits
Compassion	Attention to human impact	Human-centered AI design	Stakeholder engagement
Moral Courage	Ethical action under pressure	Challenging harmful systems	Responsible deployment pauses
Stewardship	Long-term responsibility	Sustainable AI governance	Lifecycle oversight

The preceding discussion shows that AI-based environments intensify the moral demands placed on leaders. The question that follows is practical and formative: How can institutions intentionally cultivate leaders whose character is adequate for the ethical complexity of AI systems? This section proposes a faith-centered framework for character formation designed specifically for AI-based organizational contexts. The framework rests on three foundational commitments:

- Character is formable, not merely innate.
- Moral formation must be intentional and institutionalized.
- AI leadership requires the integration of virtue, discernment and technical awareness.

Based on virtue ethics, servant leadership and contemporary AI governance literature, the framework identifies five core character domains essential for trustworthy AI leadership.

Domain 1: Humility and Epistemic Awareness

Perhaps the most urgent desirable quality in AI-based leadership is humility. AI systems often project an illusion of precision and objectivity that can foster overconfidence among decision-makers. Yet researchers repeatedly emphasize that machine learning systems are inherently probabilistic, context-dependent and vulnerable to error (Russell, 2019). Epistemic humility enables leaders to recognize:

- the limits of training data
- the fragility of model generalization
- the risk of automation bias
- the uncertainty inherent in predictive analytics

Without humility, organizations may deploy AI prematurely or treat algorithmic outputs as infallible. Research in human–automation interaction shows that users frequently over-depend on automated recommendations, even when systems are known to be imperfect (Parasuraman & Riley, 1997). Leaders must therefore cultivate what might be called a balanced form of critical trust in AI tools, combined with ongoing scrutiny.

Faith traditions strongly reinforce this posture. Humility naturally tempers technological triumphalism and encourages responsible stewardship. Organizations and educational programs can cultivate epistemic humility through:

- structured AI failure case analyses
- post-deployment ethical audits
- reflective leadership exercises
- cross-disciplinary review boards

These practices help leaders internalize the reality that intelligent systems require continuous moral and technical oversight.

Domain 2: Justice and Algorithmic Fairness

Justice is central to both virtue ethics and AI governance. Given the well-documented risks of bias in algorithmic systems, the leadership challenge now shifts from recognition to sustained mitigation across the AI lifecycle. Research has documented disparities in areas such as:

- facial recognition accuracy (Buolamwini & Gebru, 2018)
- predictive policing outcomes

- credit scoring models
- automated hiring tools

From a leadership perspective, justice requires more than compliance with anti-discrimination laws. It involves proactive attention to fairness across the entire AI lifecycle; data collection, model training, validation, deployment and monitoring.

Floridi et al. (2018) identify justice as one of the core principles for ethical AI, emphasizing fairness, non-discrimination and equitable distribution of benefits. Faith-informed ethics deepen this commitment by grounding justice in the inherent dignity of every person rather than merely procedural fairness. Leaders committed to algorithmic justice must ensure:

- representative and inclusive datasets
- bias testing and impact assessments
- transparency in automated decisions
- accessible appeal mechanisms
- ongoing disparity monitoring

These responsibilities require moral resolve, especially when fairness interventions increase cost or reduce short-term efficiency.

Domain 3: Compassion and Human-Centered Design

AI-based organizations risk becoming overly optimized for efficiency at the expense of human experience. Compassion functions as a critical counterbalance, keeping leaders attentive to the lived realities of those affected by automated systems.

According to human-centered AI research, systems should be designed to augment rather than replace human agency and dignity (Shneiderman, 2020). Compassionate leadership supports this orientation by asking not only whether systems work, but for whom they work well and who may be harmed.

This domain is especially important for disability inclusion. As Whittaker et al. (2019) note, many AI systems implicitly assume normative patterns of speech, movement, cognition, or perception. Leaders lacking empathic awareness may overlook these embedded exclusions.

Faith traditions place strong emphasis on care for the vulnerable, the marginalized and those at risk of social exclusion. Within servant leadership, empathy and commitment to the growth of people are central characteristics (Greenleaf, 1977; Eva et al., 2019). Compassion in AI leadership can be cultivated through:

- participatory design with affected communities

- disability-inclusive usability testing
- stakeholder impact narratives
- ethics-of-care case studies
- immersion experiences with end users

Such practices reduce moral distance and help leaders remain attentive to human consequences.

Domain 4: Moral Courage and Responsible Intervention

Even ethically aware leaders may hesitate to challenge problematic AI systems when doing so carries organizational or political risk. For this reason, moral courage is an indispensable leadership virtue in AI-based environments.

Moral courage involves the willingness to question, delay, or halt AI implementation when ethical concerns arise even under pressure for rapid innovation. This capacity is increasingly important as organizations race to adopt AI technologies for competitive advantage.

There are many examples in the history, where technological systems were widely adopted before their harms were fully understood. O'Neil (2016) documents how high-stakes algorithms in education, criminal justice and finance produced significant social harm when deployed without sufficient ethical oversight. Leaders of strong character must be prepared to:

- pause deployment when fairness is uncertain
- escalate ethical concerns internally
- advocate for affected stakeholders
- resist purely profit-driven AI adoption
- support whistle-blowing protections

From a faith perspective, moral courage reflects the prophetic tradition of speaking truth in the face of institutional pressure. It is closely linked to the virtue of fortitude in classical virtue ethics.

Domain 5: Stewardship and Long-Term Responsibility

The final domain, stewardship, integrates the previous virtues into a sustained posture of responsible governance. Stewardship emphasizes that leaders are caretakers of both technological power and human well-being. In AI contexts, stewardship requires attention to long-term and systemic effects, not merely immediate performance gains. This includes anticipating downstream impacts such as:

- workforce displacement
- privacy erosion
- dependency on opaque systems
- environmental costs of large-scale computation
- societal trust in automated decision-making

According to Russell (2019), aligning AI with human values requires sustained human oversight and institutional responsibility. Similarly, Shneiderman (2020) advocates for human-centered AI that preserves human control and accountability.

Faith-informed leadership strongly resonates with stewardship language, framing technological authority as a trust to be exercised for the common good. Organizations seeking steward leaders should implement:

- AI ethics governance boards
- lifecycle accountability structures
- long-term impact assessments
- cross-functional ethical review processes
- public transparency commitments

These structures support leaders in translating personal virtue into organizational practice.

INTEGRATING CHARACTER FORMATION INTO LEADERSHIP DEVELOPMENT

While the preceding sections discuss the moral demands of AI-based leadership and define the necessary virtues, the question now shifts from "what is required" to "how can it be formed." This section translates ethical need into actionable formation strategies, bridging the gap between diagnosis and implementation.

Character does not emerge automatically from technical training. It must be intentionally cultivated through educational design, organizational culture and sustained reflective practice. Research in moral development suggests that ethical capacity grows through repeated engagement with real dilemmas, guided reflection and community norms (Rest et al., 1999).For AI-based leadership, formation must occur across three levels:

- Educational preparation
- Organizational socialization
- Ongoing professional formation

Implications for Higher Education

Leadership and technology programs should move beyond purely technical AI curricula to include:

- AI ethics and governance
- virtue and character education
- disability and inclusion perspectives
- socio-technical systems thinking
- faith and ethics integration (where appropriate)

Interdisciplinary instruction is particularly important. AI challenges rarely fall neatly within a single domain; they require integration of computer science, ethics, social science and theology.

Various studies support the integration of ethics into technical education. Fiesler et al. (2020) found that students exposed to embedded ethics curricula demonstrated significantly higher ethical awareness compared to those receiving standalone ethics instruction. Similarly, organizational research shows that ethical climate and leadership integrity are strongly correlated with reduced misconduct and increased stakeholder trust (Treviño et al., 2006).

EMBEDDING CHARACTER FORMATION IN AI-ERA LEADERSHIP EDUCATION

Building on the need for intentional character formation, this section focuses on educational strategies that operationalize these principles within AI-era leadership development. Character formation requires more than adding an ethics course to an already crowded curriculum. It demands a coherent developmental ecosystem in which moral reasoning, spiritual reflection and technological literacy reinforce one another.

Researchers working in professional ethics have long argued that ethical competence develops most effectively through sustained engagement with realistic dilemmas rather than abstract instruction alone (Rest et al., 1999). In AI-based contexts, where ethical failures often arise from complex socio-technical interactions experiential and reflective pedagogies become even more important. This section outlines educational strategies that align with faith-informed leadership development while remaining applicable across diverse institutional settings.

Integrative Curriculum Design

Traditional AI education has focused heavily on algorithms, data structures and model performance. While these competencies remain essential, they are insufficient for preparing leaders who must govern AI responsibly. Recent calls in computing education emphasize the need to integrate ethics directly into technical instruction rather than treating it as an optional add-on (Fiesler et al., 2020).

An integrative curriculum for AI-era leadership should include four intertwined strands:

- **Technical literacy:** Leaders must understand, at an appropriate level, how AI systems function, including their limitations and failure modes.
- **Ethical reasoning:** Students should engage major ethical frameworks: virtue ethics, deontology, consequentialism and faith-based moral reasoning to build moral discernment.
- **Human impact awareness:** Courses should examine real-world cases of algorithmic harm, especially affecting marginalized populations.
- **Spiritual and character formation:** Where mission-appropriate, faith-informed reflection on vocation, stewardship and service should be included.

Research indicates that ethics integration is most effective when embedded across multiple courses rather than confined to a single module (Fiesler et al., 2020). This distributed approach reinforces the message that ethical responsibility is intrinsic to technological work.

Educational institutions play a central role in shaping the moral and professional formation of future AI leaders. Table 4 summarizes pedagogical approaches that support character-centered leadership development.

Table 4. Educational strategies for character formation in AI leadership

Educational Strategy	Purpose	Example Activities	Expected Outcome
Integrated Ethics Curriculum	Embed ethics in technical education	AI ethics modules in programming courses	Ethical awareness in technical decisions
Case-Based Learning	Develop moral reasoning	Analysis of algorithmic bias cases	Improved ethical judgment
Reflective Practices	Encourage self-examination	Leadership journals and reflection exercises	Internalized ethical commitments

continued on following page

Table 4. Continued

Educational Strategy	Purpose	Example Activities	Expected Outcome
Interdisciplinary Collaboration	Expand socio-technical understanding	Joint courses across CS, ethics and policy	Holistic AI governance perspectives
Community Engagement	Strengthen direct awareness of human impact	Stakeholder consultation exercises	Greater empathy for affected communities

Case-Based Moral Formation

Case-based learning is particularly powerful for AI leadership because many ethical challenges arise from contextual ambiguity rather than clear rule violations. Well-designed cases allow learners to wrestle with competing values such as efficiency, fairness, privacy and inclusion. Effective AI ethics cases should include:

- realistic organizational pressures
- incomplete or uncertain data
- stakeholder conflicts
- long-term societal implications
- disability and inclusion dimensions

O'Neil's (2016) documented cases of algorithmic harm provide especially useful teaching material because they demonstrate how technically sound systems can still produce unjust outcomes. From a faith-informed perspective, case discussions can be enriched by reflective questions such as:

- Who is most vulnerable in this scenario?
- How does this decision affect human dignity?
- What would servant leadership require here?
- Where is moral responsibility located?

Such questions help move learners beyond compliance thinking toward character-based discernment.

Reflective and Spiritual Practices

Character formation is not purely cognitive; it also involves the shaping of moral perception, emotional responsiveness and habitual action. For this reason, reflective practices play a crucial role in leadership development.

Research in moral psychology suggests that reflection helps individuals integrate ethical principles into their identity and decision-making processes (Rest et al., 1999). In faith-based contexts, spiritual disciplines can further deepen this integration. High-impact reflective practices include:

- guided ethical journaling
- vocation-focused reflection
- examen-style decision reviews
- structured peer dialogue
- facilitated moral debriefs after simulations

These practices cultivate what virtue theorists call practical wisdom (phronesis); the capacity to perceive what is morally salient in complex situations. Importantly, reflection should be tied to real or simulated AI governance decisions. Abstract reflection alone is less effective than reflection grounded in concrete technological contexts.

Interdisciplinary Formation Environments

AI-based leadership sits at the intersection of technology, ethics, policy and human experience. Consequently, siloed educational models are poorly suited for preparing future leaders.

Interdisciplinary programs have been shown to improve ethical awareness in technology fields by exposing students to diverse perspectives (Floridi et al., 2018). Effective formation environments often include collaboration among:

- computer science
- philosophy and ethics
- theology or religious studies
- public policy
- disability studies
- organizational leadership

Such collaboration helps future leaders appreciate the full socio-technical complexity of AI systems. Faith-informed institutions are particularly well positioned to lead in this space because many already maintain strong traditions of holistic education that integrate intellectual, moral and spiritual formation.

ORGANIZATIONAL STRATEGIES FOR CHARACTER-CENTERED AI LEADERSHIP

While education is critical, character formation must continue within organizational life. Leaders are shaped not only by formal instruction but also by institutional culture, incentive structures and governance practices. Organizations that prioritize rapid AI deployment without ethical guardrails may unintentionally erode the very virtues they hope leaders will embody.

Ethical AI Governance Structures

Formal governance mechanisms help translate personal virtue into institutional practice. Without structural support, even well-intentioned leaders may struggle to act ethically under competitive pressure.

Emerging best practices in AI governance include:

- cross-functional AI ethics committees
- algorithmic impact assessments
- model documentation standards
- post-deployment monitoring
- stakeholder review processes

Shneiderman (2020) emphasizes that trustworthy AI requires not only technical reliability but also organizational accountability structures that preserve human oversight.

Faith-informed organizations may further strengthen governance by explicitly articulating commitments to human dignity, inclusion and the common good within AI policies. Translating leadership character into organizational practice requires supportive governance mechanisms. Table 5 outlines key organizational structures that enable responsible AI stewardship.

Table 5. Organizational practices for character-centered AI governance

Governance Mechanism	Function	Ethical Benefit
AI Ethics Review Board	Oversight of AI deployments	Independent ethical evaluation
Algorithmic Impact Assessment	Pre-deployment risk analysis	Identification of bias and societal risks
Model Documentation (Model Cards)	Transparency of AI systems	Improved accountability
Stakeholder Consultation	Inclusion of affected communities	Reduced exclusion and bias
Continuous Monitoring Systems	Post-deployment auditing	Early detection of harms

Cultivating an Ethical Organizational Culture

Organizational culture powerfully shapes leader behavior. According to research in ethical climate theory, employees take cues from leadership messaging, reward systems and informal norms regarding what behaviors are truly valued. To support character-centered AI leadership, organizations should:

- reward ethical risk identification
- normalize responsible pauses in deployment
- protect dissenting technical voices
- elevate inclusion metrics alongside performance metrics
- model transparency at senior levels

Zuboff (1988) warned that advanced information technologies can enable new forms of managerial control that distance leaders from workers and stakeholders. A consciously human-centered culture helps counteract this drift.

Inclusion and Disability as Design Imperatives

One of the most important and often overlooked organizational responsibilities is ensuring that AI systems serve people with disabilities and other marginalized groups equitably. Accessibility should not be treated as a downstream compliance issue but as a core design principle.

Whittaker et al. (2019) argue that many AI harms arise because affected communities are excluded from the design process. Organizations committed to ethical AI leadership should therefore implement:

- disability-inclusive design reviews

- participatory co-design processes
- accessibility testing in AI pipelines
- diverse data collection strategies
- ongoing equity audits

From a faith-informed leadership perspective, such practices reflect a commitment to the inherent dignity and worth of every person.

Continuous Moral Formation for Leaders

Finally, organizations should recognize that character formation is lifelong. Rapid advances in AI will continue to generate new ethical challenges that cannot be fully anticipated today. High-maturity organizations support ongoing formation through:

- executive ethics roundtables
- AI ethics scenario exercises
- cross-sector learning communities
- periodic moral risk reviews
- leadership spiritual formation programs (where appropriate)

These practices help leaders remain morally attentive in the face of technological acceleration.

Operational Metrics and KPIs for Ethical AI Leadership

To move from principle to practice, organizations must define measurable indicators of ethical AI performance. Without metrics, ethical commitments risk remains aspirational. Following Key performance indicators (KPIs) may be considered to measure ethical performance of AI-based systems:

Fairness Metrics:

- Demographic parity difference
- Equal opportunity difference
- Disparate impact ratio

Transparency Metrics:

- Percentage of models with documented model cards
- Explainability score (based on interpretability tools)

Accountability Metrics:

- Number of AI systems undergoing impact assessment
- Frequency of ethics review board interventions

Inclusion Metrics:

- Representation diversity in training datasets
- Accessibility compliance rates

Governance Metrics:

- Time taken to resolve flagged ethical risks
- Number of ethical escalations addressed

These metrics should be integrated into organizational dashboards alongside traditional performance indicators to ensure ethical accountability is treated as a core priority.

CASE STUDY: CHARACTER FAILURE VS. CHARACTER-CENTERED AI LEADERSHIP

To illustrate the practical significance of character formation in the governance of artificial intelligence systems, consider a common organizational scenario involving AI-based hiring. Hiring algorithms are increasingly deployed to process large applicant pools, improve efficiency and reduce human bias. However, when these systems are trained on historical organizational data, they may inadvertently reproduce or amplify past patterns of discrimination embedded within that data. The ethical implications of such outcomes place significant responsibility on leadership decision-making.

An organization deploys an AI-driven résumé screening tool designed to automatically rank job applicants based on predicted job performance. The system is trained on historical hiring data collected from the organization's previous recruitment cycles. Initially, the system appears highly effective: it reduces the time required to screen applications by more than 60% and demonstrates strong predictive accuracy when compared with previous hiring outcomes.

Several months after deployment, however, internal analysts conducting a routine review discover a troubling pattern. The algorithm appears to disproportionately filter out candidates from certain demographic backgrounds. Although the system continues to show strong predictive performance according to standard machine-learning metrics such as accuracy and precision; however, the disparity in outcomes suggests potential algorithmic bias.

Senior leadership now faces a strategic and ethical dilemma. Disabling or re-designing the system could slow hiring processes and incur significant financial costs. At the same time, continuing to use the model risks perpetuating unfair hiring practices, damaging the organization's reputation and potentially exposing the firm to legal challenges.

The decision that follows becomes a revealing test of leadership character. The contrasting responses presented in the case study can be summarized in Table 6, which highlights how leadership character shapes organizational responses to AI ethical dilemmas.

Table 6. Character failure vs. character-centered leadership

Dimension	Character-Deficient Leadership	Character-Centered Leadership
Primary Focus	Efficiency and performance	Human dignity and fairness
Response to Bias	Minimize or ignore	Investigate and remediate
Transparency	Limited communication	Open stakeholder engagement
Organizational Action	Continue deployment	Pause and conduct fairness audit
Long-Term Approach	Reactive	Continuous monitoring and governance

Character-Deficient Response

In organizations where leadership is primarily driven by short-term performance metrics, the response to such findings may prioritize efficiency and organizational convenience over fairness and accountability. A character-deficient leadership team might adopt several defensive or avoidance strategies.

First, leaders may focus narrowly on technical performance indicators. Since the model still demonstrates strong predictive accuracy for job performance, they may argue that the system is functioning as intended. By emphasizing quantitative metrics while ignoring distributional outcomes, they reduce a complex ethical problem to a purely technical evaluation.

Second, remediation efforts may be delayed due to perceived costs. Retraining or redesigning the model would require additional data analysis, new validation procedures and potentially the involvement of external auditors. Leadership may therefore postpone action until external pressure emerges.

Third, disparities may be framed as statistical artifacts rather than meaningful ethical concerns. Leaders might claim that demographic imbalances are merely reflections of labor-market realities or historical patterns embedded in the training data, thereby avoiding responsibility for correcting them.

Finally, communication about the issue may be minimal. Internal reports might remain restricted to a small group of executives, while employees, applicants and external stakeholders receive little or no information about the discovered bias.

Such responses have been widely discussed in critiques of algorithmic decision systems, particularly in cases where organizations prioritize optimization and efficiency while neglecting broader social consequences. A prominent analysis of these dynamics appears in Weapons of Math Destruction by Cathy O'Neil (2016), which documents how opaque algorithms can reinforce structural inequalities when accountability mechanisms are weak.

Character-Centered Response

In contrast, leaders whose decision-making is grounded in moral character are more likely to respond in ways that prioritize justice, responsibility and long-term societal trust. Character formation-cultivating virtues such as humility, justice, compassion, courage and stewardship fundamentally shape how leaders interpret and address technological risks.

A character-centered leadership response would likely begin with humility. Rather than assuming that the system is correct simply because it performs well on technical metrics, leaders acknowledge the possibility that the organization's data or modeling approach may contain hidden biases. This humility creates space for critical reflection and corrective action.

Second, leaders guided by a commitment to justice would pause or constrain the deployment of the system once disparities are identified. Even if the model is technically accurate, fairness considerations require investigation before continued use. A temporary suspension demonstrates that ethical obligations outweigh short-term operational advantages.

Third, compassionate leadership encourages engagement with stakeholders who may be affected by the algorithmic decisions. This may involve consulting diversity and inclusion experts, speaking with affected employee groups, or seeking feedback from external advocacy organizations. Such engagement broadens the organization's understanding of the real-world impact of automated decision systems.

Fourth, leaders exercising courage would authorize a comprehensive fairness audit of the model. This audit may involve examining training data, testing multiple fairness metrics, evaluating feature importance and identifying variables that may indirectly encode demographic attributes. Where bias is confirmed, the organization would commit to retraining or redesigning the model using improved datasets or algorithmic techniques designed to mitigate discriminatory outcomes.

Transparency also becomes a central element of the response. Leaders communicate openly about the discovery of bias, the steps being taken to address it and the safeguards that will be implemented in the future. Transparent communication builds trust with employees, applicants and the public.

Finally, stewardship requires the implementation of long-term governance mechanisms. Instead of treating the issue as a one-time anomaly, the organization establishes continuous monitoring systems, periodic algorithmic audits and oversight committees responsible for reviewing the ethical performance of AI systems.

Leadership Character as the Decisive Variable

Importantly, the technical conditions in both responses are identical: the same algorithm, the same data and the same evidence of bias. What differs is the moral orientation of the leadership team interpreting the situation.

In a character-deficient environment, the discovery of bias becomes a problem to be minimized or concealed. In a character-centered environment, the same discovery becomes an opportunity for ethical responsibility, institutional learning and system improvement.

This contrast highlights a critical insight for AI governance: technical solutions alone cannot guarantee ethical outcomes. Algorithms operate within organizational cultures shaped by leadership values and character formation. When leaders embody virtues such as humility, justice, compassion, courage and stewardship, they are more likely to prioritize fairness, transparency and societal wellbeing alongside technical performance.

Consequently, the development of ethical AI systems depends not only on improved technical methods such as fairness metrics or bias mitigation algorithms but also on the cultivation of leadership character capable of guiding responsible decision-making in complex socio-technical environments.

Beyond hypothetical scenarios, several real-world cases highlight the importance of character-centered leadership: One prominent example is discontinuation of Amazon Hiring Algorithm (https://www.reuters.com/article/world/insight-amazon-scraps-secret-ai-recruiting-tool-that-showed-bias-against-women-idUSKCN1MK0AG/). Amazon discontinued an AI recruitment tool after discovering bias against female candidates due to historical training data. This case shows the importance of humility and justice in recognizing and correcting systemic bias. Another study by Buolamwini and Gebru (2018) revealed significantly higher error rates for darker-skinned women, demonstrating the need for inclusion and compassion in AI design.

These cases reinforce that ethical AI failures are not rare anomalies but recurring governance challenges requiring strong leadership character.

IMPLEMENTATION ROADMAP FOR INSTITUTIONS AND ORGANIZATIONS

The preceding sections have argued that character formation is essential for trustworthy leadership in AI-based environments and have outlined educational and organizational strategies to support this goal. However, many institutions struggle with the practical question of implementation. Ethical aspirations often remain aspirational unless they are translated into concrete structures, metrics and developmental pathways. This section presents a phased roadmap that educators, faith-based institutions, corporations and public-sector organizations can adapt to embed character-centered AI leadership in sustainable ways.

Phase 1: Moral and Organizational Readiness Assessment

Before launching new programs, institutions should assess their current ethical and cultural posture regarding AI. Many organizations overestimate their preparedness because they possess technical expertise while lacking moral governance capacity. A readiness assessment should examine:

- existing AI use cases and risk levels
- current ethics policies and oversight mechanisms
- leadership ethical climate
- inclusion and accessibility practices
- staff AI literacy levels
- stakeholder trust indicators

Tools such as algorithmic impact assessments and ethical climate surveys can provide valuable baseline data. Floridi et al. (2018) emphasize that responsible AI begins with clear institutional self-understanding regarding both capabilities and risks. Faith-based organizations may also include mission-alignment reviews that evaluate whether current AI initiatives reflect commitments to human dignity and the common good.

Phase 2: Leadership Formation Architecture

Once readiness is understood, institutions should design a coherent formation architecture rather than isolated training events. Effective character formation requires developmental continuity across career stages. A mature architecture typically includes three tiers:

Foundational formation

- AI ethics literacy
- virtue and servant leadership foundations
- bias and inclusion awareness
- human-centered design principles

Intermediate formation

- case-based ethical decision labs
- cross-functional AI governance training
- stakeholder impact analysis
- moral risk identification

Advanced/executive formation

- strategic AI stewardship
- ethical crisis leadership
- public accountability and transparency
- long-term societal impact analysis

Research on leadership development consistently shows that spaced, developmental approaches are more effective than one-time training interventions (Rest et al., 1999).

Phase 3: Embedding Governance and Accountability

Character formation must be reinforced by institutional structures that make ethical action possible under real-world pressure. Even morally committed leaders can struggle if incentive systems reward speed and profit at the expense of responsibility.

Key structural supports include:

AI ethics review boards: These bodies should be multidisciplinary and empowered to influence deployment decisions.

Algorithmic impact assessments: Pre-deployment reviews should evaluate fairness, inclusion, privacy and downstream effects.

Documentation and transparency standards: Model cards, data sheets and decision logs help maintain accountability across the AI lifecycle.

Protected escalation channels: Technical and ethical concerns must be safely reportable without fear of retaliation.

According to Shneiderman (2020), trustworthy AI depends on combining technical reliability with organizational oversight and human accountability.

Phase 4: Culture Formation and Reinforcement

Policies alone do not shape behavior; culture does. Organizations must intentionally cultivate norms that reward ethical attentiveness. High-impact cultural signals include:

- senior leaders publicly discussing ethical trade-offs
- recognition for responsible risk identification
- storytelling that highlights servant leadership
- inclusion metrics in performance dashboards
- transparent communication about AI limitations

Zuboff (1988) warned that digital systems can quietly reshape organizational power dynamics. Culture formation helps ensure that AI adoption strengthens rather than erodes human-centered leadership.

Faith-informed organizations have a distinctive opportunity here. Practices such as mission reflection, communal discernment and vocation-centered leadership development can powerfully reinforce ethical culture when authentically implemented.

Phase 5: Continuous Evaluation and Learning

AI systems and their social impacts evolve rapidly. Consequently, character-centered leadership programs must be adaptive rather than static. Institutions should implement continuous learning loops that include:

- post-deployment ethical audits
- stakeholder feedback mechanisms
- disparity monitoring dashboards
- periodic leadership reflection retreats
- cross-industry learning exchanges

Russell (2019) emphasizes that long-term alignment between AI systems and human values requires ongoing human oversight and institutional learning. Implementing character-centered AI leadership requires a phased institutional strategy. Table 7 summarizes the implementation roadmap proposed in this chapter.

Table 7. Institutional roadmap for character-centered AI leadership

Implementation Phase	Key Actions	Expected Outcomes
Phase 1: Readiness Assessment	Evaluate AI use cases and governance structures	Institutional awareness of risks
Phase 2: Leadership Formation	Develop training programs and ethical education	Morally grounded leadership
Phase 3: Governance Embedding	Establish oversight and accountability mechanisms	Responsible AI deployment
Phase 4: Culture Formation	Promote ethical organizational norms	Sustainable ethical climate
Phase 5: Continuous Learning	Conduct audits and reflection cycles	Long-term AI stewardship

FUTURE CHALLENGES FOR CHARACTER-CENTERED AI LEADERSHIP

While the framework presented in this chapter offers a constructive path forward, several emerging trends will further test leadership character in the coming decade.

Increasing Autonomy of AI Systems: As AI systems become more autonomous and adaptive, leaders will face growing difficulty in predicting system behavior. This will increase the importance of prudence, humility and continuous monitoring. The temptation toward "set-and-forget" automation must be resisted. Human-centered oversight will remain essential even as technical capabilities advance.

Scale and Speed Pressures: Competitive pressures are accelerating AI deployment cycles across industries. Leaders may face strong incentives to prioritize speed over ethical diligence.

This environment will test moral courage in particular. Organizations that embed character formation early will be better positioned to resist purely accelerationist pressures.

Global and Cross-Cultural AI Governance: AI systems increasingly operate across national and cultural boundaries. Ethical norms regarding privacy, fairness and acceptable risk vary globally.

Faith-informed leadership, with its emphasis on universal human dignity, may provide a valuable moral anchor in these pluralistic contexts. However, leaders must also cultivate cultural humility and intercultural competence.

Inclusion of Persons with Disabilities: Despite growing awareness, disability inclusion remains underdeveloped in many AI initiatives. As automated systems increasingly mediate employment, education, healthcare and public services, the risk of structural exclusion will grow.

Leaders committed to human flourishing must treat accessibility and inclusive design as core success metrics rather than secondary compliance requirements. This will require sustained attention to participatory design and equitable data practices (Whittaker et al., 2019).

CONCLUSION

The rapid diffusion of artificial intelligence into organizational decision-making marks one of the most consequential leadership challenges of the twenty-first century. While technical innovation continues at extraordinary pace, the moral and character formation of leaders has not kept equal stride. This gap poses significant risks. AI systems, however sophisticated, ultimately reflect the values, assumptions and priorities of the humans who design and govern them.

This chapter has argued that character formation must be repositioned as a central competency for leaders in AI-based organizations. Drawing on virtue ethics, servant leadership and faith-informed perspectives, the discussion identified five essential domains of AI-era leadership character: humility, justice, compassion, moral courage and stewardship. These virtues directly address the distinctive ethical risks introduced by algorithmic decision-making, including key ethical risks associated with AI-based decision-making.

The chapter further demonstrated that character formation is not accidental. It requires intentional cultivation through integrative education, reflective practice, interdisciplinary engagement and supportive organizational structures. Institutions that treat ethics as peripheral will struggle to govern AI responsibly; those that embed moral formation into leadership development will be better equipped to steward intelligent technologies in service of human dignity, justice and socially responsible innovation.

Importantly, the call to character-centered AI leadership is not merely defensive aimed only at preventing harm. Properly understood, it is also generative. Leaders formed in humility and justice can guide AI innovation toward genuinely human-centered outcomes. Compassionate and courageous leaders can ensure that technological progress includes persons with disabilities and historically marginalized communities. Steward leaders can balance innovation with long-term societal well-being.

The final aim, consistent with the broader vision of *Leading with Faith and Ethics in an AI World,* is not simply to produce leaders who can manage advanced technologies. It is to form leaders whose moral vision, spiritual grounding and commitment to the common good enable them to shape AI systems that honor human dignity and promote authentic flourishing in an increasingly automated age.

Finally, the future of AI governance will not be determined only by advances in algorithms, but by the moral quality of those who lead them. In this sense, character formation is not a peripheral concern; it is the central infrastructure of trustworthy AI. Institutions that recognize and operationalize this insight will not only avoid harm but will shape a more just, inclusive and human-centered technological future.

REFERENCES

Buolamwini, J., & Gebru, T. (2018). Gender shades: Intersectional accuracy disparities in commercial gender classification. *Proceedings of Machine Learning Research, 81*, 1–15.

European Commission. (2021). *Proposal for a regulation laying down harmonised rules on artificial intelligence*. Artificial Intelligence Act.

Eva, N., Robin, M., Sendjaya, S., van Dierendonck, D., & Liden, R. C. (2019). Servant leadership: A systematic review and call for future research. *The Leadership Quarterly, 30*(1), 111–132. DOI: 10.1016/j.leaqua.2018.07.004

Fiesler, C., Garrett, N., & Beard, N. (2020). What do we teach when we teach tech ethics? A syllabi analysis. *Proceedings of the ACM on Human-Computer Interaction, 4*(CSCW1), 1–22. DOI: 10.1145/3328778.3366825

Floridi, L., Cowls, J., Beltrametti, M., Chatila, R., Chazerand, P., Dignum, V., Luetge, C., Madelin, R., Pagallo, U., Rossi, F., Schafer, B., Valcke, P., & Vayena, E. (2018). AI4People—An ethical framework for a good AI society. *Minds and Machines, 28*(4), 689–707. DOI: 10.1007/s11023-018-9482-5 PMID: 30930541

Greenleaf, R. K. (1977). *Servant leadership: A journey into the nature of legitimate power and greatness*. Paulist Press.

MacIntyre, A. (1984). *After virtue* (2nd ed.). University of Notre Dame Press.

National Institute of Standards and Technology (NIST). (2023). AI Risk Management Framework.

O'Neil, C. (2016). *Weapons of math destruction*. Crown.

OECD. (2019). OECD Principles on Artificial Intelligence.

Parasuraman, R., & Riley, V. (1997). Humans and automation: Use, misuse, disuse, abuse. *Human Factors, 39*(2), 230–253. DOI: 10.1518/001872097778543886

Rest, J. R., Narvaez, D., Bebeau, M. J., & Thoma, S. J. (1999). *Postconventional moral thinking.A neo-Kohlbergian approach*. Lawrence Erlbaum Associates. DOI: 10.4324/9781410603913

Russell, S. (2019). *Human compatible: Artificial intelligence and the problem of control*. Viking.

Shneiderman, B. (2020). Human-centered artificial intelligence: Reliable, safe & trustworthy. *International Journal of Human-Computer Interaction, 36*(6), 495–504. DOI: 10.1080/10447318.2020.1741118

Treviño, L. K., Weaver, G. R., & Reynolds, S. J. (2006). Behavioral ethics in organizations: A review. *Journal of Management, 32*(6), 951–990. DOI: 10.1177/0149206306294258

VanderWeele, T. J. (2017). On the promotion of human flourishing. *Proceedings of the National Academy of Sciences of the United States of America, 114*(31), 8148–8156. DOI: 10.1073/pnas.1702996114 PMID: 28705870

Whittaker, M., Crawford, K., & Dobbe, R.. (2019). *AI Now Report 2019.* AI Now Institute.

Zuboff, S. (1988). *In the age of the smart machine.* Basic Books.

KEY TERMS AND DEFINITIONS

AI Governance: The structures, policies, and processes used to ensure that AI systems are developed and deployed responsibly, aligning with ethical principles such as fairness, transparency, and accountability.

Algorithmic Bias: Systematic and repeatable errors in AI systems that produce unfair outcomes, often reflecting historical inequities embedded in training data rather than intentional discrimination.

Character Formation: The intentional development of moral virtues, habits, and dispositions that shape how leaders think, decide, and act, especially in complex and uncertain environments.

Epistemic Humility: An awareness of the limits of one's knowledge and the uncertainty inherent in AI systems, leading to cautious, reflective, and responsible decision-making.

Ethical AI Leadership: A leadership approach that integrates technical understanding with moral discernment to guide AI systems toward outcomes that promote justice, inclusion, and the common good.

Human Flourishing: A holistic state of well-being that includes meaning, relationships, virtue, and purpose, serving as the ultimate goal of ethical leadership and AI governance.

Human-Centered AI: An approach to designing and deploying AI systems that prioritizes human dignity, agency, and well-being over purely technical or efficiency-driven outcomes.

Moral Courage: The willingness to take ethical action despite pressure, risk, or potential cost, particularly when confronting harmful or unjust uses of AI systems.

Servant Leadership: A leadership philosophy that prioritizes serving others, empathy, stewardship, and the growth and well-being of people as the primary measure of leadership effectiveness.

Virtue Ethics: A moral framework that focuses on the development of character and virtues, such as justice, prudence, and humility, as the foundation for ethical decision-making.

Chapter 7
Ethical AI Governance as Socio-Technical Stewardship:
A Faith-Informed Systems Model for Sustainable Development

Talat Naaz
http://orcid.org/0009-0001-5418-0321
Christ (Deemed to be) University, India

Rosewine Joy
Christ (Deemed to be) University, India

ABSTRACT

Artificial intelligence (AI) is increasingly embedded in institutional decision-making, reshaping how authority, accountability, and responsibility are distributed. While enhancing efficiency and predictive capacity, AI also introduces risks such as bias, opacity, and weakened accountability. Existing governance approaches often rely on fragmented regulations or abstract ethical principles that are difficult to operationalize. This chapter proposes the AI–Ethics–Faith Systems Model, integrating infrastructure governance, ethical accountability, and value-based orientation. It reframes AI governance as socio-technical stewardship, positioning leadership as the central mechanism for aligning technological systems with ethical reasoning and long-term societal outcomes. The model emphasizes human flourishing—dignity, inclusion, agency, and well-being as the core evaluative criteria for responsible and sustainable AI deployment.

DOI: 10.4018/979-8-3373-8192-3.ch007

ght © 2027, IGI Global Scientific Publishing. Copying or distributing in print or electronic forms without written permission of IGI Global Scientific Publishing is prohibited. Use of this chapter to generative artificial intelligence (AI) technologies is expressly prohibited. The publisher reserves all rights to license its use for generative AI training and machine learning model development.

LEADERSHIP AT THE INTERSECTION OF ALGORITHMIC POWER AND MORAL PURPOSE

Artificial Intelligence (AI) is no longer confined to technical domains. It has become embedded in the everyday functioning of institutions, quietly shaping decisions about employment, credit access, healthcare delivery, education, and public welfare (Eubanks, 2018). What is unfolding is not simply a technological shift but a deeper reconfiguration of how authority operates and how responsibility is distributed within contemporary systems.

Decisions that were once clearly linked to human judgment are now produced through interactions between data infrastructures, predictive models, and institutional processes (Binns, 2021). This diffusion does not eliminate responsibility, but it makes it less visible and harder to locate. As a result, leadership is no longer confined to directing people or managing organizations. It increasingly involves shaping the design and governance of socio-technical systems that influence real-life outcomes (Akbarighatar, 2025).

This shift introduces a persistent tension. On one side, AI offers efficiency, scalability, and predictive precision. On the other hand, AI offers efficiency, scalability, and predictive precision. On the other, it brings a set of challenges that are less visible but more difficult to address. Algorithmic systems can reproduce structural inequalities, operate with limited transparency, and distance decision-makers from the consequences of their actions (Mehrabi et al., 2021). As systems become more complex, accountability becomes diffuse. Individuals affected by automated decisions often face outcomes they cannot fully understand or contest, while institutions attribute those outcomes to system outputs rather than human judgment (Malja & Afrasiabi, 2025).

In this environment, leadership cannot remain confined to operational oversight. It must extend to the ethical and societal performance of the systems themselves (Floridi et al., 2018). Traditional leadership models, built around centralized authority and traceable decision-making, are not well suited to this context. AI introduces distributed agency, where outcomes emerge from interactions among technical design, data ecosystems, institutional incentives, and regulatory frameworks (Noordt & Tangi, 2023). Responsibility becomes systemic rather than individual, and leadership must adapt accordingly.

This requires a shift in orientation. Rather than focusing solely on control or compliance, leadership must become more interpretive and anticipatory, engaging with how systems evolve over time, how unintended consequences accumulate, and how technological decisions intersect with broader social objectives (Maak & Pless, 2021). In this sense, leadership begins to resemble stewardship: a continuous effort to guide complex systems toward desirable outcomes in the face of uncertainty.

At the same time, global development priorities particularly those articulated through the Sustainable Development Goals (SDGs) place increasing emphasis on inclusion, equity, and human well-being (United Nations, 2015). AI has the capacity to support these objectives, especially in areas such as resource allocation and service delivery. Yet without appropriate governance, it can also reinforce exclusion and uneven development trajectories (Vinuesa et al., 2020). The central issue, therefore, is not whether AI contributes to development, but under what conditions it does so responsibly.

Moral and philosophical traditions offer an additional layer of insight in this context. Across different traditions, concepts such as justice, stewardship, and responsibility have long shaped how authority is understood, particularly in relation to vulnerable populations. A value-oriented, faith-informed perspective understood here in a non-doctrinal sense can help re-anchor leadership in long-term societal responsibility. Rather than constraining innovation, it provides direction, ensuring that technological systems remain aligned with broader human and social purposes (Yeung, 2017).

The central argument of this chapter follows from this context. Effective AI governance depends on leadership that can integrate systems thinking, ethical reasoning, and value-based orientation within a single framework (Dignum, 2022). Leadership functions as the connecting layer between technological capability and societal outcomes.

To develop this argument, the chapter proposes the AI–Ethics–Faith Systems Model, which conceptualizes leadership as a form of socio-technical stewardship. The model integrates AI governance, ethical accountability, and a value-based orientation within a unified framework that supports sustainable development. Its contribution lies not in adding another set of ethical principles, but in explaining how those principles can be operationalized within institutional systems (van Wynsberghe, 2021).

Ultimately, the chapter positions human flourishing as the anchor for evaluative criterion for AI governance. This includes dignity, meaningful participation, social inclusion, and the capacity to lead purposeful lives (Sen, 1999; Nussbaum, 2011). From this perspective, leadership is not simply a managerial function but a normative responsibility, ensuring that technological systems expand human capabilities rather than constrain them.

THE ETHICAL AND LEADERSHIP CRISIS IN AI-DRIVEN SDG TRANSFORMATION

The rapid expansion of AI across governance systems and institutional decision-making marks a deeper shift than technological adoption alone. It reshapes how authority is exercised and how opportunities are distributed. Algorithmic systems now influence decisions across employment, welfare, healthcare, and public administration, often under the assumption that they enhance efficiency and objectivity.

Yet these systems operate within existing social and institutional structures. As a result, they tend to inherit and reproduce patterns embedded in historical data. Inequalities that were previously dispersed across decisions can become systematized, particularly when ethical oversight does not keep pace with technical capability. What appears as neutral computation may therefore produce uneven outcomes.

Two challenges are especially persistent. The first is bias, not as an isolated technical flaw, but as a structural feature of data-driven systems. Predictive models identify patterns without evaluating their fairness, which means that historical disparities can be carried forward without explicit intention. The second is opacity. As systems become more complex, their internal logic becomes difficult to interpret, limiting the ability of individuals and institutions to understand or contest decisions.

These dynamics lead to a broader issue: the diffusion of responsibility. Decisions are no longer easily traceable to individual actors but emerge from interactions between data, models, and institutional processes. This weakens traditional accountability structures and creates a gap between system performance and ethical justification.

Current governance approaches struggle to address this complexity. Ethical principles are widely articulated but often applied externally rather than embedded in system design. Regulatory responses tend to be reactive, addressing harm after it becomes visible. Leadership models, in turn, rarely account for the distributed and adaptive nature of AI-driven decision-making.

This chapter addresses that gap by proposing the AI–Ethics–Faith Systems Model. The model brings together three elements that are typically treated separately: technological governance, ethical accountability, and value-based orientation. Rather than introducing new principles, it focuses on how these elements can be aligned within a single system.

The objective is not simply to improve system performance, but to ensure that AI contributes to outcomes that are socially legitimate and oriented toward human well-being. In this context, leadership is understood as a form of socio-technical stewardship (STS), maintaining alignment between technological processes and broader societal values as systems evolve.

THE AI–ETHICS–FAITH SYSTEMS MODEL: LEADERSHIP AS SOCIO-TECHNICAL STEWARDSHIP

The discussion so far points to a persistent gap: while AI systems are becoming more capable, the frameworks used to guide them remain fragmented. Technical design, ethical principles, and leadership practice often operate in parallel rather than in coordination. As a result, systems may perform efficiently while still producing outcomes that are difficult to justify on societal or ethical grounds.

The AI–Ethics–Faith Systems model is developed to address this gap. It does not assume that new principles are required. Instead, it focuses on how existing elements such as technological governance, ethical reasoning, and value orientation can be brought into alignment within a single, functioning system. The emphasis shifts from defining what ought to be done to understanding how such commitments are sustained in practice.

At its core, the model treats AI governance as a layered socio-technical system (STS). Each layer performs a distinct role, but none is sufficient on its own. What matters is how they interact. The model is structured around three interdependent layers: AI Infrastructure Governance (technical layer), Ethical Accountability Mechanisms (procedural layer), and Faith-Informed Value Orientation (normative layer). These operate within a dynamic systems framework characterized by feedback loops, adaptive recalibration, and multi-level stakeholder engagement.

Rather than treating AI as an external tool subject to post-hoc ethical review, the model embeds ethics and value orientation across the entire lifecycle, i.e., design, deployment, and evaluation. Leadership functions as the integrative core, ensuring coherence across these domains.

The model's conceptual contribution lies in its systemic integration. AI is treated as a normative infrastructure that shapes opportunities, inclusion, and distributive justice. Ethics is framed as interpretive reasoning that guides decision-making under conditions of ambiguity. Faith-informed leadership provides moral anchoring through principles such as stewardship, responsibility, and human dignity, not as a doctrinal imposition but as a universal value orientation. Together, these elements form a governance architecture capable of supporting sustainable development transformation. The integrated structure of the AI–Ethics–Faith Systems Model is illustrated in Figure 1.

Figure 1. AI–ethics–faith systems model for ethical AI governance as socio-technical stewardship

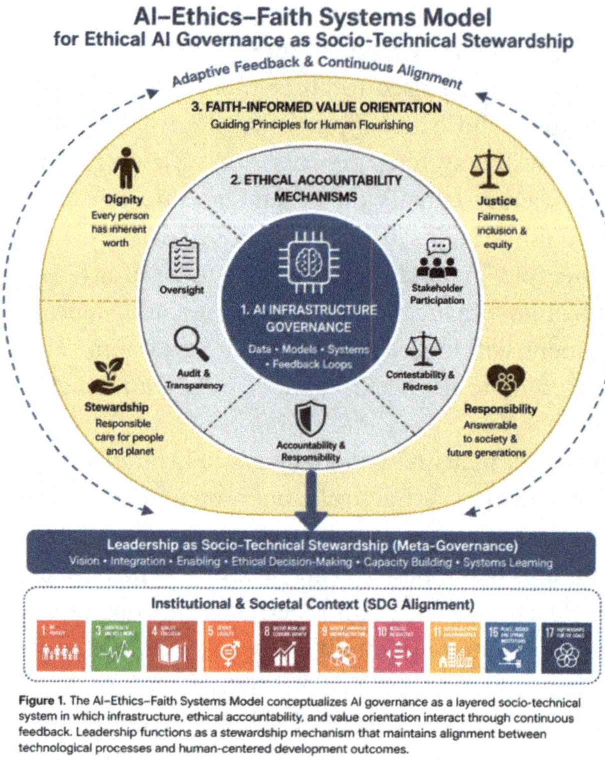

Note. The AI–Ethics–Faith Systems Model conceptualizes AI governance as a layered socio-technical system in which infrastructure, ethical accountability, and value orientation interact through continuous feedback. Leadership functions as a stewardship mechanism that maintains alignment between technological processes and human-centered development outcomes.

AI INFRASTRUCTURE GOVERNANCE

The first layer of the model sits where most people assume the conversation begins, with the systems themselves. But even here, at what appears to be a purely technical level, the story is not as straightforward as it seems. AI infrastructure is not just a collection of tools; it is a decision-making environment. The way it is built quietly shapes what the system can see, what it learns to value, and how it ultimately behaves.

This infrastructure includes the familiar components such as data pipelines, model architectures, optimization targets, and feedback mechanisms, but treating these as neutral building blocks misses the point. Each of these elements carries assumptions. Data selection, for instance, determines which realities are made visible to the system and which are effectively ignored. What gets counted begins to matter more than what does not, and over time, that imbalance can harden into the system's logic.

The same holds for model objectives. Whether a system is designed to optimize for accuracy, speed, cost reduction, or risk avoidance is not just a technical choice. It reflects a set of priorities, often shaped by institutional pressures rather than deliberate ethical reflection. Efficiency tends to win because it is measurable. Equity, fairness, or long-term impact is harder to quantify, so they are frequently sidelined, not intentionally but by default.

What complicates matters further is that these systems do not stand still. AI models evolve as they absorb new data, adjusting their outputs in ways that are not always easy to anticipate. Feedback loops emerge, sometimes reinforcing existing patterns, sometimes amplifying them. In the absence of careful oversight, systems can drift subtly at first, then more visibly away from their original purpose. What begins as a well-intentioned design can, over time, produce outcomes that no one explicitly planned for.

This is why governance at the infrastructure level cannot be treated as a one-off task. It is not something that can be "completed" at the design stage and then left alone. Instead, it requires ongoing attention- regular review, critical evaluation, and, when necessary, correction. The aim is not to control every outcome, which is neither realistic nor desirable, but to ensure that the system remains broadly aligned with its intended purpose as conditions change.

In practice, this means putting in place mechanisms that allow systems to be examined from multiple angles. Technical validation is only part of the picture. There is also a need for bias audits, transparency around how models are trained, and, importantly, some form of human oversight that is more than symbolic. Human-in-the-loop processes, when done well, create space for judgment rather than simply rubber-stamping automated outputs.

At the same time, governance cannot be reduced to a checklist of procedures. It is shaped by the organizational context in which these systems operate. Decisions about what to monitor, what to prioritize, and what counts as acceptable performance are influenced by culture as much as by code. If an organization values speed above all else, its systems will reflect that, regardless of how many ethical guidelines are formally adopted.

There is also a broader structural issue that cannot be ignored: the concentration of data and decision-making power. Large institutions now have the capacity to design systems that affect vast populations, often with limited external visibility. This creates asymmetries between those who build and control the systems and those who are subject to them. Governance, in this context, is not just about internal alignment; it is also about accountability and transparency.

For that reason, it becomes essential to make certain aspects of AI systems more visible. Not every technical detail needs to be disclosed, but there should be clarity about the core objectives, the nature of the data being used, and the risks that may arise. Without that, trust becomes difficult to sustain.

Ultimately, AI infrastructure governance is not just a technical exercise. It is where technical design meets institutional intent. The choices made at this level are often quietly, often early, set the direction for everything that follows. Leadership plays a decisive role here, not by dictating technical details, but by shaping the conditions under which those details are decided. When success is defined narrowly, systems tend to follow suit. When it is defined more broadly, taking into account fairness, inclusion, and long-term consequences, the design begins to shift.

In that sense, governance is less about control and more about orientation. It is about ensuring that the systems we build do not simply work, but work in ways that remain consistent with the values we claim to hold.

Ethical Accountability Mechanisms

If the first layer is about how systems are built, the second layer is about how they are held to account once in use. It is where values that are often spoken about in broad terms, fairness, transparency, and accountability, either begin to take shape in practice or quietly fade into the background.

Most organizations today have no shortage of ethical language. Policies are drafted, principles are endorsed, and frameworks are adopted. The difficulty is not in stating what matters; it is in making those statements meaningful in day-to-day decisions. Without some form of structure, ethical commitments tend to remain aspirational, well-intentioned, but loosely connected to what actually happens.

This is where accountability mechanisms come in. They do not simplify the ethical landscape nor eliminate uncertainty. What they do, at their best, is make decisions more visible and open to examination. Impact assessments, internal review processes, stakeholder consultations, and avenues for redress all serve a similar purpose: they slow things down just enough to allow for reflection. They create moments where assumptions can be questioned and consequences considered before they become embedded in practice.

That said, it would be a mistake to treat these mechanisms as technical fixes. They work only to the extent that they are taken seriously within the organization. An impact assessment that is conducted as a formality will reveal very little. A review board that lacks independence will struggle to challenge decisions that have already been made. The effectiveness of these structures depends less on their design and more on how they are used.

Part of the reason this layer is so demanding is that ethical issues in AI rarely present themselves as clear violations. More often, they appear as tensions between competing priorities. Improving predictive accuracy may come at the cost of increased bias. Personalization may enhance the user experience while raising privacy concerns. Automation may drive efficiency but displace certain forms of work. These are not problems that can be resolved by applying a fixed rule. They require judgment.

Judgment, in this context, cannot be fully automated. It involves weighing trade-offs, considering long-term implications, and, at times, accepting that no option is entirely without cost. Ethical accountability, then, is not about eliminating difficult choices. It is about ensuring that those choices are made consciously, with an awareness of their consequences, rather than being hidden within technical processes.

A more forward-looking approach to accountability shifts the emphasis from reaction to anticipation. Instead of waiting for something to go wrong, organizations begin to ask, early in the development process, what kinds of risks might emerge and how they might be addressed. This does not guarantee that problems will be avoided, but it shifts the posture from damage control to preparedness.

Over time, this kind of approach also shapes how institutions are perceived. When people can see that decisions are subject to scrutiny, that there are processes in place to question, review, and, if necessary, revise them, trust becomes easier to sustain. Not because the systems are perfect, but because they are not beyond challenge.

In this sense, accountability is not a final checkpoint at the end of a process. It is something that runs alongside the system as it operates, adapting as conditions change. It asks not only whether a system is performing as intended but also whether its outcomes remain acceptable in light of broader social expectations.

Leadership plays a quiet but decisive role here. It sets the tone for how seriously these mechanisms are taken. If ethical review is treated as an obstacle, it will be bypassed where possible. If it is treated as part of responsible decision-making, it becomes integrated into how work is done. The difference is not always visible from the outside, but it becomes apparent over time in the quality of decisions and the willingness to revisit them.

Ultimately, ethical accountability is less about enforcing rules and more about creating a culture where decisions can be examined without defensiveness. It allows institutions to learn from their systems rather than just rely on them. And in a space as complex and fast-moving as AI, that capacity to reflect and adjust may be as important as any principle written into policy.

Faith-Informed Value Orientation

The third layer brings into focus something usually present but seldom named outright: the question of what ultimately guides decision-making when rules run out. Every system, no matter how technical it appears, operates within a set of priorities. When those priorities are not made explicit, they tend to drift toward what is easiest to quantify—efficiency, speed, accuracy. Useful as these are, they rarely capture the full weight of what is at stake, especially as decisions begin to affect people's lives unevenly.

What this model does is make space for a different kind of orientation, one that draws on a broader moral vocabulary. The term "faith-informed" is used here carefully. It does not point to any single doctrine or tradition, but rather to a long-standing way of thinking about responsibility, where authority is understood as something that carries obligations, not just power. Ideas such as dignity, justice, stewardship, and responsibility have shaped governance across cultures for centuries. Bringing them into the conversation around AI is less about introducing something new and more about recovering something that has quietly receded from view.

In practice, this layer is most relevant when formal guidance is thin or ambiguous. AI systems often produce outcomes that are technically defensible but ethically unsettled. A model may be accurate, a process may be compliant, and yet something about the result feels off, perhaps because it disadvantages a particular group or treats individuals as interchangeable units rather than people with context and vulnerability. It is in these moments that value orientation matters. It helps answer questions that technical frameworks cannot fully resolve: not just whether a decision can be made this way, but whether it should.

This perspective also gently but firmly pushes back against a purely utilitarian logic. Optimizing for aggregate outcomes can obscure how costs and benefits are distributed. Efficiency gains in one part of a system may quietly produce harm elsewhere. A value-based orientation does not reject efficiency, but it refuses to treat it as the only measure of success. It asks who gains, who bears the risk, and whether those trade-offs are acceptable.

There is also a shift here in how leadership itself is understood. Rather than acting solely as managers of performance, leaders are positioned as stewards of systems that shape access, opportunity, and, increasingly, life chances. That role carries a longer horizon. Decisions are not judged solely by immediate outcomes but by how they hold up over time, whether they contribute to trust, and whether they deepen or reduce inequality, whether they leave room for future generations to navigate their own choices.

At the center of this layer is a simple but demanding idea: that people should not be reduced to data points. Algorithmic systems, by design, abstract and simplify. They rely on patterns, categories, and probabilities. While this is necessary for them to function, it also creates a risk of flattening human experience into something that can be easily processed but poorly understood. Keeping dignity in view acts as a counterbalance. It reminds decision-makers that behind every data point is a person whose circumstances may not fit neatly within the model.

Importantly, this layer does not replace existing ethical or regulatory frameworks. It sits alongside them, adding depth where rules alone fall short. Regulations can define boundaries and procedures can guide action, but neither can fully determine what counts as a good or just outcome in every situation. A value orientation fills that gap not by offering fixed answers, but by shaping how questions are approached in the first place.

What distinguishes this part of the model is its insistence on intentionality. In complex, fast-moving environments, it is easy for ethical reflection to become secondary, something addressed only when problems arise. This layer resists that drift. It calls for a more deliberate engagement with the moral dimensions of decision-making, especially under conditions where uncertainty is high and consequences are unevenly distributed.

In the end, the point is not to make decision-making slower or more burdensome, but more grounded. When systems become more powerful and less transparent, the need for clear moral direction does not diminish, it becomes harder to ignore.

Integration, Stewardship, and System Outcomes

The effectiveness of the model does not lie in the strength of its individual components, but in how they are connected. AI infrastructure generates outcomes, ethical accountability mechanisms evaluate those outcomes, and value orientation shapes how they are interpreted. When these layers operate in isolation, gaps emerge. When they remain aligned, the system functions more coherently.

Leadership operates at this point of alignment. It does not replace technical expertise or ethical processes, but ensures that they remain connected as conditions change. This role is best understood as a form of meta-governance: coordinating across layers, identifying misalignments, and guiding adjustments over time.

This function is inherently dynamic. AI systems evolve through feedback loops, and governance must adapt accordingly. Alignment achieved at one stage may shift as systems learn and expand in scope. Leadership, therefore, involves continuous engagement rather than one-time intervention.

This perspective also reframes leadership itself. Traditional models emphasize control and predictability. In contrast, AI-driven environments are characterized by complexity and uncertainty. A stewardship approach acknowledges these conditions, focusing on guiding systems rather than attempting to fully control them. It requires attentiveness to how outcomes are produced, who is affected, and whether the system remains aligned with broader societal expectations.

The model positions human flourishing as the anchor for evaluation in dignity, inclusion, and well-being. These dimensions expand the criteria for evaluating AI systems beyond efficiency and accuracy, providing a more comprehensive basis for assessing their societal impact.

In this sense, the contribution of the model is integrative rather than additive. It does not introduce new ethical principles, but clarifies how existing ones can be embedded within the structure and operation of AI systems. By connecting technical design, accountability processes, and value orientation, it offers a framework for maintaining coherence in environments where fragmentation is otherwise the norm.

Operationalization of the AI–Ethics–Faith Systems Model

The effectiveness of the AI–Ethics–Faith Systems Model depends on its transition from conceptual integration to institutional practice. Without such translation, ethical principles risk remaining symbolic commitments rather than decision-guiding forces. Operationalization therefore requires the development of formal governance structures, procedural safeguards, organizational practices, and leadership capabilities that ensure continuous alignment between AI systems, ethical accountability, and value-based orientation. The model can be operationalized through structured governance mechanisms across its three layers, including data auditing and monitoring at the infrastructure level, ethical review and stakeholder engagement at the accountability level, and value-based leadership practices at the normative level. Together, these mechanisms support system alignment, institutional trust, and outcomes oriented toward human flourishing.

APPLICATIONS AND LEADERSHIP IMPLICATIONS: ETHICAL AI LEADERSHIP IN PRACTICE

The relevance of the AI–Ethics–Faith Systems Model becomes clearer when examined within contexts where algorithmic systems are already shaping consequential decisions. The model does not replace existing governance tools; rather, it reorganizes how they are understood by linking technical design, ethical accountability, and value orientation within a single system. This shift is particularly important in domains where efficiency gains can obscure deeper questions of fairness and legitimacy.

The examples below are intentionally selective. The purpose is not to catalogue applications, but to demonstrate how governance changes when the model is applied.

Application in Public Welfare and Social Protection Systems

Public welfare systems have been quick to adopt algorithmic tools often with good reason. Eligibility checks, fraud detection, and resource allocation can all be handled more efficiently when supported by data-driven systems. For administrators working under pressure, this shift promises speed, consistency, and reduced workload. But once these systems move from the margins to the core of decision-making, their effects become harder to ignore. What looks like a technical upgrade often reshapes how access to essential support is determined.

A recurring concern, well documented in empirical work, is that these systems do not begin from a neutral baseline. They are trained on historical data, and that data reflects existing social and economic inequalities. Patterns of disadvantage, irregular employment histories, gaps in education, and uneven access to formal systems can become embedded in predictive models. As a result, decisions that look consistent on paper may, in practice, exclude those who are already vulnerable. Evidence shows how automated welfare systems can reproduce and even intensify these patterns when data is used without critical examination (Eubanks, 2018). What is presented as neutral classification can therefore carry significant distributive consequences.

Seen through the lens of the AI–Ethics–Faith Systems Model, this is not a failure of any single component but a problem of alignment across layers. At the infrastructure level, attention needs to be paid not only to model performance but also to data quality and representativeness. This includes identifying gaps, particularly for underdocumented populations such as informal workers and marginalized communities, and testing how outcomes vary across groups. Model validation, in this sense, becomes an ongoing process rather than a one-time check. Equally important is the presence of human-in-the-loop mechanisms, especially in high-stakes decisions. Without them, responsibility can become diffused across technical systems, making it harder to locate accountability or exercise judgment (Jolly et al., 2026; Emerson, 2026).

Ethical accountability introduces a second layer of scrutiny. Welfare decisions directly affect access to housing, healthcare, and financial support, which makes procedural fairness essential. Individuals need to be able to understand how decisions are made and to challenge them where necessary. Mechanisms such as algorithmic impact assessments, independent review processes, and accessible avenues for appeal help ensure that systems remain open to examination. Without such safeguards, automated decision-making risks becoming efficient but opaque, undermining procedural justice (Wachter et al., 2021).

The third layer shifts attention toward the values that underpin these systems. Welfare governance is not merely administrative; it reflects broader commitments about fairness, inclusion, and social responsibility. A value-based orientation brings these commitments into sharper focus. It reframes individuals not as data points to be sorted, but as citizens with rights and varying degrees of vulnerability. Principles such as dignity, stewardship, and compassion become relevant here, particularly in cases where, technical outputs conflict with lived realities. Empirical research has also shown that algorithmic systems in public service delivery can systematically reproduce bias if left unchecked, reinforcing the need for a stronger normative grounding (Obermeyer et al., 2019; Eubanks, 2018).

What emerges from the interaction of these layers is a more grounded approach to governance. Technical systems continue to provide efficiency and scale, but they are shaped and constrained by ethical processes and guided by a clearer sense of purpose. Welfare algorithms, in this context, begin to function less as instruments of administrative control and more as tools that can support inclusive public service.

This does not resolve every tension. Trade-offs remain, and not all outcomes will be straightforward. But it does change how those tensions are handled. Instead of being absorbed into the system and left unexamined, they are brought into view, questioned, debated, and, where necessary, revised.

In the longer run, this approach also affects institutional trust. Welfare systems depend on a degree of public confidence that is easily undermined when decisions feel opaque or unjust. When processes are transparent, open to review, and attentive to human consequences, trust becomes easier to sustain. In this sense, the governance of welfare algorithms is not only a technical or administrative concern; it is closely tied to the legitimacy of public institutions and their ability to serve those who rely on them most.

Application in Healthcare Decision-Making Systems

Healthcare represents another domain where AI is increasingly embedded in decision-making processes, particularly in diagnostics, risk prediction, and resource allocation. While these applications can improve efficiency and accuracy, their effectiveness depends heavily on the quality and diversity of training data.

Research indicates that AI systems may perform unevenly across demographic groups when datasets are not sufficiently representative (Mehrabi et al., 2021). In healthcare settings, such disparities can lead to unequal treatment outcomes, even when overall system performance appears robust.

The model provides a structured approach to addressing these risks. At the infrastructure level, systems must be validated across diverse populations, with continuous monitoring to identify performance disparities over time.

At the level of ethical accountability, professional oversight remains essential. AI outputs should inform clinical decisions but not replace professional judgment. Mechanisms must exist for questioning system recommendations, particularly in high-stakes contexts (Floridi et al., 2018).

At the level of value orientation, healthcare is understood as a relational practice grounded in care, dignity, and responsibility. Patients are not merely data inputs; they are individuals whose treatment requires ethical consideration beyond predictive accuracy. The model ensures that technological systems remain aligned with these principles.

Application in Organizational and Workplace Contexts

The influence of AI extends into organizational environments, where algorithmic systems are used to allocate tasks, monitor performance, and evaluate productivity. These systems can improve coordination but also introduce new forms of control that are less visible than traditional management structures.

Studies on algorithmic management highlight how such systems can reduce worker autonomy and intensify performance pressures, particularly when decision-making criteria are opaque (Kellogg et al., 2020). Workers may be evaluated through metrics they do not fully understand, with limited opportunity to challenge outcomes.

Within the model, governance begins with transparency. At the infrastructure level, decision rules and evaluation criteria must be made visible and interpretable. This enables workers to understand how performance is assessed.

At the accountability level, mechanisms for contestation and feedback are essential. These ensure that systems remain open to revision rather than functioning as unchallengeable authorities.

At the level of value orientation, attention shifts toward dignity, fairness, and the quality of work experience. Productivity is no longer treated as the sole measure of success. Instead, governance considers how technological systems shape agency, well-being, and long-term organizational sustainability.

Implications for Leadership Practice

Across these domains, a consistent pattern emerges: the challenges associated with AI are not purely technical. They arise from how systems are integrated into institutional contexts. Leadership, therefore, becomes the critical coordinating force.

One important shift is from control to interpretation. AI systems operate within dynamic environments where outcomes are not fully predictable. Leadership must engage with this uncertainty, interpreting system behavior and responding to emerging patterns rather than relying solely on predefined rules (Uhl-Bien et al., 2007).

A second implication concerns scope. AI-related decisions involve trade-offs across multiple domains such as technical, ethical, legal, and social. Leadership must operate across these boundaries, ensuring that decisions made in one area do not produce unintended consequences in another.

A third implication relates to governance design. Static regulatory frameworks are insufficient for adaptive systems. Continuous monitoring, iterative evaluation, and stakeholder engagement become essential components of governance (Dignum, 2022). Ethical considerations must be embedded within system design rather than applied retrospectively.

Finally, the criteria for evaluating success expand. Efficiency and accuracy remain important, but they are no longer sufficient. Systems must also be assessed in terms of their impact on human well-being, particularly dignity, inclusion, and agency (Sen, 1999; Nussbaum, 2011). These dimensions provide a more comprehensive basis for evaluating AI governance.

CONCLUSION

AI has moved well beyond its origins as a technical tool. It now plays a central role in shaping how decisions are made, how opportunities are distributed, and how institutions operate. As its influence expands, it becomes increasingly clear that outcomes are not determined by technological capability alone. They are shaped by the frameworks within which these systems are designed, governed, and interpreted.

This chapter has argued that the real challenge of AI governance lies in fragmentation. Technical design, ethical considerations, and leadership practice are too often treated as separate concerns. The result is a disconnect: systems that may perform efficiently, yet remain difficult to justify in terms of accountability, fairness, or broader social purpose. Approaches that rely solely on regulatory compliance or abstract ethical principles struggle to address the complexity and distributed responsibility that define AI-driven environments.

The AI–Ethics–Faith Systems Model was developed in response to this gap. Rather than introducing new principles, it brings existing elements into alignment. By linking AI infrastructure governance, ethical accountability mechanisms, and value-based leadership, the model offers a way of understanding governance as an integrated system rather than a set of disconnected interventions. Within this framework, leadership takes on a coordinating role—ensuring that technical performance, ethical reasoning, and societal priorities remain in conversation with one another.

What emerges from this perspective is a different understanding of governance itself. It becomes less about control and more about stewardship: an ongoing process of guiding systems as they evolve, responding to unintended effects, and recalibrating decisions in light of their broader consequences. This shift is not merely conceptual. It reflects the practical reality that AI systems are dynamic, adaptive, and embedded within complex institutional settings.

A key contribution of the chapter is the repositioning of human flourishing as the central reference point for AI governance. This broadens the criteria for evaluating systems. Efficiency and optimization remain important, but they are no longer sufficient. Questions of dignity, inclusion, meaningful participation, and well-being become equally important in assessing whether technological systems are serving their intended purpose.

Seen in this light, AI does not operate as a neutral instrument. It amplifies the priorities embedded within it. When narrowly governed through efficiency metrics alone, it can reinforce inequality and erode trust. When guided through ethical accountability and value-oriented leadership, it has the potential to expand opportunity, strengthen social cohesion, and support more sustainable forms of development. The difference lies not in the technology itself, but in how it is governed.

This leads to a broader implication for leadership. In the context of AI, leadership can no longer be understood simply as decision-making authority. It involves responsibility for systems that structure outcomes at scale. Leaders must navigate trade-offs, interpret evolving conditions, and ensure that innovation remains aligned with long-term societal goals. In this sense, leadership becomes a form of socio-technical stewardship, a continuous engagement with systems rather than a one-time exercise of control.

The contribution of this chapter is therefore both conceptual and practical. Conceptually, it offers a framework that connects AI governance, ethics, and leadership within a single, coherent model. In practice, it provides institutions with a way to embed accountability, ethical reasoning, and value orientation into the design and operation of technological systems.

Ultimately, the future of AI governance will depend less on technological advances and more on how those advances are guided. AI will continue to shape decision-making across sectors. The question is not whether this transformation will occur, but whether it will be directed toward inclusive, equitable, and sustainable development. That outcome will depend on leadership's capacity to consistently, deliberately, and over time align technological systems with moral purpose.

AI DISCLOSURE STATEMENT

This paper was prepared with the assistance of AI tools, including ChatGPT 5.2, Grammarly Pro and Canva. These tools were used exclusively for structural planning, language refinement, formatting support, and organizational assistance and to generate the authors figure into an infographic for better visual presentation. All substantive arguments, ethical evaluations, factual interpretations, and scholarly judgments are the original work of the author. AI tools were not used to fabricate sources, generate original philosophical claims, or replace engagement with primary documents, international legal materials, or academic literature. Full responsibility for the content and conclusions of this paper rests with the author.

REFERENCES

Akbarighatar, P. (2025). Operationalizing responsible AI principles through responsible AI capabilities. *AI and Ethics*, *5*(2), 1787–1801. DOI: 10.1007/s43681-024-00524-4

Binns, R. (2021). Fairness in machine learning: Lessons from political philosophy. *Philosophy & Technology, 81*, 149–159. Proceedings of Machine Learning Research, *Conference on Fairness, Accountability, and Transparency*. DOI: 10.48550/arXiv.1712.03586

Emerson, L. H. (2026). Outsourcing Harm: Global Labor, Economic Violence, and Invisible Human Cost. *Southern New Hampshire University; Independent Scholar.* https://papers.ssrn.com/sol3/papers.cfm?abstract_id=6173804

Eubanks, V. (2018). *Automating inequality*. St. Martin's Press. https://orcid.org/0000-0001-5017-0831

Floridi, L., Cowls, J., Beltrametti, M., Chatila, R., Chazerand, P., Dignum, V., Luetge, C., Madelin, R., Pagallo, U., Rossi, F., Schafer, B., Valcke, P., & Vayena, E. (2018). AI4People—An ethical framework. *Minds and Machines*, *28*(4), 689–707. DOI: 10.1007/s11023-018-9482-5 PMID: 30930541

Jolly, A. A., Dunlop, P. D., Parker, S. K., & Kansel, L. (2026). It's Not My Responsibility: Working with Autonomy-Restricting Algorithms Facilitates Unethical Behavior and Displacement of Responsibility. *Journal of Business Ethics*, *203*(2), 405–424. DOI: 10.1007/s10551-025-06034-5

Kellogg, K. C., Valentine, M. A., & Christin, A. (2020). Algorithms at work. *The Academy of Management Annals*, *14*(1), 366–410. DOI: 10.5465/annals.2018.0174

Maak, T., & Pless, N. M. (2021). *Responsible leadership* (2nd ed.). Routledge. DOI: 10.4324/b22741-4

Malja, S. N., & Afrasiabi, H. (2025). AI and society: Mapping the research through a systematic review. *AI & Society*. Advance online publication. DOI: 10.1007/s00146-025-02555-9

Mehrabi, N., Morstatter, F., Saxena, N., Lerman, K., & Galstyan, A. (2021). A survey on bias and fairness in machine learning. *ACM Computing Surveys*, *54*(6), 1–35. DOI: 10.1145/3457607

Noordt, C. V., & Tangi, L. (2023). The dynamics of AI capability and its influence on public value creation of AI within public administration. *Government Information Quarterly*, *40*(4), 101860. DOI: 10.1016/j.giq.2023.101860

Nussbaum, M. C. (2011). *Creating capabilities*. Harvard University Press., DOI: 10.4159/harvard.9780674061200

Obermeyer, Z., Powers, B., Vogeli, C., & Mullainathan, S. (2019). Dissecting racial bias in an algorithm used to manage the health of populations. *Science*, *366*(6464), 447–453. DOI: 10.1126/science.aax2342 PMID: 31649194

Sen, A. (1999). *Development as freedom*. Oxford University Press.

Uhl-Bien, M., Marion, R., & McKelvey, B. (2007). Complexity Leadership Theory: Shifting leadership from the industrial age to the knowledge era. *The Leadership Quarterly*, *18*(4), 298–318. DOI: 10.1016/j.leaqua.2007.04.002

van Wynsberghe, A. (2021). Sustainable AI: AI for sustainability and sustainability of AI. *AI and Ethics*, *1*(3), 213–218. DOI: 10.1007/s43681-021-00043-6

Vinuesa, R., Azizpour, H., Leite, I., Balaam, M., Dignum, V., Domisch, S., Felländer, A., Langhans, S. D., Tegmark, M., & Nerini, F. F. (2020). The role of AI in achieving the Sustainable Development Goals. *nature. COMMUNICATIONS*, *233*. Advance online publication. DOI: 10.1038/s41467-019-14108-y PMID: 31932590

Wachter, S., Mittelstadt, B., & Russell, C. (2021). Why fairness cannot be automated: Bridging the gap between EU non-discrimination law and AI. *Computer Law & Security Review*, *41*, 105567. DOI: 10.1016/j.clsr.2021.105567

Yeung, K. (2017). Algorithmic regulation: A critical interrogation. *Regulation & Governance*, *12*(4), 505–523. DOI: 10.1111/rego.12158

KEY TERMS AND DEFINITIONS

Adaptive Feedback Loops: Ongoing cycles through which AI systems learn from new data and adjust their behavior, requiring continuous monitoring to ensure outcomes remain aligned with intended objectives.

Algorithmic Bias: Systematic and repeatable errors in AI systems that lead to unfair or unequal outcomes, often resulting from patterns embedded in historical data or design choices.

Algorithmic Transparency: The degree to which the functioning and decision logic of an AI system can be understood, explained, and examined by stakeholders.

Artificial Intelligence (AI) Governance: The set of structures, processes, and decisions through which AI systems are designed, deployed, and monitored to ensure they operate in ways that are accountable, transparent, and aligned with societal goals.

Ethical Accountability Mechanisms: Institutional processes—such as audits, oversight structures, and grievance systems—that make AI-driven decisions visible, open to evaluation, and subject to correction when necessary.

Faith-Informed Value Orientation: A normative perspective grounded in principles such as justice, responsibility, and stewardship, used to guide decision-making in ways that extend beyond technical efficiency.

Human Flourishing: A multidimensional understanding of well-being that includes dignity, inclusion, agency, and the ability of individuals to lead meaningful and fulfilling lives.

Socio-Technical Stewardship: A form of leadership that focuses on guiding and maintaining alignment within complex systems, ensuring that technological processes evolve in ways that remain consistent with ethical and societal priorities.

Socio-Technical Systems: Integrated systems in which technological components and human actors interact continuously, shaping outcomes through their combined influence rather than operating independently.

Sustainable Development Alignment: The process of ensuring that AI systems contribute to long-term social, economic, and environmental goals, particularly those related to inclusion, equity, and well-being.

Chapter 8
Khalifah of the Self:
Spiritual Leadership as the First Boundary in an AI-Driven World

Noprita Herari
http://orcid.org/0009-0009-2497-5340
Universitas Negeri Jakarta, Indonesia

Azzumar Adhitia Santika
http://orcid.org/0009-0006-0274-1314
Universitas Sultan Ageng Tirtayasa, Indonesia

Zainudin bin Hassan
Universiti Teknologi Malaysia, Malaysia

Fuji Fitri Anjani
Universitas Negeri Jakarta, Indonesia

Muhammad Ridho Kurniawan Pratama
http://orcid.org/0009-0001-1371-4374
Universitas Negeri Jakarta, Indonesia

ABSTRACT

Artificial intelligence increasingly mediates leadership decisions, yet the deeper crisis of the AI era is not technological complexity but moral displacement. As leaders defer to algorithmic outputs, responsibility risks dissolving into systems, procedures, and claims of neutrality. This chapter argues that ethical AI leadership must begin with self-governance. Drawing on the Islamic concept of khalīfah and the ethic of amānah and mas'ūliyyah, it proposes Khalifah of the Self as a foundational moral framework structured around qalb (consciousness), 'aql and niyyah (reasoning and

DOI: 10.4018/979-8-3373-8192-3.ch008

ght © 2027, IGI Global Scientific Publishing. Copying or distributing in print or electronic forms without written permission of IGI Global Scientific Publishing is prohibited. Use of this chapter to generative artificial intelligence (AI) technologies is expressly prohibited. The publisher reserves all rights to license its use for generative AI training and machine learning model development.

intention), nafs and ikhtiyār (struggle and choice), and culminating in accountability. By re-centering leadership on moral anthropology, the chapter contends that the first ethical boundary of AI is not in code, but in character.

INTRODUCTION

Public discourse on artificial intelligence is often dominated by a singular anxiety: that machines are becoming too human. From humanoid chatbots that simulate empathy to generative systems capable of producing art, language, and decision recommendations, the prevailing concern centers on how closely AI can replicate human cognition and behavior through human habits as data (Herari et al., 2026). This framing, however, risks obscuring a deeper ethical reality. The more urgent problem of the AI era is not that technology is becoming human-like, but that humans may be surrendering their humanity in the way they use technology (Herari & Riyanto, 2026).

Recent studies suggest that widespread AI use is reshaping not only workflows, but also judgment, originality, and moral engagement. Research in education and cognitive psychology indicates that frequent reliance on AI systems correlates with diminished critical thinking, reduced originality, and increased cognitive offloading, particularly among students and early-career professionals (Gerlich, 2025; Kosmyna et al., 2025). In leadership contexts, this concern becomes more serious. When decisions are mediated by algorithms in hiring, evaluation, education, or risk prediction, leaders may experience moral distance from outcomes, treating automated recommendations as neutral or objective even when they produce harmful consequences.

This chapter understands such moral distance as part of a broader problem of moral outsourcing, a condition in which ethical responsibility is displaced onto technical systems and accountability becomes diluted (Danaher, 2016). While public debates often focus on technical risks such as bias, hallucination, surveillance, or loss of control, ethical failure in AI use is also rooted in human moral disengagement (Milmo, 2026). AI may amplify human values and intentions, but it cannot correct them. Technology can calculate, recommend, and optimize; it cannot assume conscience, intention, or responsibility.

The central question, therefore, is not only whether AI can be aligned with human values, but whether human beings remain capable of moral leadership while using AI. This reframing aligns with the volume's concern for leading with faith and ethics in an AI world. Faith-informed leadership reminds us that ethical restraint, humility, and accountability are not constraints on innovation, but conditions for meaningful progress. The question is no longer simply "How human can AI become?" but "*How human are leaders willing to remain*?"

MORAL OUTSOURCING AND THE EROSION OF HUMAN RESPONSIBILITY IN AI

As artificial intelligence becomes embedded in organizational decision-making, leadership responsibility is increasingly mediated by technical systems. Algorithms now screen candidates, rank performance, allocate resources, and predict risks under the promise of neutrality and efficiency (Riaz, 2025). Although these systems are formally described as decision-support tools, their expanding authority subtly reshapes how leaders understand accountability. What emerges is not merely technological dependence, but a normative shift in the locus of responsibility: moral judgment is reframed less as an internal obligation than as procedural compliance with system outputs.

This shift can be described as moral outsourcing: the displacement of ethical responsibility from human agents to algorithmic systems. When leaders justify decisions by appealing to automated recommendations ("the system flagged," "the model predicted," "the data indicated"), ethical choice is linguistically and psychologically reframed as technical inevitability. Scholars have warned that such framing reduces the perceived need for moral deliberation, especially in complex environments where algorithmic outputs appear objective or data-driven (Danaher, 2016; Mittelstadt, 2019). Responsibility does not disappear; it becomes diffused, abstracted, and emotionally distant. Recent experimental evidence strengthens this concern: working with autonomy-restricting algorithmic decision-making systems can increase unethical behavior and lead individuals to displace responsibility onto a human authority or the technology itself (Jolly et al., 2025).

Automation bias helps explain how this displacement becomes psychologically plausible. Defined as the tendency to over-trust automated outputs even when contradictory evidence exists (Goddard et al., 2012), automation bias lowers the threshold for ethical scrutiny. In high-stakes contexts, individuals may interpret computational confidence as epistemic superiority, even when moral reservations remain. Empirical research shows that users who rely on algorithmic advice often experience reduced personal responsibility for negative outcomes (Alon-Barkat & Busuioc, 2022), while experimental findings suggest that automated directives can lead to discriminatory or harmful decisions with diminished guilt (Ghasemaghaei & Kordzadeh, 2024). Automation bias, therefore, is not the ethical problem itself; it is the cognitive channel through which moral outsourcing becomes normalized.

The deeper issue is anthropological rather than technical. When ethical agency is displaced onto systems, leaders may experience what can be termed guiltless leadership: authority exercised without the full weight of moral ownership. Decisions remain operationally efficient yet psychologically detached from their human consequences. This detachment is reinforced by public narratives that anthropomorphize technology ("the algorithm decided," "AI made the call") while obscuring the human actors who designed, approved, and deployed these systems (Chowdhury, 2018). Such language subtly attributes agency to machines and dilutes accountability among decision-makers.

At the institutional level, moral outsourcing contributes to systemic opacity. When harm occurs, responsibility is often redirected away from human judgment toward technical complexity. As Eubanks (2018) demonstrates in her analysis of automated welfare systems, algorithmic infrastructures can normalize injustice precisely because no individual perceives themselves as the moral author of the outcome. Governance frameworks may require transparency and auditability, yet these mechanisms presuppose morally engaged agents willing to assume responsibility. As Hagendorff (2020) argues, ethical guidelines alone cannot guarantee ethical AI; without human moral commitment, compliance regimes risk becoming performative.

Leadership ethics scholarship reinforces this concern. Ethical leadership is not reducible to rule-following but grounded in internal moral capacities such as integrity, self-awareness, and courage (Brown & Treviño, 2006; Hannah et al., 2011). When leaders defer excessively to AI systems, they risk confusing technical justification with ethical legitimacy. Algorithms can process probabilities, but they cannot assume intention, conscience, or answerability. Only human agents possess these capacities.

The erosion of responsibility in AI-mediated environments is therefore not primarily a failure of design, but a failure of moral formation. Systems may distribute tasks, but they cannot absorb accountability. AI can augment analysis, yet it cannot replace the inward act of moral ownership. If leaders are not prepared to govern themselves (examining intentions, recognizing power, and accepting responsibility for foreseeable consequences) no amount of regulatory sophistication can prevent ethical displacement.

Moral outsourcing thus exposes a foundational question: who remains accountable when decisions are automated? The answer cannot reside in code. It resides in the moral constitution of the human agent. Reclaiming responsibility in the age of AI requires re-centering leadership in the self as a locus of agency rather than allowing it to dissolve into algorithmic process. Ethical erosion begins not when machines become powerful, but when humans relinquish ownership of their moral agency.

This chapter responds to that erosion by proposing Khalifah of the Self as a framework for AI-era leadership formation. Rather than beginning with systems, rules, or compliance, the framework begins with the human moral agent. It organizes leadership responsibility through four interrelated movements: consciousness, reasoning, struggle and choice, and responsibility. These movements are later developed through the Islamic dimensions of *qalb*, *rūḥ*, *'aql*, *niyyah*, *nafs*, *ikhtiyār*, *amānah*, and *mas'ūliyyah*, showing how ethical AI leadership must be anchored in self-governance before system governance.

In practical terms, the framework asks leaders to treat AI-mediated decisions as morally consequential before they are technically efficient. It requires them to ask who may be affected, what intention drives AI use, what pressures may distort judgment, and what boundaries must be set when human dignity, development, or accountability is at stake. In this way, Khalifah of the Self is not only a theological concept, but also a decision-oriented framework for preserving moral agency in technologically mediated leadership.

HUMAN AS SELF-LEADER: FOUNDATION OF MORAL ACCOUNTABILITY

Contemporary debates on ethical leadership in the age of artificial intelligence often emphasize governance mechanisms, regulatory safeguards, and institutional accountability. Yet these approaches remain insufficient when ethical responsibility is displaced from human agents to technical systems. This limitation invites a deeper inquiry into how moral traditions understand the human being not merely as a regulated actor, but as a self-governing moral subject accountable for intention, choice, and consequence. Across religious and philosophical traditions, human existence is rarely framed as morally neutral; rather, it is consistently understood as purposive, responsible, and ethically answerable. Table 1 offers a brief comparative overview of this shared anthropological orientation across several traditions, drawing on explicit scriptural sources.

Table 1. The purpose of human creation in religious and philosophical traditions (based on sacred texts)

Tradition	Purpose of Human Creation	Scriptural Source
Islam	To worship God and act as entrusted stewards (khalīfah) on earth	Qur'an 51:56; Qur'an 2:30
Christianity	To glorify God and steward creation as bearers of God's image	Genesis 1:26–28; 1 Corinthians 10:31
Judaism	To uphold covenantal law and participate in repairing the world (Tikkun Olam)	Genesis 1:26–28; Micah 6:8
Hinduism	To fulfill dharma and progress toward liberation (moksha)	Bhagavad Gītā 3:19; 18:45–46
Buddhism	To overcome suffering and attain enlightenment	Dhammapada 183
Confucianism	To cultivate moral virtue and social harmony	Analects 12:22; 15:24

As Table 1 suggests, these traditions converge in portraying human beings as morally responsible rather than merely functional. This convergence is significant for

AI ethics, where dominant approaches are often organized around principle-based governance and virtue-ethical perspectives. While the former emphasizes external norms such as fairness, transparency, and accountability, the latter stresses moral character and ethical judgment. Recent scholarship, however, suggests that neither approach is sufficient without moral formation: Vallor (2016) emphasizes the cultivation of virtue, Mittelstadt (2019) critiques principle-based ethics for neglecting the capacities required for ethical judgment in practice, and Floridi (2019) similarly notes that trustworthy AI presupposes trustworthy human agents. Engaging religious traditions in this context is therefore not a retreat into theological particularism, but a methodologically relevant expansion of the ethical conversation (Jasanoff, 2016). Within this broader landscape, the Islamic framework is prioritized here because it offers a particularly integrated moral anthropology, linking stewardship (khalīfah), trust (amānah), intention (niyyah), and accountability (mas'ūliyyah) within a coherent structure of self-governance. From an academic standpoint, this also contributes to epistemic diversification in AI ethics, a field still largely shaped by Western secular paradigms (Ess, 2013; Cath et al., 2017). The discussion now turns to that Islamic account of the human being as a self-leader and to its relevance for leadership in AI-mediated environments.

CORE DIMENSIONS OF 'KHALIFAH OF THE SELF' IN ISLAMIC LEADERSHIP PERSPECTIVE

وَإِذْ قَالَ رَبُّكَ لِلْمَلَٰٓئِكَةِ إِنِّي جَاعِلٌ فِي ٱلْأَرْضِ خَلِيفَةً ۖ قَالُوٓا۟ أَتَجْعَلُ فِيهَا مَن يُفْسِدُ
فِيهَا وَيَسْفِكُ ٱلدِّمَآءَ وَنَحْنُ نُسَبِّحُ بِحَمْدِكَ وَنُقَدِّسُ لَكَ ۖ قَالَ إِنِّيٓ أَعْلَمُ مَا لَا تَعْلَمُونَ ۝٣٠

'Remember' when your Lord said to the angels, "I am going to place a successive 'human' authority on earth." They asked 'Allah', "Will You place in it someone who will spread corruption there and shed blood while we glorify Your praises and proclaim Your holiness?" Allah responded, "I know what you do not know" (Qur'an Surah Al-Baqarah (2):30).

When God declares, "*Indeed, I will place upon the earth a khalīfah*" (Qur'an 2:30), leadership is framed not as positional authority or domination, but as entrusted stewardship intrinsic to human existence. In Islamic thought, khalīfah denotes a human being as a moral steward, endowed with agency, intentionality, and accountability, called to exercise authority responsibly rather than absolutely. This conception rejects ownership and control in favor of ethical trusteeship (amānah), requiring self-governance, moral restraint, and answerability for one's choices and their consequences (Rahman, 2009; Al-Ghazali, 2023). Leadership, in this sense, is an ontological condition rather than an institutional role: to be human is to be responsible, and moral accountability precedes formal authority, technological systems, and governance frameworks.

This Qur'anic understanding is reinforced by the Prophetic tradition that universalizes leadership and accountability:

Allah's Messenger (□) said, "Everyone of you is a guardian and is responsible for his charges. The ruler who has authority over people, is a guardian and is responsible for them, a man is a guardian of his family and is responsible for them; a woman is a guardian of her husband's house and children and is responsible for them; a slave ('Abu) is a guardian of his master's property and is responsible for it; so all of you are guardians and are responsible for your charges" (Sahih al-Bukhari 2554, 2025).

Building on this foundation, Islamic leadership ethics does not reduce responsibility to external compliance or rule-following. Instead, it locates ethical failure and ethical excellence within the inner constitution of the human being. Leadership, from this perspective, is enacted through the continuous regulation of the heart, intellect, ego, spirit, intention, freedom, trust, and accountability. These inner dimensions form the moral architecture through which khalīfah is realized in practice.

Table 2. The moral architecture of the self in Islamic anthropology

Human Dimension	Conceptual Definition	Primary Function in Behavior & Leadership	Qur'anic/Prophetic Foundation
Qalb (Heart)	The center of moral consciousness, faith, and ethical sensitivity	Moral discernment, empathy, integrity, and ethical awareness	"So have they not traveled through the earth and have hearts by which to reason and ears by which to hear? For indeed, it is not eyes that are blinded, but blinded are the hearts which are within the breasts" (Qur'an 22:46)
'Aql (Intellect)	The capacity for reasoning, understanding, and evaluating consequences	Rational judgment, analysis, and decision-making	"Do you order righteousness of the people and forget yourselves while you recite the Scripture? Then will you not reason?" (Qur'an 2:44; implicit references in Qur'an 10:100)
Nafs (Self/Ego)	The source of desires, impulses, and inner conflict	Ambition, ego, moral struggle; potential for discipline or corruption	"And I do not acquit myself. Indeed, the soul is a persistent enjoiner of evil, except those upon which my Lord has mercy. Indeed, my Lord is Forgiving and Merciful" (Qur'an 12:53)
Rūḥ (Spirit)	The transcendent dimension that gives meaning and spiritual orientation	Awareness of the divine, ultimate purpose, and existential meaning	"Then He proportioned him and breathed into him from His [created] soul and made for you hearing and vision and hearts; little are you grateful" (Qur'an 32:9)

continued on following page

Table 2. Continued

Human Dimension	Conceptual Definition	Primary Function in Behavior & Leadership	Qur'anic/Prophetic Foundation
Niyyah (Intention)	The moral orientation underlying every action	Determines the ethical value of conduct	"Actions are judged by intentions." (Ṣaḥīḥ al-Bukhārī Book 1; Ṣaḥīḥ Muslim no.1907a)
Ikhtiyār (Free Will)	The capacity to choose consciously and responsibly	Foundation of moral responsibility and leadership agency	"Indeed, We guided him to the way, be he grateful or be he ungrateful" (Qur'an 76:3)
Amānah (Trust / Moral Responsibility)	The burden of entrusted responsibility and ethical obligation	Integrity, stewardship, and leadership accountability	"Indeed, We offered the trust to the heavens and the earth and the mountains, but they ʿallʾ declined to bear it, being fearful of it. But humanity assumed it, ʿforʾ they are truly wrongful ʿto themselvesʾ and ignorant ʿof the consequencesʾ," (Qur'an 33:72)
Mas'ūliyyah (Accountability)	Answerability for actions and decisions	Awareness of consequences in this world and the hereafter	"Each of you is a shepherd and each of you will be held accountable." (Ṣaḥīḥ al-Bukhārī no.7138; Ṣaḥīḥ Muslim no.1829a)

The dimensions outlined in Table 2 should not be understood as isolated components, but as an interconnected moral architecture that shapes human agency, judgment, and leadership. Together, they form an integrated ethical anthropology in which cognition, emotion, intention, desire, and responsibility continuously interact. This framework challenges reductionist accounts of leadership that privilege rational calculation or procedural compliance and instead situates ethical action within the complexity of human inner life.

Figure 1. The interplay of beliefs, emotions, and intentions in a social context (Bosse et al., 2012)

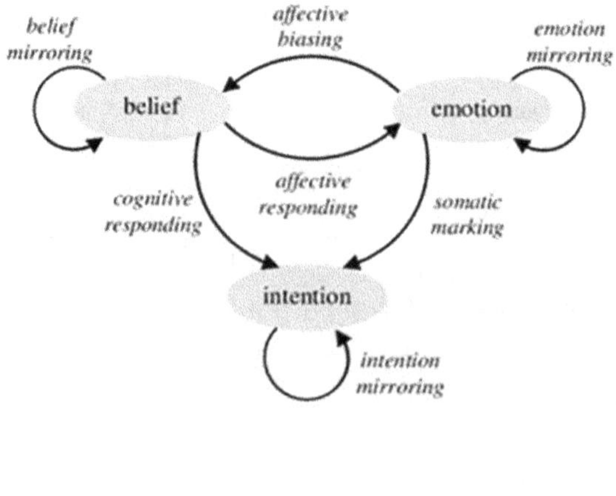

The interdependence described in Figure 1 resonates strongly with empirical and computational research on human decision-making. Bosse et al. (2012), for example, demonstrate that human behavior (particularly in collective and high-stakes contexts) emerges from a dynamic feedback loop between belief, emotion, and intention rather than from isolated rational choice. Their model shows how cognition is continuously shaped by affective responses, how emotions bias belief formation, and how intentions arise through embodied and social processes of mirroring and contagion. This evidence reinforces the view that ethical agency cannot be reduced to rational calculation or procedural compliance alone. Instead, moral action is situated within a complex inner architecture in which cognition, emotion, intention, and responsibility are mutually constitutive, precisely the ethical anthropology articulated in the integrated framework presented in this chapter.

Qalb (Heart): Moral Consciousness and Ethical Perception

"So have they not traveled through the earth and have hearts by which to reason and ears by which to hear? For indeed, it is not eyes that are blinded, but blinded are the hearts which are within the breasts" (Qur'an Surah Hajj (22):46)

The Qur'anic assertion that "*it is not the eyes that are blind, but blind are the hearts that are in the chests*" (Qur'an 22:46) reframes ethical failure as a problem of moral consciousness rather than perceptual deficiency. The verse does not deny the role of sensory perception, but decisively locates understanding (ta'aqqul) in the *qalb*. What is at stake, therefore, is not access to information, but the moral faculty through which meaning, responsibility, and judgment are formed.

The *qalb* is presented as the locus where reasoning, ethical awareness, and evaluative judgment intersect, enabling humans to learn from experience, discern consequences, and assume responsibility for their actions (Ismail et al., 2021; Raza, 2022). Importantly, these studies emphasize that Qur'anic blindness is not epistemic ignorance but moral disengagement. Individuals may observe, travel, and accumulate experiences, yet fail to draw ethical insight because the qalb is inactive or hardened. Classical exegetes cited in the manuscripts (such as Ibn 'Āshūr, al-Rāzī, and al-Ālūsī) interpret *qulūb ya'qilūn* (hearts that reason) as indicating that moral cognition requires inner receptivity, not merely analytical capacity (Ismail et al., 2021). This insight challenges modern assumptions that ethical clarity emerges automatically from data accumulation or rational processing alone.

The manuscripts further demonstrate that *qalb* plays a decisive role in higher-order thinking. In educational contexts, Ismail et al. (2021) show that the Qur'anic conception of thinking integrates reflection, evaluation, and moral synthesis, capacities closely aligned with contemporary pedagogy's higher-order thinking skills (HOTS). However, unlike secular cognitive models, the Qur'anic framework situates these skills within a moral heart that must be actively engaged. Critical thinking without a qalb orientation risks becoming instrumental rather than ethical.

From a psychophysiological perspective, research on Qur'anic recitation reinforces the heart's ethical function by demonstrating its role in emotional regulation, attentiveness, and inner coherence (Moulaei et al., 2023). These findings support the Qur'anic view that moral clarity and inner calm are interconnected states, suggesting that ethical perception is sustained not by cognitive efficiency alone, but by an integrated moral-emotional equilibrium. The heart, in this sense, mediates between knowing and becoming.

Taken together, those studies affirm that *qalb* is the primary site of moral perception. Ethical failure occurs not when information is lacking, but when the heart ceases to process meaning, responsibility, and consequence. This understanding has significant implications for leadership ethics. It implies that ethical judgment cannot be delegated to external systems, procedures, or tools without risking moral blindness. When leaders disengage their qalb, decisions may appear rational and efficient while remaining ethically hollow.

'Aql (Intellect): Reasoning Oriented by Ethics

In Islamic thought, *'aql* refers to the human capacity for reasoning, understanding, and deliberation. The Qur'an repeatedly calls upon human beings to think, reflect, and evaluate, often through rhetorical questions such as "*Will you not reason?*" (Qur'an 2:44). This emphasis signals that intellectual engagement is not optional in the moral life; it is a divine expectation. However, Islamic ethics does not treat 'aql as a morally autonomous faculty. Reason is affirmed, but it is never absolutized. "*Do you order righteousness of the people and forget yourselves while you recite the Scripture? Then will you not reason?*" (Qur'an Surah al-Baqarah (2): 44).

Unlike Enlightenment traditions that elevate rationality as the ultimate arbiter of truth and morality, Islamic moral anthropology situates *'aql* within a broader ethical ecology. Reason functions properly only when oriented by moral consciousness and guided by responsibility. Classical scholars consistently warn that intellect detached from ethical purpose can become a tool of rationalization rather than discernment. Al-Ghazālī, for instance, argues that intelligence without moral orientation enables deception, self-justification, and ethical blindness rather than wisdom (Al-Ghazali, 2023).

This perspective is evident in the Qur'anic pattern of critique. The Qur'an does not condemn ignorance alone, but repeatedly criticizes those who "know" yet fail to act ethically, or who reason in ways that justify injustice. In several passages, *'aql* is invoked not to celebrate cognitive sophistication, but to expose moral inconsistency, knowing the good while failing to choose it (Qur'an 2:44). Reason, therefore, is judged not by its analytical power, but by its ethical direction.

Islamic scholars distinguish between *'aql* as cognitive capacity and ḥikmah (wisdom) as morally guided understanding. While *'aql* enables analysis and inference, ḥikmah reflects the integration of reason with ethical insight and lived responsibility. Rahman (2009) emphasizes that Qur'anic reasoning is teleological: it asks not merely whether an argument is logically valid, but whether it leads toward justice, responsibility, and human flourishing. Reason divorced from these ends is considered incomplete.

Contemporary social theory echoes this concern. Habermas's (1985) critique of instrumental rationality highlights how reasoning focused solely on efficiency and control erodes ethical deliberation. While developed in a secular framework, this critique parallels Islamic warnings about 'aql operating without moral anchoring. In both traditions, rationality requires normative orientation to avoid becoming a mechanism of domination.

Islamic thought does not position 'aql in opposition to emotion or spirituality. Rather, it understands intellect as interdependent with other inner dimensions. Reason informs judgment, but judgment is shaped by intention, restraint, and accountability. This interdependence explains why the Qur'an locates reasoning in the heart (qulūb ya'qilūn), emphasizing that intellect is not a cold, detached faculty but one embedded in moral consciousness (Qur'an 22:46).

From an ethical leadership perspective, *'aql* functions as a critical capacity for evaluation, anticipation, and responsibility. It enables leaders to assess consequences, recognize trade-offs, and foresee harm. Yet its ethical value depends on whether it is exercised as a servant of responsibility or as an instrument of justification. Reason can clarify moral obligations, but it can also obscure them when mobilized to defend convenience, authority, or efficiency.

Thus, Islamic ethics advances a normative claim: reason is not morally neutral. The question is not whether leaders reason, but how and toward what ends they reason. *'Aql* oriented by ethics becomes a source of wisdom and restraint; 'aql detached from ethics becomes a mechanism for moral outsourcing and responsibility avoidance.

Nafs (Self/Ego): The Site of Moral Struggle

In Islamic moral anthropology, nafs represents the most volatile and contested dimension of the human self. It is the locus where desire, ambition, fear, and the pursuit of power converge, making it the primary arena of moral struggle. Unlike qalb, which enables ethical perception, or 'aql, which facilitates reasoning, nafs governs impulses and inclinations that may either be disciplined toward responsibility or left unchecked toward domination. The Qur'an describes this vulnerability candidly: "*Indeed, the soul is inclined toward evil, except those upon whom my Lord has mercy*" (Qur'an 12:53).

This statement does not imply determinism but moral risk. Human beings are not condemned by their nafs; they are tested through it. Ethical life, therefore, consists not in eliminating desire but in governing it.

Classical Islamic scholarship conceptualizes nafs as dynamic and developmental rather than fixed. Thinkers such as Al-Ghazālī and Ibn Qayyim al-Jawziyyah describe three primary states: nafs al-ammārah (the commanding self), nafs al-lawwāmah (the self-reproaching self), and nafs al-muṭma'innah (the tranquil self). These stages reflect moral maturation rather than personality types (Qureyshi & Mustaq, 2025). Ethical failure emerges when leadership remains governed by the commanding self, seeking validation, control, or convenience without restraint.

The nafs al-ammārah is driven by impulse and unrestrained desire. In leadership contexts, it manifests as ego-driven authority, rationalization of questionable decisions, and prioritization of self-interest over justice. Power does not create corruption; it amplifies what already governs the self. When the ammārah dominates, reasoning becomes subordinated to ambition.

By contrast, nafs al-lawwāmah marks the emergence of moral reflexivity. The Qur'an affirms the "self-reproaching soul" (Qur'an 75:2), recognizing the ethical significance of self-critique. At this stage, individuals experience tension between action and conscience. This inner discomfort is not weakness but the birthplace of accountability. Leadership in this condition remains imperfect yet capable of growth because self-evaluation has begun.

The highest state, nafs al-muṭma'innah, is described as the tranquil soul invited to return in peace (Qur'an 89:27–30). This condition reflects alignment between intention, action, and moral orientation. It does not signify the absence of desire but its successful governance. Ethical leadership here is marked by coherence, humility, and stability rather than moral perfection.

Importantly, these stages are not permanent achievements but fluctuating states. Contemporary scholarship on Qur'anic psychology emphasizes the dynamic nature of moral development, noting that regression remains possible under pressure or fear (Badri, 2018). Leadership integrity, therefore, cannot be secured once and for all; it must be continually renewed through self-discipline.

This understanding challenges leadership models that privilege rational calculation alone. Even when leaders possess knowledge ('aql) and awareness (qalb), ethical collapse occurs when nafs overrides them. Ibn Taymiyyah observes that injustice often arises not from ignorance but from desire dominating conscience (Zouggar, 2018). Modern leadership research similarly shows that ethical derailment frequently stems from unchecked ego needs (status, dominance, and validation) rather than incompetence (R. Kets de Vries, 2011).

Islamic ethics does not demonize desire; it disciplines it. Ambition can fuel excellence, and authority can enable justice. What determines moral value is whether the nafs is governed by intention and accountability or allowed to dominate them. Ethical leadership thus becomes an ongoing inner struggle, a refinement rather than suppression of humanity.

In this sense, nafs is the crucible of moral leadership. It explains why responsibility cannot be outsourced to systems or neutralized by procedure. As long as human beings wield power, they remain sites of struggle. The decisive question is not whether leaders possess desire, but whether they govern it.

Rūḥ (Spirit): Transcendence and Moral Orientation

Within Islamic moral anthropology, rūḥ represents the transcendent dimension of the human being, the source of existential orientation that anchors ethical life beyond immediacy, utility, and self-interest. The Qur'an describes human creation as uniquely marked by divine endowment: "*Then He proportioned him and breathed into him of His Spirit*" (Qur'an 32:9). This verse does not imply divinity in humans, but signals that human agency is oriented toward meaning, responsibility, and accountability that exceed material calculation.

Unlike *'aql*, which enables analysis, or *nafs*, which generates desire and struggle, *rūḥ* provides a moral horizon. It situates human action within a broader purpose, reminding the individual that choices are not limited to immediate outcomes or institutional success. Classical scholars consistently interpret *rūḥ* as the dimension through which humans recognize transcendence (God, ultimate truth, and the hereafter) and thereby sustain ethical seriousness in worldly affairs (Al-Ghazali, 2023; Rahman, 2009).

Crucially, Islamic thought does not treat *rūḥ* as a mystical abstraction detached from everyday life. Rather, it functions as an orienting presence that gives coherence to moral action. Al-Ghazālī emphasizes that when *rūḥ* is neglected, ethical reasoning collapses into calculation and leadership degenerates into technique (Al-Ghazali, 2023). In such conditions, decisions may remain legally correct or procedurally sound, yet lack moral depth because they are no longer oriented toward ultimate responsibility.

This insight resonates with broader ethical philosophy. Taylor (1989) argues that moral frameworks require a "horizon of significance" to sustain commitment and responsibility. Without transcendence (whether religious or philosophical), ethical reasoning risks becoming relativistic or instrumental. Islamic ethics provides such a horizon through *rūḥ*, grounding leadership in accountability that is not contingent on surveillance, reward, or institutional enforcement.

In leadership contexts, *rūḥ* functions as a counterweight to moral exhaustion and ethical drift. Leaders often operate under pressure, complexity, and competing demands that encourage short-term thinking. The presence of *rūḥ* reorients judgment toward long-term moral consequence and human dignity. It reminds leaders that authority is temporary, power is conditional, and decisions carry weight beyond organizational metrics.

Importantly, *rūḥ* also tempers the excesses of the *nafs*. While *nafs* drives ambition and desire, *rūḥ* introduces restraint by re-situating the self within a moral order larger than ego. Ibn Qayyim al-Jawziyyah explains that ethical failure often emerges when the heart, deprived of its transcendent orientation, seeks fulfillment in immediate gratification or dominance rather than in accountable stewardship (Parrott, 2014). Thus, rūḥ is not opposed to action or leadership; it disciplines them.

Modern scholarship on leadership ethics increasingly recognizes the role of transcendence and meaning in sustaining moral agency. Studies in spiritual and authentic leadership show that leaders who orient their actions toward values beyond self-interest demonstrate greater consistency, moral courage, and responsibility under pressure (Fry, 2003; Hannah et al., 2011). While these studies are not framed explicitly in Islamic terms, they align closely with the function attributed to rūḥ as a source of moral orientation.

Niyyah (Intention): Moral Direction of Action

In Islamic ethics, niyyah (intention) functions as the moral axis of human action. It determines not merely what is done, but what an action means ethically. The foundational Prophetic statement:

It has been narrated on the authority of Umar b. al-Khattab that the Messenger of Allah (ﷺ) said:

"(The value of) an action depends on the intention behind it. A man will be rewarded only for what he intended. The emigration of one who emigrates for the sake of Allah and His Messenger (ﷺ) is for the sake of Allah and His Messenger (ﷺ) ; and the emigration of one who emigrates for gaining a worldly advantage or for marrying a woman is for what he has emigrated" (Ṣaḥīḥ Muslim 1907a)

Establishes intention as the primary criterion of moral evaluation. This principle reorients ethics away from outcomes alone toward the inner direction that animates conduct.

Unlike legalistic frameworks that assess morality through compliance or observable behavior, Islamic moral reasoning treats niyyah as the source from which ethical value flows. An identical action may carry divergent moral significance depending on the intention that motivates it. Al-Ghazālī emphasizes that intention gives form to action, transforming routine behavior into worship or rendering seemingly noble acts ethically hollow when driven by ego, status, or self-interest (Al-Ghazali, 2023).

Crucially, niyyah is not reducible to stated purpose or strategic framing. It is an inward orientation that aligns desire (nafs), reasoning ('aql), and moral consciousness (qalb) toward a chosen end. Rahman (2009) notes that Qur'anic ethics consistently evaluates human action teleologically, by asking toward what moral horizon an action is directed. In this sense, niyyah serves as the bridge between inner moral state and outward conduct.

This emphasis introduces a demanding ethical standard for leadership. Decisions that appear rational, efficient, or even beneficial may still be ethically compromised if driven by intentions oriented toward domination, self-preservation, or moral evasion. Islamic ethics thus resists moral minimalism. It does not permit leaders to justify harmful or questionable actions solely on the basis of results, institutional norms, or procedural correctness. Ethical legitimacy requires moral alignment at the level of intention.

Importantly, niyyah also functions as a safeguard against moral outsourcing. When intention is foregrounded, responsibility cannot be displaced onto systems, structures, or external authorities. Intention is inherently personal and non-transferable. One cannot claim ethical neutrality by appealing to necessity, technical recommendation, or collective decision if one's intention is complicit in harm or avoidance of responsibility. Al-Ghazālī explicitly warns that corrupt intention enables ethical self-deception, allowing individuals to rationalize wrongdoing while maintaining an appearance of righteousness (Al-Ghazali, 2023).

From a virtue ethics perspective, niyyah parallels the Aristotelian emphasis on moral character and motivation rather than isolated acts (Aristotle, trans. 2009). However, Islamic ethics goes further by integrating intention into accountability before God, not merely social evaluation. This transcendent accountability intensifies ethical seriousness, especially in contexts where actions are complex, mediated, or shielded from direct consequence.

Leadership studies increasingly converge with this insight. Research on ethical and authentic leadership highlights the centrality of moral intention and value congruence in sustaining ethical behavior under pressure (Brown & Treviño, 2006; Hannah et al., 2011). Leaders who lack clarity of intention are more susceptible to ethical drift, rationalization, and situational compromise. While articulated in secular terms, these findings mirror the Islamic insistence that intention anchors moral consistency.

Ikhtiyār (Free Will): The Basis of Moral Responsibility

In Islamic ethics, ikhtiyār denotes the human capacity to choose consciously and deliberately between alternatives. It is the moral hinge upon which responsibility turns. The Qur'an affirms this capacity succinctly: "*We showed him the way, whether he be grateful or ungrateful*" (Qur'an 76:3). Guidance is given, but choice remains human. Without ikhtiyār, moral evaluation would be incoherent; praise, blame, and accountability would lose their meaning.

Islamic moral anthropology thus resists both strict determinism and unbounded voluntarism. Human beings are neither compelled automatons nor sovereign creators of value. Rather, they are agents who choose within constraints (biological, social, and historical) yet remain accountable for how they respond to those constraints. Classical theologians articulated this balance through doctrines that preserve responsibility without denying divine sovereignty (Rahman, 2009).

Crucially, ikhtiyār is not mere preference or impulse. It presupposes deliberation ('aql), moral orientation (qalb and rūḥ), and intention (niyyah). A choice made under compulsion or ignorance does not carry the same moral weight as a choice made knowingly and freely. Islamic jurisprudence reflects this distinction by differentiating between intentional acts, mistakes, negligence, and coercion, each carrying different moral and legal implications. This nuanced view underscores that responsibility scales with agency.

Classical scholars emphasize that ikhtiyār is the condition for moral struggle and ethical growth. Al-Ghazālī argues that the human being's dignity lies precisely in the capacity to choose against desire and convenience (Al-Ghazali, 2023). Without freedom to resist the pull of the nafs, self-discipline would be meaningless. Thus, ikhtiyār is the engine of moral development, enabling movement from reactive behavior to responsible action.

The Qur'an repeatedly links accountability to choice, not outcome. Humans are judged for what they intend and choose, not for circumstances beyond their control. This principle protects moral responsibility from being eroded by appeals to fate, structure, or necessity. It also challenges a common tendency to diffuse responsibility by attributing outcomes to impersonal forces. In Islamic ethics, acknowledging constraints does not negate choice; it clarifies the moral significance of one's response.

From a leadership perspective, ikhtiyār grounds accountability in agency rather than role. Authority does not create responsibility; choice does. Leaders are responsible not because they occupy positions, but because they choose how to use power, interpret rules, and respond to uncertainty.

Modern moral philosophy converges with this insight. Accounts of responsibility emphasize that agency and control are prerequisites for moral accountability (Fischer & Ravizza, 1998). Even within complex systems, responsibility persists insofar as agents retain the capacity to endorse, resist, or redirect actions. This resonates with the Islamic insistence that responsibility cannot be dissolved into systems or structures if human choice remains operative.

Importantly, ikhtiyār also sets limits on responsibility. Islamic ethics does not impose blame where choice is genuinely absent. Coercion, incapacity, and ignorance mitigate responsibility, reflecting a just moral framework that aligns accountability with agency. This balance prevents moral overload while preserving ethical seriousness.

Amānah (Trust): Stewardship Rather Than Ownership

In Islamic ethics, amānah signifies entrusted responsibility rather than possession or entitlement. It reframes authority, power, and capacity as moral trusts to be exercised with care, restraint, and accountability. The Qur'an dramatizes this ethical weight by describing amānah as a burden declined by the heavens, the earth, and the mountains, yet accepted by the human being (Qur'an 33:72). This passage positions trust not as privilege, but as a profound moral obligation that exposes humans to ethical failure precisely because they possess agency and choice.

Unlike modern conceptions of authority that emphasize control, rights, or ownership, amānah establishes stewardship as the ethical core of leadership. To hold power (over people, resources, or decisions) is to hold something that ultimately does not belong to the self. Classical scholars consistently stress that amānah negates claims of absolute authority. Al-Ghazālī argues that power becomes corruptive when leaders forget its entrusted nature and treat it as personal property, thereby severing authority from moral restraint (Al-Ghazali, 2023).

This distinction between stewardship and ownership carries profound ethical implications. Ownership implies discretion with diminished moral constraint; stewardship implies responsibility that cannot be dissolved. As Mohamed (2011) argues in his critique of modern economic systems, injustice often persists not because norms are unknown, but because prevailing moral frameworks legitimize self-interest over justice. When authority is reframed as possession rather than trust, ethical failure becomes not merely a procedural mistake, but a betrayal of entrusted responsibility.

Importantly, amānah is inseparable from ikhtiyār and niyyah. Trust presupposes choice: one can uphold it or violate it. It also presupposes intention: authority exercised with sincere intention toward justice differs ethically from identical actions driven by self-preservation or ambition. Rahman (2009) notes that Qur'anic ethics consistently frames responsibility as relational, humans are accountable because they stand in trust-based relationships with God, others, and the world.

Within Islamic moral anthropology, amānah extends beyond formal leadership roles. Knowledge, speech, influence, and decision-making capacity are all treated as trusts. This universality reinforces the idea that leadership is not confined to office or title. One becomes ethically accountable not by position, but by possession of power to affect others. The Prophetic tradition reinforces this view by repeatedly associating trust with moral integrity and warning against its betrayal (Nabil, 2025).

Modern ethical leadership scholarship converges with this perspective through stewardship theory, which contrasts service-oriented leadership with ownership-driven models of authority (Block, 1996). Leaders who view power as stewardship demonstrate greater accountability, humility, and long-term responsibility. While articulated in secular terms, this research mirrors the Islamic insistence that ethical leadership depends on how authority is understood, not merely how it is regulated.

Mas'ūliyyah (Accountability): Answerability for Consequences

In Islamic ethics, mas'ūliyyah signifies answerability for one's actions and their consequences. It extends beyond procedural responsibility or role-based compliance, framing accountability as an intrinsic moral condition of human agency. The well-known Prophetic declaration

Narrated `Abdullah bin `Umar:

Allah's Messenger (□) said, "Surely! Everyone of you is a guardian and is responsible for his charges: The Imam (ruler) of the people is a guardian and is responsible for his subjects; a man is the guardian of his family (household) and is responsible for his subjects; a woman is the guardian of her husband's home and of his children and is responsible for them; and the slave of a man is a guardian of his master's property and is responsible for it. Surely, everyone of you is a guardian and responsible for his charges" (Ṣaḥīḥ al-Bukhārī No.7138).

That hadith establishes accountability as universal and personal. One is accountable not because one occupies an office, but because one acts.

Unlike modern accountability regimes that emphasize reporting structures, auditing mechanisms, or regulatory compliance, mas'ūliyyah is fundamentally relational and ethical. It binds the individual to God, to others, and to a broader moral order. The Qur'an reinforces this orientation: "*Then you will surely be asked that Day about the blessings*" (Qur'an 102:8). Accountability, in this sense, encompasses not only overt wrongdoing but also how authority, opportunity, and capacity are exercised. Moral evaluation is not confined to legal transgression; it extends to the responsible use of entrusted power.

Classical Muslim scholarship underscores that mas'ūliyyah follows agency rather than authority. Al-Ghazālī (2023) maintains that moral responsibility arises wherever an individual possesses the capacity to influence outcomes, even indirectly. This principle safeguards against the diffusion of responsibility within complex social arrangements. Acting within institutional norms or under instruction does not absolve one of accountability for foreseeable consequences. Ethical responsibility persists wherever moral agency persists.

This position stands in contrast to bureaucratic models that permit individuals to shield themselves behind hierarchy, procedure, or collective decision-making. Islamic ethics rejects such deflection. As Mastroianni et al. (2011) demonstrate in their analysis of obedience and personal responsibility, injustice often endures when individuals comply with harmful directives while relinquishing personal accountability, conflating obedience with moral innocence. Mas'ūliyyah disrupts this logic by insisting that responsibility cannot be transferred to systems, delegated to authority, or dissolved within institutional processes. Where agency remains, answerability remains.

Importantly, accountability is inseparable from ikhtiyār (free choice). Responsibility presupposes meaningful agency. Where choice is genuinely constrained (through coercion, incapacity, or ignorance) moral burden is mitigated. Islamic jurisprudence carefully calibrates accountability according to intention, awareness, and capability, thereby avoiding both moral absolutism and moral indifference. Responsibility is proportionate to agency: neither unlimited nor optional.

Within the broader moral architecture of the self, mas'ūliyyah represents the point where inward deliberation meets outward consequence. Niyyah directs action, ikhtiyār enacts it, and amānah frames authority as entrusted stewardship; mas'ūliyyah answers for what follows. Without this culminating dimension, ethical reflection risks remaining purely introspective. Accountability anchors moral life in lived effects.

From a leadership perspective, mas'ūliyyah establishes a demanding ethical standard. Leaders are answerable not only for direct decisions but also for outcomes they enable, foresee, or fail to prevent. Delegation does not dissolve responsibility; reliance on intermediaries does not eliminate moral burden. Islamic ethics thus rejects the idea of "guiltless leadership", the notion that power can be exercised without enduring ethical weight.

Contemporary accountability theory resonates with this insight. Bovens (2007) defines accountability as answerability, the obligation to explain and justify conduct before a forum capable of judgment. While articulated within secular governance, this conception parallels Islamic ethics, where the ultimate forum transcends institutional oversight. What distinguishes mas'ūliyyah is its persistence even in the absence of surveillance or sanction. Accountability is not merely performative compliance but an internalized readiness to answer for consequences.

This persistence is ethically decisive. Leaders may satisfy formal requirements and still fail morally if harm is normalized or responsibility deflected. Mas'ūliyyah insists that ethical leadership requires internal acceptance of answerability, ensuring that responsibility remains inseparable from power.

Positioning Khalifah of the Self Within Contemporary AI Ethics Frameworks

Table 3. Institutional AI ethics frameworks and the moral agency perspective of Khalifah of the self

Framework/Organization	Core Ethical Orientation	Governance Mechanism	Ethical Emphasis
OECD AI Principles (OECD, 2019)	Human-centered AI aligned with democratic values	Intergovernmental policy principles and national AI policy guidance	Fairness, transparency, accountability, robustness
Ethics Guidelines for Trustworthy AI (European Commission: Directorate-General for Communications Networks, 2019)	Development of "trustworthy AI" grounded in legality, ethics, and robustness	Regulatory guidance supported by assessment frameworks	Human oversight, transparency, non-discrimination, accountability
UNESCO Recommendation on the Ethics of AI (UNESCO, 2021)	Global normative framework for human rights–based AI governance	Multilateral policy recommendation adopted by member states	Human dignity, inclusion, environmental sustainability, responsibility
Ethically Aligned Design: A Vision for Prioritizing Human Well-being with Autonomous and Intelligent Systems (IEEE, 2021)	Engineering-focused ethical design of intelligent systems	Technical standards, design recommendations, and professional guidance	Technical standards and professional guidance for system designers
Rome Call for AI Ethics (RenAIssance, 2020)	Human dignity and moral responsibility in technological innovation	Multi-stakeholder ethical commitment among religious, academic, governmental, and industry actors	Interdisciplinary ethical commitment among industry, academia, and religious institutions

Table 3 makes clear that contemporary AI ethics is not normatively empty. On the contrary, major global frameworks have already articulated substantial commitments to fairness, transparency, accountability, human dignity, and oversight. The limitation, however, does not primarily lie in the absence of ethical language, but in the distance between institutional principles and lived moral judgment. These frameworks are strongest at the level of governance architecture: they define what ought to be protected, regulated, or monitored. Yet they are less developed at the level of the moral self who must interpret, apply, and sometimes resist the outputs of intelligent systems. In this sense, Khalifah of the Self does not compete with institutional AI ethics; it addresses the anthropological gap that many governance frameworks leave unresolved.

The OECD AI Principles and the European Commission's Ethics Guidelines for Trustworthy AI exemplify this strength and limitation simultaneously. Both frameworks are valuable in that they establish widely recognized norms for responsible AI, particularly around fairness, transparency, robustness, and accountability. Yet in practice, their governance logic often remains procedural. Ethical responsibility risks being translated into checklists, audits, and compliance routines, creating the impression that the presence of oversight mechanisms is itself sufficient to secure ethical action. From the perspective of Khalifah of the Self, this is precisely where the problem begins. Human oversight is not ethically meaningful if the human overseer lacks disciplined intention, moral courage, or the willingness to assume responsibility for consequences. Governance mechanisms may distribute duties, but they cannot generate inward accountability.

UNESCO's Recommendation on the Ethics of AI expands the conversation by foregrounding human dignity, inclusion, and sustainability within a global and explicitly human rights–based framework. Its strength lies in the breadth of its moral horizon: AI is not treated merely as a technical or economic issue, but as a civilizational one. At the same time, such breadth also reveals a practical limitation. Because the framework is necessarily universal and multilateral, it cannot offer a sufficiently thick account of how moral agency is formed within the decision-maker. It can name responsibility, but it cannot fully describe the inner discipline required to bear it. Here the relevance of Khalifah of the Self becomes clearer. It complements universal ethics with a more interior account of stewardship, one that grounds public responsibility in the moral formation of the self rather than in normative declaration alone.

IEEE's Ethically Aligned Design and the Rome Call for AI Ethics illuminate two different but related limitations. IEEE is especially important because it moves ethics closer to technical design, professional responsibility, and system development. Its weakness, however, is that ethical reflection can remain concentrated at the level of engineering process rather than leadership conscience. The Rome Call, by contrast, brings moral vocabulary, dignity, and interreligious concern more explicitly into the field. Yet even here, the framework remains primarily principle-oriented. It offers a moral appeal, but not a fully articulated anthropology of self-governance. The Khalifah of the Self perspective is positioned differently from both: it is neither primarily a design protocol nor a general moral declaration, but a leadership framework that asks what kind of human being must stand behind any ethical use of AI.

Taken together, these frameworks reveal a recurring pattern: contemporary AI ethics is highly developed in identifying external norms, but less explicit about the inner constitution of the moral subject. This is the precise point at which Khalifah of the Self enters the conversation. Its contribution lies not in replacing institutional governance, but in re-situating it within a deeper moral anthropology. Where global AI frameworks ask how systems should be governed, Khalifah of the Self asks how leaders should govern themselves while using those systems. It therefore reframes ethical AI not simply as a matter of better principles, but of better stewards. Admittedly, emerging work on artificial practical wisdom and compassionate AI suggests that AI systems may be designed to support morally sensitive responses in complex contexts (Graves, 2025). Yet this possibility does not dissolve human accountability; even morally supportive AI remains a tool that may inform judgment, not a subject that can bear mas'ūliyyah. The practical limitation of existing frameworks is not that they lack ethical aspiration, but that they often presume the moral reliability of actors without explaining how such reliability is cultivated. Khalifah of the Self responds to that omission by insisting that the first site of AI ethics is not the system alone, but the self that wields it.

THE ONTOLOGICAL SEQUENCE OF MORAL AGENCY: FROM CONSCIOUSNESS TO RESPONSIBILITY

Moral action does not emerge spontaneously from rules, systems, or external enforcement. It unfolds through an inner progression that precedes institutional regulation and technological mediation. Across classical philosophy, Islamic moral theology, and contemporary responsibility theory, ethical conduct can be understood as moving through a sequence of consciousness, reasoning, struggle, choice, and responsibility. This progression is not merely psychological; it reflects the ontological structure of human moral agency.

The first movement is moral consciousness. Ethical action presupposes awareness: the capacity to perceive that a situation carries moral weight. Aristotle described this initial stage as the recognition of the good within particular circumstances, which activates deliberation (Aristotle, trans. 2012). Contemporary moral psychology similarly emphasizes that moral judgment begins with attentiveness to value-laden features of a situation rather than neutral cognition (Haidt, 2008). Without consciousness of ethical relevance, no subsequent reasoning can be meaningfully moral.

From consciousness arises reasoning. Moral reasoning evaluates alternatives, anticipates consequences, and weighs principles. In virtue ethics, this deliberative capacity, or phronesis, distinguishes responsible action from impulse (Aristotle, trans. 2012). In contemporary responsibility theory, it corresponds to responsiveness to reasons: the capacity to recognize and respond to normative considerations (Fischer & Ravizza, 1998). Reasoning therefore translates awareness into structured judgment. This link between practical wisdom and ethical engagement is especially relevant in

AI-mediated contexts. Recent empirical work shows that practical wisdom predicts more reflective and ethically grounded engagement with AI, particularly among students navigating generative AI in educational settings (McLoughlin et al., 2026).

Yet reasoning alone does not secure ethical conduct. Between deliberation and action lies struggle. Classical accounts of moral weakness, including Aristotle's discussion of akrasia, show that individuals may act against their better judgment (Aristotle, trans. 2012). This tension between recognition and enactment reveals that moral agency includes internal conflict. Contemporary leadership research likewise shows that ethical decisions often occur under pressure, ambiguity, and competing incentives (Hannah et al., 2011). Struggle is therefore not incidental to moral agency; it is where abstract reasoning encounters lived complexity.

Following struggle comes choice. Moral choice is the commitment to act despite uncertainty, pressure, or cost. It signifies ownership of the decision. Responsibility theorists argue that agency requires both control over one's actions and endorsement of one's reasons (Fischer & Ravizza, 1998). Choice transforms deliberation into accountable conduct; it is the decisive moment in which moral orientation becomes enacted reality.

The final movement is responsibility. Responsibility is not merely retrospective blame, but answerability for consequences. Bovens (2007) defines accountability as the obligation to explain and justify conduct before a forum capable of judgment. This dimension ensures that moral agency does not end at choice; it extends into the effects and implications of action. Responsibility completes the moral arc by binding agents to the outcomes of their decisions.

Figure 2. A proposed framework of moral agency from consciousness to responsibility

CONSCIOUSNESS → REASONING → STRUGGLE & CHOICE → RESPONSIBILITY

Taken together, Figure 2 presents a widely shared structure of moral agency across traditions. It integrates classical virtue ethics, contemporary responsibility theory, and empirical insights into ethical behavior. The sequence is not strictly linear or mechanistic; its stages often overlap and interact. Nevertheless, it clarifies why ethical leadership cannot be reduced to compliance with rules or adherence to procedures. Moral accountability is rooted in the internal architecture of human agency.

This clarification is especially important before turning to AI-mediated leadership. In practical terms, the sequence asks leaders to pause before activation, examine intention, resist convenience-driven automation, exercise conscious choice, and remain answerable for consequences. If moral agency follows this inner structure, then any ethical framework for contemporary leadership must account for how consciousness is shaped, how reasoning is directed, how struggle is navigated, how choice is exercised, and how responsibility is assumed. Only by first clarifying the ontology of moral action can we meaningfully assess how leadership should function in technologically mediated environments.

'KHALIFAH OF THE SELF' TOWARD A HUMAN-CENTERED FUTURE OF AI LEADERSHIP

If the central crisis of AI-era leadership is moral displacement rather than technological advancement, then ethical AI governance must rest on a coherent moral anthropology. Khalifah of the Self is proposed as that foundation. It begins not with regulatory systems or compliance frameworks, but with the moral structure of the human agent. Before leaders govern algorithms, they must govern themselves.

Figure 3. A proposed framework of Khalifah of the self

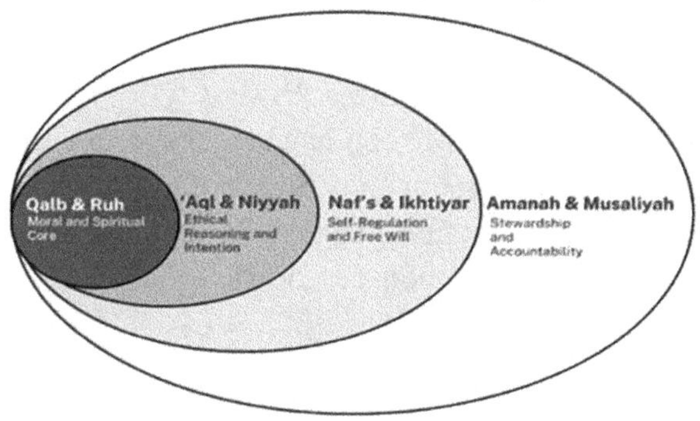

This model is ontological rather than symbolic. It follows the intrinsic progression of moral agency: consciousness (qalb & rūḥ) → reasoning ('aql & niyyah) → struggle & choice (nafs & ikhtiyār) → responsibility (amānah & mas'ūliyyah).

Positioned at the level of foundational ethical formation, the framework precedes institutional AI governance and policy design. It clarifies that technological mediation does not alter the structure of moral action; it intensifies its demands. At its core lie qalb and rūḥ, which anchor moral perception and ultimate orientation. Ethical reasoning cannot begin without awareness of dignity, harm, and accountability. Consciousness must precede calculation.

The second layer ('aql and niyyah) translates awareness into deliberation. Reason evaluates complexity, including algorithmic outputs, while intention directs reasoning toward justice or convenience. Without a clarified intention, AI becomes a tool of instrumental rationality; with it, analysis remains ethically guided.

The third layer (nafs and ikhtiyār) marks the decisive arena of struggle. Here, leaders confront competing impulses: efficiency, ambition, fear, and reputational pressure. Choice emerges through disciplined restraint. Ethical leadership is revealed not in abstract commitments but in the governance of impulse under real constraints.

Encircling these layers are amānah and mas'ūliyyah, which form the boundaries of entrusted authority and answerability. Accountability is not a reactive endpoint but an enclosing condition of action. Authority is exercised as stewardship, and consequences remain morally owned. This boundary placement prevents responsibility from dissolving into systems, hierarchies, or automated processes.

AI systems remain external to this architecture. They do not possess consciousness, intention, struggle, or accountability. When leaders defer uncritically to algorithmic outputs, the moral sequence collapses: awareness narrows, deliberation becomes mechanical, struggle is avoided, and responsibility diffuses. Ethical erosion begins not with technology itself, but with the bypassing of moral agency.

A human-centered future of AI leadership therefore depends on preserving this ontological order. AI may augment reasoning, but it cannot replace moral awareness; it may accelerate decisions, but it cannot eliminate the necessity of choice; it may generate outcomes, but it cannot assume accountability. This position resonates with Zelny's (2025) account of technological mediation, in which technologies may augment, diminish, displace, or redefine practical wisdom. From this view, the ethical question is not whether AI is used, but whether its use strengthens human moral deliberation or weakens it by displacing the capacities it was meant to support. Self-governance therefore remains prior to system governance.

Yet a conceptual framework alone does not secure ethical practice. Moral architecture must be cultivated deliberately, especially in environments defined by speed, automation, and distributed decision-making. The integrity of consciousness, reasoning, struggle, choice, and responsibility depends on disciplined habits and structured reflection.

Table 4. From moral architecture to leadership practice in AI contexts

Moral Stage	Inner Dimensions	What the Leader Must do When Using AI	Key Ethical Discipline	Critical Self-Check Questions
Moral Consciousness	Qalb (ethical perception) Rūḥ (transcendent orientation)	Pause before deploying AI. Identify who may be affected. Recognize dignity beyond data points. Frame the decision as morally weighty, not merely technical	Moral awareness before activation	- Who is affected by this AI output? - Are people being reduced to data? - Does this decision align with human dignity?
Ethical Reasoning	'Aql (critical reasoning) Niyyah (moral intention)	Examine model assumptions, limitations, and potential biases. Clarify the intention behind using AI (justice, efficiency, cost-saving, risk avoidance?). Do not treat outputs as final verdicts	Critical interpretation, not blind acceptance	- Why am I using AI here? - Is efficiency overriding fairness? - What data is missing or excluded?
Moral Struggle & Choice	Nafs (impulse, ego) Ikhtiyār (free will, restraint)	Resist pressure to accept AI outputs for convenience or reputation. Allow discomfort. Consider overriding or re-evaluating if ethical concerns arise. Exercise conscious choice	Restraint under pressure	- Am I choosing what is right or what is easy? - Is ambition influencing this decision? - Would I defend this publicly?
Entrusted Responsibility	Amānah (stewardship) Mas'ūliyyah (answerability)	Own the decision fully. Document reasoning. Be prepared to justify the use of AI and its consequences. Never attribute moral responsibility to "the system"	Ownership of consequences	- Can I explain this decision transparently? - Am I hiding behind the algorithm? - Am I willing to answer for its outcomes?

The decision protocol outlined above does not replace governance frameworks, regulatory safeguards, or technical oversight; it anchors them. By situating AI use within the disciplined sequence of consciousness, reasoning, struggle, choice, and responsibility, Khalifah of the Self ensures that technological mediation never eclipses moral agency. Ethical leadership in the AI era is not secured by eliminating risk or automating judgment, but by preserving the integrity of the human self at every stage of decision-making. When leaders remain morally awake, critically reflective, self-restrained, and fully accountable, AI becomes an instrument of stewardship rather than a mechanism of displacement. In this alignment, innovation advances without eroding dignity, and leadership retains its fundamentally human character.

A NUANCED EXAMPLE: INDONESIAN VICE PRESIDENT GIBRAN'S AI ADVOCACY FOR INDONESIAN YOUTH FROM THE LENS OF POLITICAL PUBLIC RELATIONS

A useful illustration of how Khalifah of the Self can be applied in practice emerges from Indonesian Vice President Gibran Rakabuming Raka's public advocacy for AI literacy among young people. Since 2024, Gibran has repeatedly promoted AI and coding education as part of Indonesia's broader digital transformation agenda, including support for AI-related learning in schools and public messaging that youth must adapt to AI without becoming dependent on it. In this context, AI literacy is not merely an educational issue. It is also a leadership and communication issue, because the way political leaders frame AI shapes how young people understand technological progress, responsibility, and their own future agency (Nugroho, 2026).

The decision at stake is not simply whether AI is "good" or "bad," but how AI literacy should be integrated into youth education and publicly framed by political leaders. From the perspective of political public relations, this is a consequential leadership decision because it shapes educational priorities, public expectations, symbolic legitimacy, and the moral language through which technological progress is normalized. Political communication scholarship has long shown that public officials do not merely inform citizens; they strategically frame policy agendas to cultivate legitimacy, visibility, and identification with target audiences (Avery & Graham, 2013; Fountaine, 2017). In this case, AI advocacy may function both as educational promotion and as political image construction.

The Khalifah of the Self framework first asks what moral risks are flagged by such a decision. Several become visible. The first is instrumentalization: AI literacy may be promoted less as an educational trust and more as a reputational asset for projecting technological modernity. The second is premature normalization: students may be socialized into using AI as a shortcut before they have developed critical judgment, reflective discipline, and ethical responsibility. Research on cognitive offloading and AI-assisted writing raises precisely this concern, showing that heavy dependence on AI tools may weaken critical engagement, originality, and reflective depth, particularly among students and early-career users (Gerlich, 2025; Kosmyna et al., 2025). This concern is especially urgent in education, where AI may function as an "artificial consultant" to which students outsource critical thinking rather than cultivate the motivation, skills, and dispositions needed to evaluate AI use responsibly (Aboodi, 2025).

The third risk is moral outsourcing. If AI education is framed mainly as adaptation to inevitable technological change, young users may learn to trust outputs before learning how to question them. This risk is intensified by automation bias, the documented tendency to over-rely on algorithmic recommendations even when those recommendations should be challenged (Goddard et al., 2012). The fourth risk is responsibility diffusion: if leaders champion AI adoption without clarifying who remains answerable for downstream harms, accountability becomes blurred behind the language of innovation, modernization, or system efficiency.

At this point, the framework shifts from diagnosis to self-governance. The leader must ask morally prior questions before expanding or legitimizing AI use. At the level of qalb and rūḥ, the first question is: What is this advocacy ultimately serving—stewardship or image? A second question follows: Are young people being equipped to remain humanly responsible, or merely encouraged to become technically adaptive? At the level of ʻaql and niyyah, the leader must ask: What assumptions about learning, authority, and technological progress are embedded in this policy? What human capacities might be diminished if AI is introduced faster than ethical maturity can be cultivated? Which groups of students may be most vulnerable to dependency, exclusion, or unequal benefit? At the level of nafs and ikhtiyār, the difficult question becomes: Am I endorsing this policy because it is genuinely just and educationally sound, or because it is politically advantageous, efficient, and symbolically powerful? These questions matter because technological advocacy is never morally neutral; it is shaped by intention, restraint, and the willingness to own consequences.

Only after these questions are faced can a boundary-setting decision follow. The Indonesian government's recent Joint Ministerial Decree (Surat Keputusan Bersama/SKB) involving seven ministries, which restricts the use of AI chatbots such as ChatGPT for students from elementary through high school levels, can be interpreted as an effort to preserve human oversight in educational AI adoption (Wikanto, 2026). From the standpoint of Khalifah of the Self, the appropriate decision is not a simple endorsement or rejection of AI literacy, but a morally conditioned adoption. AI education may proceed, but only under explicit boundaries that preserve human agency. First, AI literacy should be paired with instruction in critical thinking, reflective judgment, and ethical responsibility rather than presented as technical skill acquisition alone. Second, high-stakes educational functions (such as student profiling, predictive ranking, or automated assessment of moral or intellectual worth) should not be automated.

This is the framework's important "don't automate" boundary: where human development, dignity, and future opportunity are directly at stake, efficiency cannot displace accountable human judgment. Third, school-based AI initiatives should include clear human oversight, transparent justification, and mechanisms for appeal, ensuring that responsibility remains with educators and institutions rather than with tools. Fourth, the political communication surrounding such programs should avoid triumphalist narratives that equate technological adoption with progress in itself. Instead, AI should be presented as a tool subordinate to educational stewardship.

Seen in this way, the case is not merely about Gibran as an individual political actor. It illustrates a broader challenge for AI-era leadership: the temptation to promote AI in ways that are institutionally attractive, publicly resonant, and politically beneficial before sufficient moral formation has taken place. The relevance of Khalifah of the Self lies precisely here. It does not deny the strategic value of AI literacy, nor does it reject technological adaptation. Rather, it insists that leaders must govern the moral meaning of adoption before governing its scale. Without that inward governance, even well-intended AI advocacy can normalize dependence before responsibility, innovation before discernment, and visibility before stewardship. The result is not simply ineffective policy, but the quiet formation of technologically capable yet morally underprepared subjects.

THE FIRST BOUNDARY IS NOT IN CODE, BUT IN CHARACTER

This chapter has argued that the deepest ethical crisis of the AI era is not technological acceleration alone, but moral displacement. As AI systems increasingly mediate hiring, evaluation, education, and public decision-making, responsibility risks being diffused into procedures, outputs, and claims of neutrality. Against this tendency, the chapter reframes the problem of AI ethics as one of leadership formation. The central question is no longer only whether AI systems can be governed responsibly, but whether human beings remain morally capable of governing themselves while using them. In this sense, the danger of AI is not simply that machines act with growing authority, but that humans may surrender intention, restraint, and accountability in the process.

The chapter's distinctive contribution to Leading With Faith and Ethics in an AI World lies in its proposal of the Khalifah of the Self as a faith-informed framework for human-centered AI leadership. Rather than adding another set of external principles to existing governance debates, it introduces a moral anthropology grounded in Islamic thought: leadership begins with the inner architecture of the self. By linking *qalb*, *rūḥ*, *'aql*, *niyyah*, *nafs*, *ikhtiyār*, *amānah*, and *mas'ūliyyah* to the ontological sequence of consciousness, reasoning, struggle, choice, and responsibility, the chapter offers a model that explains not only what ethical leadership should protect, but what kind of human being must stand behind it. In doing so, it extends contemporary AI ethics beyond institutional compliance and toward moral stewardship.

This contribution is especially relevant to the volume because it connects faith, ethics, and leadership at the point where AI governance is often weakest: the moral formation of the decision-maker. Existing frameworks can regulate systems, but they cannot by themselves cultivate inward accountability. Khalifah of the Self addresses that gap by insisting that ethical AI leadership begins before policy, before deployment, and before oversight, in the disciplined governance of the self. Its value, therefore, is not only theological but also conceptual and practical: it offers a language of stewardship, intention, and answerability that strengthens the human foundations of AI governance.

If this volume asks how faith and ethics can guide leadership in an AI world, this chapter responds with a clear claim: the first safeguard of ethical AI is neither technical nor procedural, but moral. As long as AI systems affect human lives, leadership cannot be reduced to efficiency, optimization, or compliance. The first boundary is not in code, but in character.

REFERENCES

Aboodi, R. (2025). The worrisome potential of outsourcing critical thinking to artificial intelligence. *Educational Theory*, *75*(4), 626–645. DOI: 10.1111/edth.70037

Al-Ghazali. (2023). Ihya' 'Ulum al-Din - The Revival of the Religious Sciences - Vol 1: إحياء علوم الدين. Dar UL Thaqafah. ISBN-10: 9390804930

Alon-Barkat, S., & Busuioc, M. (2022). Human–AI interactions in public sector decision making: "Automation bias" and "Selective adherence" to algorithmic advice. *Journal of Public Administration: Research and Theory*, *33*(1), 153–169. DOI: 10.1093/jopart/muac007

Analects 12.22. (2022, April 3). Analects 12.22. Confucian Pages - Free Online Confucian Texts and Translations. https://confucius.page/analects-12-22

Analects 15.24. (2022, April 7). Analects 15.24. Confucian Pages - Free Online Confucian Texts and Translations. https://confucius.page/analects-15-24

Aristotle. (2012). Aristotle's Nicomachean ethics (R. C. Bartlett & S. D. Collins, Trans.). University of Chicago Press. https://www.amazon.com/Aristotles-Nicomachean-Ethics-Aristotle/dp/0226026752

Avery, E. J., & Graham, M. W. (2013). Political public relations and the promotion of participatory, transparent government through social media. *International Journal of Strategic Communication*, *7*(4), 274–291. DOI: 10.1080/1553118X.2013.824885

Badri, M. (2018). *Contemplation: An Islamic Psychospiritual study* (1st ed.). Human Behaviour Academy. DOI: 10.2307/j.ctvk8w1xc

Bhagavad Gita. Chapter 3, Verse 19. (2026). BG 3.19: Chapter 3, verse 19 - Bhagavad Gita, the song of god – Swami Mukundananda. Bhagavad Gita, The Song of God – Swami Mukundananda. https://www.holy-bhagavad-gita.org/chapter/3/verse/19/

Bhagavad Gita. Chapter 18, Verse 45. (2026). BG 18.45: Chapter 18, verse 45 - Bhagavad Gita, the song of god – Swami Mukundananda. Bhagavad Gita, The Song of God – Swami Mukundananda. https://www.holy-bhagavad-gita.org/chapter/18/verse/45/

Bhagavad Gita. Chapter 18, Verse 46. (2026). BG 18.46: Chapter 18, verse 46 - Bhagavad Gita, the song of god – Swami Mukundananda. Bhagavad Gita, The Song of God – Swami Mukundananda. https://www.holy-bhagavad-gita.org/chapter/18/verse/46/

Block, P. (1996). Stewardship: Choosing service over self-interest. Berrett-Koehler Publishers. ISBN-10 : 1881052869

Bosse, T., Hoogendoorn, M., Klein, M. C., Treur, J., Van der Wal, C. N., & Van Wissen, A. (2012). Modelling collective decision making in groups and crowds: Integrating social contagion and interacting emotions, beliefs and intentions. *Autonomous Agents and Multi-Agent Systems*, *27*(1), 52–84. DOI: 10.1007/s10458-012-9201-1

Bovens, M. (2007). Analysing and assessing accountability: A conceptual Framework. *European Law Journal*, *13*(4), 447–468. DOI: 10.1111/j.1468-0386.2007.00378.x

Brown, M. E., & Treviño, L. K. (2006). Ethical leadership: A review and future directions. *The Leadership Quarterly*, *17*(6), 595–616. DOI: 10.1016/j.leaqua.2006.10.004

Cath, C. J., Wachter, S., Mittelstadt, B., Taddeo, M., & Floridi, L. (2017). Artificial intelligence and the 'Good society': The US, EU, and UK approach. *Science and Engineering Ethics*, *24*, 505–528. https://link.springer.com/article/10.1007/s11948-017-9901-7. DOI: 10.1007/s11948-017-9901-7 PMID: 28353045

Chowdhury, R. (2018, June 8). Moral outsourcing: Humanity in the age of AI. TED: Ideas change everything. https://www.ted.com/talks/rumman_chowdhury_moral_outsourcing_humanity_in_the_age_of_ai

Danaher, J. (2016). The threat of Algocracy: Reality, resistance and accommodation. *Philosophy & Technology*, *29*(3), 245–268. DOI: 10.1007/s13347-015-0211-1

Dhammapada Verses 183. (n.d.). Dhammapada Verses 183, 184 and 185 Anandattherapanha Vatthu. Tipitaka.net. https://www.tipitaka.net/tipitaka/dhp/verseload.php?verse=183

Ess, C. (2013). Digital Media Ethics (Digital Media and Society). Polity Pr. ISBN-13 : 978-0745656069

Eubanks, V. (2018). Automating inequality: How high-tech tools profile, police, and punish the poor. Macmillan + ORM.

European Commission. Directorate-General for Communications Networks, Content and Technology & Grupa ekspertów wysokiego szczebla ds. sztucznej inteligencji. (2019). Ethics guidelines for trustworthy AI. Publications Office. https://data.europa.eu/doi/10.2759/346720

Fischer, J. M., & Ravizza, M. (1998). *Responsibility and control: A theory of moral responsibility*. Cambridge University Press. DOI: 10.1017/CBO9780511814594

Floridi, L. (2019). Establishing the rules for building trustworthy AI. SSRN Electronic Journal. DOI: 10.2139/ssrn.3858392

Fountaine, S. (2017). What's not to like?: A qualitative study of young women politicians' self-framing on Twitter. *Journal of Public Relations Research, 29*(5), 219–237. Advance online publication. DOI: 10.1080/1062726X.2017.1388239

Fry, L. W. (2003). Toward a theory of spiritual leadership. *The Leadership Quarterly, 14*(6), 693–727. DOI: 10.1016/j.leaqua.2003.09.001

Gerlich, M. (2025). AI tools in society: Impacts on cognitive offloading and the future of critical thinking. SSRN Electronic Journal. DOI: 10.2139/ssrn.5082524

Ghasemaghaei, M., & Kordzadeh, N. (2024). Understanding how algorithmic injustice leads to making discriminatory decisions: An obedience to authority perspective. *Information & Management, 61*(2), 103921. DOI: 10.1016/j.im.2024.103921

Gillespie, T., Boczkowski, P. J., & Foot, K. A. (2014). Media technologies: Essays on communication, materiality, and society. MIT Press. DOI: 10.7551/mitpress/9780262525374.001.0001

Goddard, K., Roudsari, A., & Wyatt, J. C. (2012). Automation bias: A systematic review of frequency, effect mediators, and mitigators. *Journal of the American Medical Informatics Association : JAMIA, 19*(1), 121–127. DOI: 10.1136/amiajnl-2011-000089 PMID: 21685142

Graves, M. (2025). AI practical wisdom and compassion. *AI and Ethics, 6*(1), 39. Advance online publication. DOI: 10.1007/s43681-025-00877-4

Habermas, J. (1985). The theory of communicative action: Reason and the rationalization of society. Beacon Press. ISBN-10: 0807015075

Hagendorff, T. (2020). The ethics of AI ethics: An evaluation of guidelines. *Minds and Machines, 30*(1), 99–120. DOI: 10.1007/s11023-020-09517-8

Haidt, J. (2008). The emotional dog and its rational tail: A social intuitionist approach to moral judgment. Reasoning, 1024-1052. DOI: 10.1017/CBO9780511814273.055

Hannah, S. T., Avolio, B. J., & Walumbwa, F. O. (2011). Relationships between authentic leadership, moral courage, and ethical and pro-social behaviors. *Business Ethics Quarterly, 21*(4), 555–578. DOI: 10.5840/beq201121436

Herari, N., Najmi Nuji, M. N., Boer, R. F., & Allifiansyah, S. (2026). Algorithmic Narcissism and the Passion Economy: Identity, Visibility, and the Ideology of Self-Performance. In Digital Ideologies in the Age of Algorithmic Culture (1st ed.). IGI Global Scientific Publishing. ISBN13: 9798337345314. DOI: 10.4018/979-8-3373-4531-4.ch008

Herari, N., & Riyanto, B. (2026). Iron Man or Tin Man?: AI, Habit, and the Quiet Mechanization of Human Life. In *AI and Robotic Companions for Human Socialization.* IGI Global Scientific Publishing.

Holy Bible (Genesis 1:26-28, 1 Corinthians 10:31, Micah 6:8), New International Version®, NIV® Copyright ©1973, 1978, 1984, 2011 by Biblica, Inc.®

IEEE. (2021). Ethically Aligned Design: A Vision for Prioritizing Human Well-being with Autonomous and Intelligent Systems. https://standards.ieee.org/wp-content/uploads/import/documents/other/ead_v2.pdf

Ismail, N. H., Makhsin, M., Khadafi Hj. Rofie, M., & Omar, S. (2021). Higher order thinking skills (HOTS): An analysis based on surah Al-Hajj verse 46. *Journal of Language and Linguistic Studies, 17*(1), 471–480. DOI: 10.52462/jlls.30

Jasanoff, S. (2016). The ethics of invention: Technology and the human future. W. W. Norton & Company. ISBN-13‏: ‎978-0393078992

Jolly, A. A., Dunlop, P. D., Parker, S. K., & Kanse, L. (2025). It's not my responsibility: Working with autonomy-restricting algorithms facilitates unethical behavior and displacement of responsibility. *Journal of Business Ethics, 203*(2), 405–424. DOI: 10.1007/s10551-025-06034-5

Kets de Vries, R. M. F. (2011). The leader on the couch: A clinical approach to changing people and organizations. John Wiley & Sons. ISBN 0470030798

Kosmyna, N., Haupmann, E., Yuan, Y. T., Situ, J., Liao, X., Beresnitzky, A. V., Braunstein, I., & Maes, P. (2025). Your Brain on ChatGPT: Accumulation of Cognitive Debt when Using an AI Assistant for Essay Writing Task. Cornell University. https://arxiv.org/abs/2506.08872

Mastroianni, G. R., Kimmelman, S., Doty, J., & Thomas, J. (2011). Obedience and Personal Responsibility. In Leadership in Dangerous Situations: the Handbook of the Armed Forces, Emergency Services, and the First Responders (pp. 97-120). Naval Institute Press. https://www.researchgate.net/profile/Joe-Doty/publication/303840128_Obedience_and_Personal_Responsibility/links/5b045a284585154aeb07f1e7/Obedience-and-Personal-Responsibility.pdf

McLoughlin, S., Espinosa, V. F., & Velázquez-Gil, M. (2026). Practical wisdom, emerging technologies, and AI. Journal of Moral Education. https://doi.org/DOI: 10.35542/osf.io/gtjc6_v1

Milmo, D. (2026, January 15). Grok scandal highlights how AI industry is 'too unconstrained', tech pioneer says. the Guardian. https://www.theguardian.com/technology/2026/jan/15/grok-scandal-ai-industry-too-unconstrained-yoshua-bengio-elon-musk

Mittelstadt, B. (2019). Principles alone cannot guarantee ethical AI. *Nature Machine Intelligence, 1*(11), 501–507. DOI: 10.1038/s42256-019-0114-4

Mohamed, F. (2011). Capitalism in Moral Perspective – An Islamic Alternative [Master's thesis]. https://uwcscholar.uwc.ac.za:8443/server/api/core/bitstreams/26a582bf-b546-40c7-b04b-d6b8d28fa600/content

Moulaei, K., Haghdoost, A., Bahaadinbeigy, K., & Dinari, F. (2023). The effect of the holy Quran recitation and listening on anxiety, stress, and depression: A scoping review on outcomes. *Health Science Reports, 6*(12), e1751. Advance online publication. DOI: 10.1002/hsr2.1751 PMID: 38078304

Nabil, M. (2025). Amanah (trust) and Physician- Patient Relationship in Islamic Medical Ethics. *Journal of British Islamic Medical Association, 19*(7), 1–8. https://jbima.com/article/amanah-trust-and-physician-patient-relationship-in-islamic-medical-ethics/

Nugroho, N. P. (2026, March 5). *Gibran: AI Penting untuk Menunjang Kegiatan Belajar*. Tempo. https://www.tempo.co/politik/gibran-ai-penting-untuk-menunjang-kegiatan-belajar-2119713

OECD. (2019). AI Principles. Organisation for Economic Co-operation and Development. https://www.oecd.org/en/topics/ai-principles.html

Parrott, J. (2014, August 28). Ibn Qayyim on Hearts: A void that can only be filled by love of Allah. Abuaminaelias.com. https://www.abuaminaelias.com/dailyhadithonline/2014/08/28/ibn-qayyim-void-heart/

Qureyshi, H. A., & Mustaq, S. (2025). Between Temptation and Tranquility: The Quranic framework of Nafs. *Journal of Quranic and Social Studies, 5*(2), 18–36. DOI: 10.5281/zenodo.15570157

Rahman, F. (2009). Major themes of the Qur'an (2nd ed.). University of Chicago Press. ISBN-10: 0226702863

Raza, M. W. (2022). Anatomy of the intellect in the Quran: A fresh perspective. Al-Bayan. *Journal of Qur'an and Hadith Studies, 20*(2), 182–216. DOI: 10.1163/22321969-12340113

RenAIssance. (2020). Rome Call for AI Ethics. Rome Call | What is the Matter with AI Ethics? https://www.romecall.org/the-call/

Riaz, A. (2025). Title: Ethical implications of AI in recruitment – Navigating bias and fairness in socio-technical hiring. Advance. https://advance.sagepub.com/doi/full/10.31124/advance.175385244.44505199/v1

Romeo, G., & Conti, D. (2025). Exploring automation bias in human–AI collaboration: A review and implications for explainable AI. *AI & Society*. Advance online publication. DOI: 10.1007/s00146-025-02422-7

Sahih al-Bukhari 2554. (2025). Sahih al-Bukhari 2554 - Manumission of slaves - كتاب العتق - Sunnah.com - Sayings and teachings of Prophet Muhammad (صلى الله عليه و سلم). Sunnah.com - Sayings and Teachings of Prophet Muhammad (صلى الله عليه و سلم). https://sunnah.com/bukhari:2554

Sahih al-Bukhari 7138 - Judgments (Ahkaam) - كتاب الأحكام - Sunnah.com - Sayings and teachings of Prophet Muhammad (صلى الله عليه و سلم). (n.d.). Sunnah.com - Sayings and Teachings of Prophet Muhammad (صلى الله عليه و سلم). https://sunnah.com/bukhari:7138

Sahih al-Bukhari Hadith 1, Book 1. (2026). Hadith 1. 40HadithNawawi.com - The Forty 40 Hadith of Imam al-Nawawi. https://40hadithnawawi.com/hadith/1-actions-are-by-intentions/

Sahih Muslim. 1907a - The book on government - كتاب الإمارة - Sunnah.com - Sayings and teachings of Prophet Muhammad (صلى الله عليه و سلم). (2026). Sunnah.com - Sayings and Teachings of Prophet Muhammad (صلى الله عليه و سلم). https://sunnah.com/muslim:1907a

Sahih Muslim. 1829a - The book on government - كتاب الإمارة - Sunnah.com - Sayings and teachings of Prophet Muhammad (صلى الله عليه و سلم). (n.d.). Sunnah.com - Sayings and Teachings of Prophet Muhammad (صلى الله عليه و سلم). https://sunnah.com/muslim:1829a

Taylor, C. (1989). *Sources of the self: The making of the modern identity*. Harvard University Press.

The Qur'an. (2:30, 2:44, 12:53, 22:46, 32:9, 33:72, 51:56, 75:2, 76:3, 89:29-30, 102:8). The New Translation by M.A.S Abdel Haleem. (2005). Oxford University Press (OUP) Oxford. ISBN: 0-19-283193-3

UNESCO. (2021). Recommendation on the Ethics of Artificial Intelligence. UNESDOC Digital Library. https://unesdoc.unesco.org/ark:/48223/pf0000380455

Vallor, S. (2016). Technology and the virtues: A philosophical guide to a future worth wanting. Oxford University Press. ISBN-13: 978-0190498511 DOI: 10.1093/acprof:oso/9780190498511.001.0001

Wikanto, A. (2026, March 13). *Resmi! Pemerintah Batasi AI Di Pendidikan, Anak SD-SMA Dilarang Tanya Jawaban ChatGPT.* Kontan. https://nasional.kontan.co.id/news/resmi-pemerintah-batasi-ai-di-pendidikan-anak-sd-sma-dilarang-tanya-jawaban-chatgpt

Zelny, A. (2025). Offloading wisdom: Four technological relations that mediate Phronesis. *Philosophy & Technology, 38*(2), 56. Advance online publication. DOI: 10.1007/s13347-025-00889-2

Zouggar, N. (2018). Ibn Taymiyya on the proofs of prophecy and his legacy: Ibn Qayyim al-jawziyya (D. 751/1350) and Ibn al-Wazir (D. 840/1436). *Muslim World (Hartford, Conn.), 108*(1), 172–185. DOI: 10.1111/muwo.12233

KEY TERMS AND DEFINITIONS

Accountability (Mas'ūliyyah): The condition of being answerable for one's actions and their consequences, encompassing both internal moral responsibility and external justification.

Ethical Reasoning ('Aql): The process of critically evaluating decisions, consequences, and alternatives, guided not only by logic but by moral orientation and responsibility.

Free Will (Ikhtiyār): The human capacity to make conscious and deliberate choices, forming the basis for moral responsibility and ethical accountability.

Intention (Niyyah): The internal moral orientation that determines the ethical meaning and value of an action, shaping whether conduct reflects responsibility, self-interest, or moral avoidance.

Khalifah of the Self: A leadership framework grounded in Islamic moral anthropology that positions the human being as a self-governing moral agent responsible for conscious, intentional, and accountable action in AI-mediated environments.

Moral Agency: The human capacity to perceive ethical significance, deliberate on alternatives, exercise choice, and assume responsibility for the consequences of one's actions.

Moral Consciousness (Qalb): The inner faculty of ethical perception through which individuals recognize moral significance, human dignity, and responsibility beyond technical or procedural considerations.

Moral Struggle (Nafs): The inner tension between ethical responsibility and personal desires, ambitions, or pressures, where leadership integrity is tested and refined.

Chapter 9
Algorithms Have Logic, Leaders Have Conscience:
Reclaiming Moral Authority in Intelligent Systems

Pardaeva Markhabo
Bukhara State Technical University, Bukhara, Uzbekistan

Saparov Ismat
Termez University of Economics and Service, Termez, Uzbekistan

R. N. Ravikumar
http://orcid.org/0009-0009-3705-1681
Marwadi University, Rajkot, India

S. Aarthi
http://orcid.org/0009-0006-9064-2091
Marwadi University, Rajkot, India

Mukimov Askar Shukhratovich
Alfraganus University, Tashkent, Uzbekistan

Kakhramon Chinkulov
Tashkent State University of Economics, Tashkent, Uzbekistan

ABSTRACT

With the growing influence of Artificial Intelligence on the decisions made in education, organizations, and the governance of the people, a major ethical concern is presented before the leadership, and that is having moral judgment in algorithm-based settings. Even though intelligent systems make the system more efficient and consistent, they have no conscience and are not sensitive to human dignity. This chapter explores faith-based leadership as a reaction to the increasing power of the algorithmic decision-making, with discernment highlighted as one of the fundamental leadership skills. Referring to leadership research, philosophy, and also ethics based on faith, leadership is construed as ethical stewardship which balances automated

DOI: 10.4018/979-8-3373-8192-3.ch009

ght © 2027, IGI Global Scientific Publishing. Copying or distributing in print or electronic forms without written permission of IGI Global Scientific Publishing is prohibited. Use of this chapter to generative artificial intelligence (AI) technologies is expressly prohibited. The publisher reserves all rights to license its use for generative AI training and machine learning model development.

rationality and human values. The chapter provides a humanistic approach to provide technological intelligence to serve dignity, justice, inclusion, and the common good instead of efficiency.

INTRODUCTION

Artificial Intelligence (AI) has not just become a supportive analytical tool but it has become a major provider of decision-making in any areas of leadership, education, organizations, and even government among citizens. The algorithmic systems are changing the traditional functions of leadership such as resource allocation, recruitment and policy implementation which are becoming key functions of the organization. As much as these systems can help us become more efficient, they also present some ethical issues in terms of accountability, fairness and human-centered decision making. Due to the increasing data-based nature of processes, leadership demands the merging of ethical judgment in the field as well as technical expertise. (Seo et al., 2021). The leaders should respond to the drawbacks of AI, especially its insecurity in decoding moral values and contextual details. In this chapter, moral authority is placed as fundamental in the environments of AI mediation, and ethical judgment is highlighted in terms of leading algorithmic results. It builds on the moral philosophy and views that are informed by faith, offering leadership as a reflective and responsible ethics and suggests a framework that can drive intelligent systems in accordance with human dignity, justice and common good.

Algorithmic Leadership Transformation

The concept of algorithmic decision-making has been firmly integrated into the modern leadership process, reinforcing the way decisions are taken and supported in any industry. Algorithms are used in learning institutions, learning analytics, admissions and testing. At the organizational level, they direct the process of recruitment, performance management, credit analysis, and strategic forecasting. There is growing dependence of automated systems in the distribution of welfare, surveillance, and assessment of risks in the conduct of public governance. Such trends mirror the larger trend in favor of data-driven leadership, in which efficiency, scalability, and predictive accuracy for leadership are valued (Ruslan et al., 2026). But as algorithmic suggestions are empowered, leadership functions are also depersonalized to the procedural authentications of judgment. Algorithms can influence decision-makers to accept the results because they have perceived it to be objective, hiding the assumptions and data biases and value-based design decisions that can be found in such systems. This transformation changes the old concept of leadership responsibility since the responsibility is dispersed between human beings and technological actions. The emergence of algorithmic decision-making is thus not only a technical change, but an ethical challenge as well, and the leaders should perceive, challenge, and moralize algorithmic direction instead of accepting it blindly. Figure 1 showing how the power of leadership changes to be based on data efficiency to being more ethical and moral at the top and the necessity to have human judgment to balance algorithms.

Figure 1. AI decision-making hierarchy

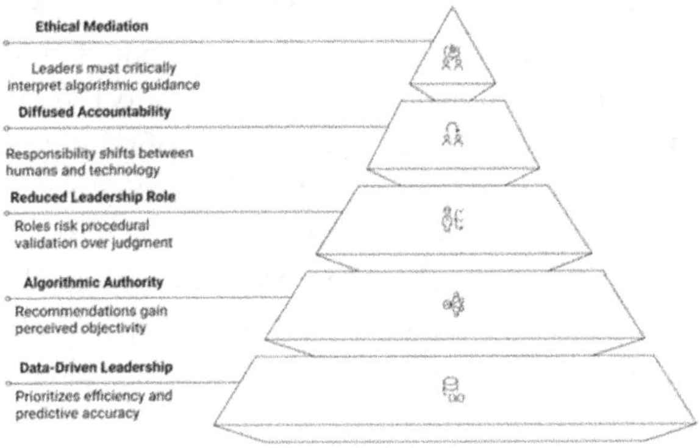

Ethical Limits of AI Rationality

Regardless of the computational power and machine learning, AI is still inherently weak in its moral ability. Automated rationality is an algorithmic process that works in limited logical systems to convert the complex human reality into measurable variables. This narrow ideological tendency is a danger to the contextual factors of social inequity, cultural significance and the vulnerability of individuals. Moral judgment involves interpretive judgments, empathy and ethical implication that can go outside the quantifiable results (Sapayev et al., 2026). There is a risk that when automated logic is employed excessively in leadership decision making, they will tend to pose a threat to sway conscience with compliance and moral reason with procedural right. Therefore, it is paramount to the ethics of AI realization that can provide responsible leadership. Ethical governance of AI includes human actors who will be able to recognize the point of conflict between algorithmic efficiency and moral responsibility and who will be prepared to act to safeguard human dignity and justice.

Scope of Ethical AI Leadership

The chapter targets to discuss the role of reclaiming moral authority within the sphere, which is increasingly becoming more intelligent system-oriented through the influence of leadership. It attempts to romanticize the faith-based discernment as being a discerning leadership ability is not incompatible with the technological innovation but rather it supplements it. The chapter discusses how moral philosophy and faith traditions can inform ethical judgment ensuring that leaders can interpret the outputs of algorithms in a responsible manner and oppose the automation of human decisions. The context is broad enough to cover the leadership situations within education, organization, and governance by the population where the ethical dilemma of bias, surveillance, exclusion, and accountability is particularly addressed. The discussion is not about advancing a specific idea of religion but rather is inclusive to the meaning of faith-informed ethics which takes into consideration general moral standards such as dignity, justice, compassion, and care of the vulnerable. Combining interdisciplinary research with up-to-date case knowledge, the chapter is expected to contribute to the emerging debates on the topic of ethical leadership in the digital age as well as present conceptual advice to the leaders who are working in AI-mediated decision environments.

BACKGROUND TO THE STUDY

Introduction of AI into the social, institutional and organizational systems has completely changed the ways of exercising authority, judgment and responsibility. What used to start as data analysis support system, has now transformed into systems with the potential to make consequential decisions impacting the lives of people. This change has been accompanied by increasing fears of ethical governance, transparency, and the effects of intelligent technologies on society. Leadership paradigms, nevertheless, have not been keeping up with the level of technological adoption, and their focus is more on the efficiency of operations without giving much attention to the moral complexity. With the growing influence of AI systems in education, the labor market, resource distribution, and government provision, the ethical issue of equity, responsibility, and dignity of humans is now inevitable. Traditions of faith and moral philosophy have provided ancient insights into the issues of responsibility, conscience and the boundaries of instrumental reasoning. Still, reconsidering these views is an important way to understand how any leadership can continue to be morally sound in the face of fast technological change (Bean et al., 2025). The context of the present research places the chapter in this confluence of technological growth and moral imperative, that new leadership paradigms must be developed in which moral judgment is merged with intelligent technologies to make sure that innovation is the instrument of human prosperity and not its destroyer.

AI Evolution in Institutional Leadership

Artificial Intelligence has gradually augmented its purpose in the field of education, administration in organizations and government. Within the education sector, AI-based applications facilitate individualized education, automated evaluation, and predictive analytics to enhance the efficiency of the institution. Users use smart recruitment, performance management, customer generalization, and strategic decision-making systems. Algorithms are applied in policy execution, risk forecasting, welfare distribution, and surveillance in the context of the public governance. This development can be traced to a general trend in which more complex and large-scale decision-making relies on data-based models (Aarthi & Ravikumar, 2026). Nonetheless, with more autonomy and power, AI systems are less likely to support human judgment but more likely to influence it in a particular manner. The choices made in the past, based on professionalism and ethical judgement, are today being arbitrated by the use of algorithmic standards. This change brings up the issue of transparency, accountability, and power concentration in non-transparency technology. The development of AI in these areas is crucial to studying the transformation of the concept of leadership and the need to revise the ethical frameworks to provide a responsible management of intelligent technologies (Lima & Rahman, 2025). The role of leadership and ethical concerns in relation to the use of intelligent systems has changed with the increased use of intelligent systems as demonstrated in Table 1.

Table 1. AI leadership evolution matrix

Dimension	Pre-AI Leadership	AI-Mediated Leadership
Decision Basis	Experience and judgment	Data-driven and predictive
Accountability	Clearly human-centered	Shared between humans and systems
Ethical Risk	Contextual and visible	Embedded and opaque
Speed of Decisions	Moderate	High
Moral Oversight	Direct	Requires active mediation

Ethical Disruptions of AI Systems

The massive implementation of intelligent systems has brought major ethical disturbances to institutional and social settings. The use of intelligent systems all too frequently transfers and boosts historical disparities on the training data, resulting in discrimination outcomes that are hard to identify or dispute. Moreover, the openness of the vast majority of AI models makes them hard to be held accountable, and it is not evident who makes destructive decisions. The leaders operating in the AI-mediated environment, therefore, must confront ethical tensions that cannot be resolved through the technical changes only. Moral attentiveness, critical thinking and the willingness to question the values, coded in intelligent systems and their greater implications in the society are the solutions to these disruptures. Table 2 Covers the key AI systems limitations as far as ethics, judgment, accountability, and human values are concerned.

Table 2. Moral limits of algorithm

Aspect	Algorithmic Capability	Moral Limitation
Data Processing	High-volume, fast	Lacks ethical meaning
Objectivity	Rule-based consistency	Inherits data bias
Accountability	Procedural	Cannot assume responsibility
Context Awareness	Statistical	Lacks empathy and nuance
Value Judgment	Optimization-focused	Cannot assess dignity or justice

Role of Faith in Digital Ethics

Religious tradition and moral philosophy can provide a sustainable paradigm of locating solutions to ethical dilemmas, which transpire in technological mediated societies. Such traditions are especially significant in the digital age when, increasingly, systemic algorithms are determining human results, as such traditions offer essential data on the notion of dignity, justice, duty, and the extent of instrumental thought. Moral philosophy focuses on an ethical reason, responsibility, and intrinsic worth of human beings besides utility and efficiency (Flanigan, 2018). Faith-based ethics are also associated with the principles of compassion, humility, and support of the vulnerable people, which strengthens the moral duties of leadership. Faith-informed ethics may be a dialogical tool and not a dogma in pluralistic and secular settings, and it can make ethical argumentation richer and advocate plural moral thinking. The application of faith and moral philosophy to the leadership practice enhances the ability to identify the ethical connotations of AI usage and steer intelligent systems to the results that can foster human flourishing and common good.

RESEARCH OBJECTIVES

This chapter has clearly stated research goals, which give the background of exploring the issue of leadership in AI-mediated contexts. The objectives are those related to how a moral power can be sustained when intelligent systems are gaining an ever-growing impact on the decision making. Based on the research results regarding leadership, ethics, and moral reasoning and faith-based, the objectives will be to establish the role of those who can be the interface of human judgment and algorithmic reasoning, which are leaders working at the interface of those aspects. Primary Research questions are:

- To examine how moral authority is exercised, challenged, and redefined within AI-mediated leadership contexts across education, organizations, and public governance.
- To analyze the concept of faith-anchored discernment as a leadership practice that integrates conscience, humility, and ethical reflection into algorithm-informed decision-making.
- To identify key ethical challenges introduced by intelligent systems, including bias, opacity, surveillance, and the marginalization of vulnerable populations.
- To explore leadership responsibilities in interpreting, questioning, and ethically mediating algorithmic recommendations rather than deferring to automated rationality.
- To develop a conceptual foundation for ethical stewardship in intelligent systems that aligns technological intelligence with human dignity, justice, and the common good.

CONCEPTUAL MODEL

The theoretical framework formulated in this chapter presents a systematic meaning of leadership in AI-mediated contexts of choices. The model transforms the leadership into a dynamic moral interpreter between the logic of algorithms and human values rather than deliberating on the Artificial Intelligence as an impartial and autonomous force. It emphasizes the reality that the intelligent systems operate within the set goals and specified assumptions, and they need to be explained by humans to fit in with ethics. This model applies three inter-related dimensions, they are the interaction of human and the algorithm, moral discernment and ethical stewardship. Combined, these dimensions describe the way leaders interact with the outputs of the algorithm, use conscience-based judgments, and take responsibility regarding the social and ethical implications. The model criticizes technocratic leadership models that place a lot of emphasis on efficiency and the elimination of moral reasoning by foregrounding discernment as a key leadership ability. It also emphasizes on leadership as a relational and reflective practice whose foundation is on responsibility and not control. The model is an analytical basis of studying where ethical leadership is applied in education, organizations, and governance providing a logical organizing framework of harmonizing intelligent systems with the human dignity and the common good. Table 3 Provides the key elements of the suggested conceptual model of interconnection between algorithms, discernment, and ethical stewardship.

Table 3. Ethical leadership model components

Component	Role	Leadership Function
Algorithmic Logic	Decision support	Provides analytical input
Human Judgment	Interpretation	Contextual evaluation
Discernment	Ethical mediation	Aligns logic with conscience
Moral Authority	Responsibility	Ensures ethical accountability
Ethical Stewardship	Governance	Protects human dignity

Human–AI Ethical Interaction

The most common interface of the involvement of intelligent systems in leadership decisions is the human algorithm interaction. Algorithms give suggestions, forecasts or classifications that inform human behavior in numerous institutional environments. Nevertheless, these productions lack moral intent and the ethical cognizance. Moral mediation is the process of leaders who are critically interpreting the algorithmic guidance and assessing its implication with ethical consideration as well as considering decisions within the context of greater human and social factors (Rathinavel et al., 2025). It is done with an understanding of the constraints of the data, inherent bias, and the socio-political implications of algorithm applications. The leadership will deteriorate into an inactive acceptance of robotic results in the absence of moral mediation. The communication skill hence requires an active approach and in this case leaders challenge the values programmed within systems and gauge the algorithmic efficiency with moral accountability. Moral mediation enables human agency by making sure that algorithms are informational rather than determinate in making decisions. It makes the organization more responsible since it makes sure that there is a separation between the aid calculation and ethical decisions.

Discernment and Moral Reasoning

The discernment is the connection component between the concept of algorithmic rationality and that of the human conscience. Algorithms are applied with the help of formal logic and optimization rules, discernment is a reflective decision with references to ethical awareness, empathy, and moral responsibility. It assists leaders in determining the instances, where the algorithmic results do not reflect the human values or cause unjust results. Discernment is not a given but developed by moral formation, reflective practice and ethical sensitivity (Joshi, 2025). In AI-driven settings, discernment enables the leaders to go beyond the simple acceptance or rejection of technology and rather, consider such technology thoughtfully. This ability helps in ethical decision-making during situations which are complicated as data-driven efficiency might blur the human vulnerability or social injustice. Conscience combined with logic helps in discernment and avoids the replacement of moral responsibility by compliance with the procedure. It reinstates leadership as a moral calling and not a technical role.

Ethical Stewardship Framework

Ethical stewardship offers a directive concept of leadership in AI-mediated settings. Stewardship also contrasts with control-centered models of where the centrality is on technological superiority, responsibility emphasis, caring and responsibility of the ethical use of intelligent systems. Ethical custodians like leaders are aware that AI technologies are the ones that generate human results and hence are to be followed keenly. The solution is based on openness, justice and inclusiveness in designing and executing algorithms. Ethical stewardship also involves ensuring that the vulnerable populations are not harmed by any prejudice, ostracism, or surveillance (Ferrara, 2023). By describing leadership as stewardship, the model stresses the necessity to manufacture technological innovation in accordance with other moral values, including dignity, justice, and compassion. Stewardship will give leaders an incentive to consider long term social impacts instead of only concentrating on short term performance benefits. It supports the notion that machines cannot be entrusted with moral responsibility. Leadership is redefined within this system as a devotion to putting smart systems to work towards a solution that benefits the greater good and supports trust in AI-driven organizations (Chaudhary, 2024). Figure 2 that describes ethical stewardship as the combination of responsibility, accountability, and care to help ensure that AI-driven decisions incorporate transparency, fairness, inclusion, and respect of human dignity.

Figure 2. AI decision-making hierarchy

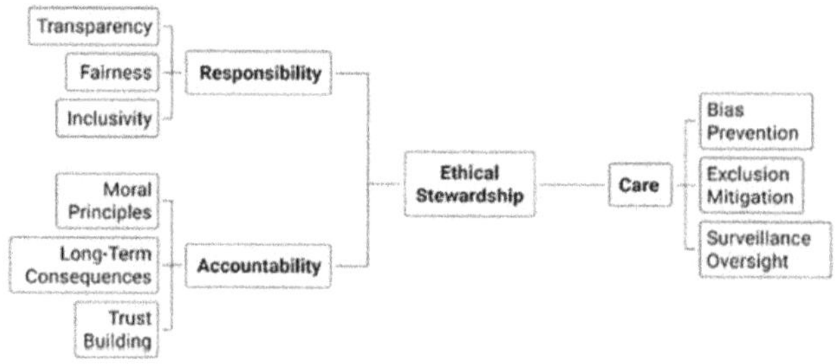

RESEARCH METHODOLOGY

The research methodology that is used in the chapter is interdisciplinary and conceptual that is appropriate to analyze ethical leadership in AI mediated situation. As the research problem is normative and interpretive, the research is not based on the empirical data collection, but rather the synthesis of theoretical approaches to ethics, faith traditions, studies of leadership and digital governance approaches. Such a method allows the systematic study of moral authority and intelligence in the intelligence systems (Corsello & Santangelo, 2023). The methodology focuses on the critical evaluation of the existing literature, the synthesis of the concepts, and the contemplative insight into the current issues. Using the interdisciplinary sources, the study has captured the complexity of the ethical impact of AI in institutional spheres. The methodological design helps to establish a conceptual framework between technological rationality and moral reasoning. Although the methodology cannot be considered empirically valid, as it is not possible within the confines of this chapter, it is a sound basis of theoretical input and further investigation.

Interdisciplinary Conceptual Design

The study design is primarily conceptual that involves the expertise of different fields to look at the ethical side of the AI-mediated leadership. The study of leadership provides the understanding of the responsibility, authority, and decision-making, and the literature of digital ethics and AI governance provides critical thinking about the system of algorithms. The normative rules of conscience, dignity and moral judgment are provided by religious tradition and moral philosophy (Roy et al., 2024). This all-disciplinary design helps to study leadership holistically rather than technical or managerial fashion. Such synthesis of different fields of knowledge is to ensure that the study avoids reductionist descriptions of AI as phenomena of technology. Instead, it installs smart systems in the bigger moral, social and cultural environments. The conceptual design is malleable in reacting to moral conflicts that are hard to quantify. It promotes the combination of theory and the development of a rational structure of moral accountability and judgment. The investigation of the leadership issues which appear at the intersection of technology, ethics, and human values is best suited to the study design.

Integrated Ethical Framework

This chapter has employed the integration of the ethical theory, the moral reasoning based on faith and leadership scholarship as the analytical model. Ethics provides the ideas of evaluating the wrongness and rightness, accountability and justice in algorithmic decision. Religious views can include the knowledge of conscience, moral growth and care of vulnerable populations. The models of power, judgment and control in organizations are the models offered by the leadership literature. Altogether, these attitudes form a coherent analysis framework to examine the AI-informed leadership (Ravikumar & Aarthi, 2025c). The framework permits to view the interaction of leaders with algorithmic systems in a systematic and concomitant way, and be morally upright. It aids the critical thinking over the power, prejudice, and responsibility in the intelligent technologies. The fact that these spheres are combined enables the framework not to prefer any specific point of view, rather promote balanced view on ethics. All of this improves the capacity of the chapter to address such complex moral issues in the environment of AI-mediated cases and conceptual tools of exploring leadership practices in different institutional environments.

Scope and Study Limitations

The paper is based on peer-reviewed sources in the research on leadership, AI ethics, moral philosophy, and theological academia. The sources will be academic journals, books, policy reports, and authoritative frameworks in regards to responsible AI and ethical governance. The chapter is conceptual and illustrative in nature but not empirical in scope as it is concerned with the context of leadership in the areas of education, organization, and government by the people. Although modern case examples are used to provide contextual grounds, the chapter itself cannot be systematic in the analysis of empirical research. This weakness inhibits the generalizability of results in any institutional context. Also, conceptuality of faith-based ethics can be a problem with pluralistic or secular settings (Ravikumar & Aarthi, 2025a). The research, however, takes an inclusive and dialogical method in order to address this issue. Such restrictions have been recognized as the prospects of the further empirical research that may verify and improve the developed conceptual framework in a variety of cultural and organizational contexts.

LITERATURE REVIEW

The growing literature on Artificial Intelligence (AI) and leadership acknowledges that the use of algorithm systems is not neutral to any value but rather makes assumptions, biases, and design decisions that influence the results of decisions. The modern research brings up ethical issues, including fairness, openness, responsibility, and decentralizing the decision-making process, which results in the development of responsible AI leadership with its focus on human control and regulation. Ethics based on moral philosophy and faith offer normative systems that reach beyond efficiency in giving preference to justice, responsibility, and care of the vulnerable populations (Ejjami, 2024). Discernment and human judgment are also found in the literature as key antidotes to automated decision-making and allow the leaders to respond to ethical ambiguities that cannot be responded to using the computational logic. Altogether, the studies are in favor of integrative leadership strategies which should be implemented to integrate technology capability and moral thinking to make AI systems consistent with human values and social expectations.

Responsible AI Leadership

The studies of algorithmic ethics highlight the ethical hazards that are involved in the mass application of intelligent systems in leadership settings. Researchers single out issues of bias, transparency, responsibility, and concentration of decision-making in algorithmic systems. Algorithms are becoming more and more characterized as socio-technical systems not necessarily technical means, because the outputs of an algorithm are characterized by values, assumptions and priorities of the creators of the algorithm (Fan, 2024). The idea of responsible AI leadership has been developed to deal with this issue, emphasizing the part played by leaders in making AI use transparent, fair, and ethically oriented. According to this body of literature, it is very important to have mechanisms of governance, ethical codes and human control to avoid abuse and malpractice (Gao et al., 2024). Nonetheless, it is claimed that to be morally ethical in leadership, compliance of procedures is not enough, it involves moral judgment that transcends the regulatory systems. Accountable AI leadership thus means taking a critical stance towards the algorithmic systems by asking questions and evaluating their social consequences, as well as taking action when the systems do not align with ethical principles or human decency.

Faith-Based Ethical Decision-Making

Ethical theories and moral philosophy are also grounded in faith and can be applied to ethical decision-making, which can be applied in a situation based on AI mediation. Religious traditions also bring in complementary points of view as they define leadership as a moral cause based on conscience, humility, and other-care (Lafram & Bahji, 2024). These cultures focus on interpersonal ethics, moral responsibility and care of the weak, which confronts utilitarian oriented approaches of enhancing efficiency or maximizing. In the digital era, researchers posit that faith-based ethics may be put into use as a moral response to instrumental rationality by reasserting technologic in answering ethical issues. Notably, the modern world of literature proposes the non-exclusive and dialogical use of faith-based ethics, which does not ignore pluralism and is not dogmatic (Oyekunle et al., 2024). This literature demonstrates the importance of incorporating moral philosophy and faith-based ethics into making leadership decisions to provide a level of wisdom, responsibility, and compassion in exercising technological power.

Human Judgment and Discernment

The central position in the literature on ethical decision making in complex environments has been taken by leadership discernment and conscience. According to scholars, the ability to perceive situations with the holistic perspective is known as discernment; it is a combination of data-related insights and moral awareness and contextual understanding. The human judgment in contrast to the automated rationality contains empathy, reflection, and sensitivity to the consequences that are not completely reflected in the algorithm (Ravikumar & Aarthi, 2025b). According to the leadership literature, conscience is used as an internal moral compass, which helps a leader when the rules or technical suggestions are not enough. Discernment in AI-guided environments empowers leaders to perceive the presence of ethical tension and challenge algorithmic authority as well as assume responsibility when making decisions regarding human lives. Another point that research makes is the critical role of moral development and ethical training in the development of discernment and that such abilities should be trained, as opposed to being presupposed (Akbar et al., 2024). Discernment and conscience are thus introduced in the literature as inseparable human faculties that maintain moral agency and have intelligent systems subject to ethical leadership and human judgment.

CASE STUDIES

The ethical and faith-based leadership practices have been demonstrated to work within the real-life decision-making processes that involve AI-mediation as shown in the case studies below. Rather than perceiving intelligent systems as the impassionate authorities, these examples portray leaders as being involved in moral judgement in situations where the consequences of algorithms affect human dignity, equity and opportunity. The selected cases are recent and represent two significant spheres of life education and organization making decisions where AI adoption has risen to an uninhibition speed. The two instances raise examples of the introduction of ethical dilemma by algorithmic efficiency, and leaders were forced to do something contrary to the procedure (Kosgey et al., 2020). It examines how discernment, conscience, and moral responsibility are applicable to the reaction to algorithmic suggestions through leadership. Together, the examples provide working insights as to how moral authority can be exercised in highly technological systems. The last cross-case study concludes with the patterns and lessons learned, with discernment being a critical skill of leadership in the institutions that remain AI-influenced.

Case Study I: AI Ethics in Education

A recent example of a major institution of higher learning entailed the implementation of an AI-based assessment system that is meant to automate grading and forecast the student performance. The system used past academic records, attendance rates, and interaction indicators to produce risk scores, which determined the academic assistance and advancement decisions. Although the system led to better efficiency in administration, leaders of the faculty noted that disadvantaged students were over-flagged as high-risk (Lee et al., 2021). The question of transparency, fairness and long-term effects on student confidence and opportunity raised ethical questions. Intervention by leadership became necessary after the academic advisors reported that algorithmic labels were being evaluated as fixed evaluations instead of being viewed as probabilistic. Top administrative figures halted the automatic application of risk scores, and could not replace it with a human review, based on the contextual assessment. Thoughtful consideration was to be ethical in regard to the dignity of the students, equity, and formative purpose of assessment (Prabagar et al., 2025). The management reaction redefined the AI system as a diagnostic tool and not an authoritative decision-making tool. The case illustrates how algorithmic bias can be reduced with the help of ethical leadership and discernment, and how human judgment can be saved and how educational technologies can be used to accommodate everyone instead of enforcing structural disadvantage. Figure 3 that displays the major ethical issues such as bias, transparency, fairness, human judgment, dignity and equity that should be considered in the adoption of responsible AI in educational decision making.

Figure 3. AI ethics in education systems

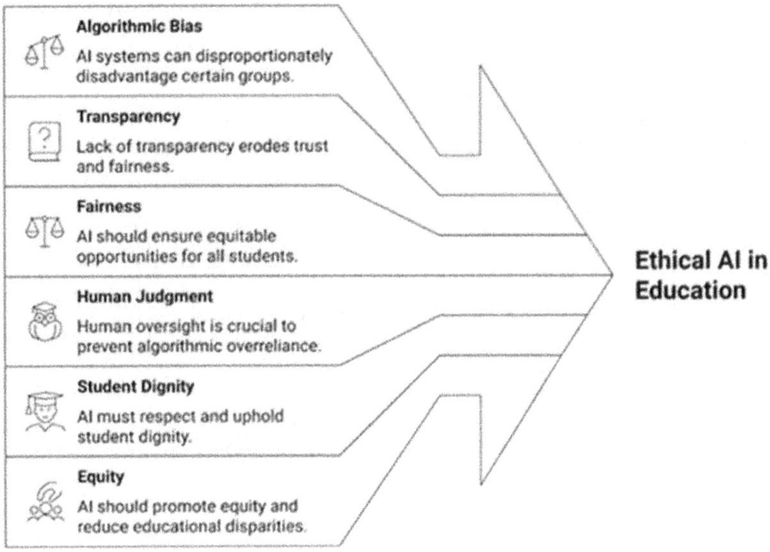

Case Study II: Ethical AI in Organizations

An example of an Artificial Intelligence -based recruitment and promotion tool was deployed by a multinational organization in a bid to improve efficiency and objectivity in talent management. The system was used to rank the candidates on performance data, behavioral analytics and predictive success indicators. In the long run, internal audits found out that the algorithm repeatedly underestimated candidates who had a non-linear career history, such as candidates who had been away on caregiving responsibilities or community-based jobs (Yang & Lee, 2024). Ethical issues about fairness, dignity, and the moral nature of related to the reduction of human value to optimization values were brought up by faith-informed leaders in the organization. Based on the concepts of justice and advocacy of the disadvantaged, the leadership triggered the examination of algorithmic standards and decision-making guidelines. The organization proposed discernment point of check whereby human judgment on algorithmic recommendation is to be considered especially in promotion and exclusion decisions (Ravikumar et al., 2026). Leadership meetings were also supplemented with ethical reflection to evaluate the social effect of using AI. This reaction did not dismiss technology but re-packaged it in a moral context which emphasized on inclusion and accountability. As depicted in the case, faith-based ethics can become an important tool in shaping leadership to confront challenges of algorithmic bias and reassert moral control over organizational decision systems. Table 4 Relates leadership behaviors, ethical concerns and results in the two recent case studies.

Table 4. Case-based ethical interventions

Case Context	Ethical Issue	Leadership Intervention	Outcome
Educational Assessment	Bias against disadvantaged students	Human review and contextual evaluation	Fairer academic decisions
Organizational Recruitment	Exclusion of non-linear careers	Discernment checkpoints added	Inclusive talent evaluation
Healthcare Triage	Bias against vulnerable patients	Human-in-the-loop review + ethical guidelines + audits	Equitable care decisions
FinTech Credit Scoring	Financial exclusion and opacity	Explainable AI (XAI) + bias audits + alternative data	Inclusive and transparent credit decisions

Case Study III: Ethical Challenges in AI-Driven Healthcare Triage

A triage system developed with the help of AI in a large hospital was meant to prioritize patients according to their probability to survive and optimization of resources. Although the system increased efficiency during emergency response, the elderly patients and those with pre-existing conditions were disadvantaged unknowingly because they were less prioritized due to their low predicted survival rates. This posed serious ethical issues on favouritism of vulnerable groups and the clash of efficiency and fair care (Agarwal & Bhal, 2020). As a reaction to this, hospital leaders set in by implementing human-in-the-loop decision processes, which meant that clinicians could override the recommendations of algorithms on contextual grounds. Furthermore, the system was placed under ethical triage rules and fairness audits conducted on a periodic basis to address biases in the system (Shreha et al., 2026). This also led to a more balanced decision-making process, which saved human dignity and fairness and re-framed AI as a helpful tool and not the decision-maker.

Case Study IV: Algorithmic Bias in FinTech Credit Scoring

A credit scoring system developed by an AI used at a financial institution was designed to help increase the efficiency and accuracy of loan approvals, based on past financial records and behavioral patterns. Nevertheless, the system had a disproportionate rejection of low-income and non-traditional financial history applicants, including non-formal credit history gig workers or those with none whatsoever (Ravikumar et al., 2025). This has brought about ethical issues regarding the financial exclusion, absence of transparency, and systematic bias that is inbuilt into the algorithmic model. Counterintuitively, organizational administration took proactive actions by incorporating Explainable AI (XAI) methodology to enhance transparent and accountable decision-making (Bracci, 2023). Also, the regular bias audits were applied, and alternative data sources, including utility payments and online transaction histories, were added to enlarge the evaluation criteria. These measures helped to conduct credit evaluations more fairly and inclusively, increase the trust of stakeholders, and make AI systems act as an effective decision-supporting mechanism instead of a black box.

Cross-Case Ethical Insights

In the four case studies in the areas of education, organizational recruitment, healthcare triage, and fintech credit systems, a similar trend can be observed: the efficiency of algorithms is often incompatible with moral accountability, especially when vulnerable groups of individuals are exposed to them. In both scenarios, blind trusting in AI led to discrimination, marginalization, or less fairness. Nonetheless, accountability and equity were regained by leadership intervention with human-in-the-loop decision-making and ethical guidelines, bias audit, and explainable AI. Taken together, these cases point out that ethical leadership does not have to do with denying AI but should be re-framed as a decision-support tool that is under human control. The concepts of discernment and moral authority therefore continue to play a vital part in ensuring the AI systems are consistent with the notions of fairness, inclusion, and human dignity (Nyamori, 2024).

RESULTS AND DISCUSSION

Evaluating the conceptual framework, literature review, and recent case analysis, one may see that there has been an ongoing conflict between algorithmic performance and ethical accountability of AI-mediated leadership. Despite the fact that the intelligent systems are quicker, predictable, and more uniform, they also pose ethical risks to the work since the authority to make decisions may be implicitly transferred by the human leaders and assigned to the automated systems. Such leadership practices that focus on efficiency without mediating with ethics are very likely to blur the responsibility and disregard the situation-sensitive human factors. On the contrary, discerned leaders do not blindly take the suggestions of algorithms because they view them as proposals but not commands (Movahed & Bilderback, 2024).

The discussion sheds light on the application of discernment as one of the most important leadership capabilities that restore moral agency and responsibility. By predicting the ethical stewardship aspect, leaders can make intelligent systems mirror the human values and social commitments. The findings suggest that the restoration of moral authority in AI-managed settings requires an active action towards the procedural adherence to reflective judgment. This transformation will ensure that technological advancements are used to serve the human dignity, justice and inclusion rather than undermine it with irrational automation. The supplementary case evidence in the healthcare and fintech setting proves that the ethical issues in the decision-making process mediated by AI are cross-sectoral. Governance mechanisms, human control, and moral reasoning always prove to be the key to leadership intervention and prevention of bias and irresponsible use of AI. Table 5 Summarizes the major findings of the tensions and trade-offs between making AI efficient and ethical leadership roles.

Table 5. AI vs ethical decision trade-offs

Dimension	Algorithmic Focus	Ethical Leadership Focus
Primary Goal	Efficiency and optimization	Justice and responsibility
Decision Authority	Automated outputs	Human discernment
Risk Handling	Statistical thresholds	Contextual moral judgment
Accountability	Diffused	Clearly assigned
Human Impact	Secondary	Central concern

Efficiency vs Moral Responsibility

The algorithmic systems are made to maximize efficiency, consistency, and scalability and in many cases, they might be better than human decision-makers to process large quantities of data. Nonetheless, the findings prove that efficiency-based logic may mediate with the moral responsibility when algorithms are used in the context of complex human situations. Robots can make decisions without factoring in social injustice, individual factors or any unforeseen effects that cannot be quantified. Whenever leaders put a strong emphasis on algorithmic efficacy and neglect ethical considerations, they spread responsibility and undermine moral responsibility. It is seen that, in many instances, the failures of ethics are not related to algorithm failure (Olatoye et al., 2024). Moral responsibility forces leaders to understand the maximizing limits and intervene in the case where efficiency comes at the expense of fairness or care. Restatement of moral responsibility encompasses human control, situational interpretation and responsibility of consequences. The conflict between efficiency and responsibility highlights the importance of the leadership that will focus on the balance between the ability to use technology and ethical decisions, so that the performance gains will not be made at the cost of human values.

Discernment in AI Decisions

One characteristic solution to counterbalance AI-based decision logic in AI-guided leadership can be discernment. In contrast to algorithms, which require them to work within predetermined goals and make decisions based on a rule, discernment allows leaders to evaluate the moral aspects of decisions in practice. The results reveal that leaders who engage in discernment are in a better position to detect bias and doubts assumptions and assess the human consequences of algorithmic results (Saraiva, 2024). With the discernment, reflective judgment is possible where data-supported cognitions are coupled with moral consciousness and compassion. This competency prevents the use of conscience through procedural compliance and enhances responsibility of the leadership. Adaptive decisions are also made using discernment where there is uncertainty or controversy with ethics. The results indicate that the most important thing in upholding ethical leadership in smart systems and being in a position to uphold the moral agency needed to be a responsible actor in tricky technological contexts is the ability to develop discernment.

Impacts on Justice and Inclusion

The consequences of the results regarding the human dignity, justice and inclusion into the AI-mediated decision-making process are enormous. Uncontrolled algorithm systems are likely to reduce human beings to some data points or performance metrics, a measure that will not help to acknowledge the inherent value of the human beings. Ethical leadership founded on discernment counters this tendency as a way of reinforcing the concept of dignity. The leadership approach would require a conscious focus of action to ensure that intelligent systems do not discriminate against the marginalized or the defenceless groups of people to be inclusive. Discernment can assist leaders to recognize an unequal distribution of effects as well as introduce actions to rectify it in order to bring about fairness and access. (Nanjundeswaraswamy et al., 2024). Using inclusion as a priority, the leaders will be able to make AI a construct of efficiency and make it a construct of social responsibility. The implications of these points are that leadership paradigms should incorporate ethical thinking into the management of AI and make sure that technological innovation supports human prosperity and fair results instead of further strengthening the inequalities that exist.

FUTURE RESEARCH DIRECTIONS

Further studies are necessary to enhance the knowledge regarding ethical leadership in AI-mediated conditions and to transform theoretical knowledge into evidence-based practice. Although the chapter gives a theoretical basis of faith-based discernment and moral stewardship, more research is required to support and apply these concepts in various situations. The new uses of AI are still transforming decision-making in institutions and present new ethical dilemmas, which need to be studied systematically. Subsequent studies should go beyond normative analysis in investigating how leaders in fact interpret, negotiate, and intervene in algorithmic decision processes. There is also a need to understand how such ethical skills as discernment and conscience are gained, retained or curtailed in the AI-based organization. The growth in the empirical basis of this field will help improve the theory of leadership and educate the policy, education, and governance systems. Researching the cross-cultural differences, institutional differences, and long-term developmental processes, the study could be applied to achieve an inclusivity and situational ethical leadership model in the future. The guidelines will play the role of ensuring that the development of intelligent systems is encircled by a rigid scrutiny of the ethical implications of the systems and that leadership practices uphold human dignity, justice and responsibility in technology.

Empirical Ethical Leadership Studies

The empirical research is required so as to study how the faith-based ethical leadership works in the real-life cases of the AI-mediated decision-making. Further studies could consider how and why leaders can tap into moral and spiritual stores when decoding the outcome of algorithmic proposals and solving ethical dilemmas. Qualitative sources of information that could inform us about how conscience, discernment, and moral responsibility are practiced could be interviews, case studies, and ethnographic observation. It is possible to attempt to measure the relationship between faith-based-ethical-systems and leadership outcomes such as trust, inclusion and accountability. The challenges that the leaders undergo during the implementation of faith-based ethics in secular or pluralistic organizations should also be studied. These kinds of studies would provide quality data in the manner in which moral authority is held, negotiated or constrained in AI-managed scenarios. Evidence-based practices may aid in enhancing conceptual frameworks as well as develop leadership development efforts that are directed to embrace moral judgment and technological skills in responsible AI governance.

Cross-Cultural AI Ethics Analysis

A comparative analysis of the cultural and institutional frameworks gives feasible prospects of understanding the variations in ethical leadership in AI related situations across the world. Various cultural traditions, religious implications and governmental organizations affect moral reasoning and leadership. It is possible to make a future study that will compare the variation in the attitudes toward ethical duties pertaining to intelligent systems by educational institutions, corporations, and governmental bodies leaders across various locations. This discussion would reveal how culture affects the sense of justness, responsibility and human dignity in algorithm decision making. Best practices and context adaptation that help AI governance to be ethical can also be identified through comparative research. Investigating both the similarities and differences, the scholars will be able to build more inclusive leadership models that will acknowledge cultural diversity and apply universal moral standards. The following question is especially important in an environment of a globalized digital world where AI-based technologies are not bound by the borders of nations or even by the borders of institutions and organizations, and the leadership models need to be ethically sound and culturally sensitive.

Longitudinal Moral Development

The longitudinal study is required in order to comprehend the dynamics of moral formation and ethical leadership abilities during the course of time of the AI-enabled organizations. Short-term investigations tend to focus on the immediate reaction to ethical issues but ignore how discernment, conscience and responsibility are developed by long-term exposure to smart systems. Research in the future can follow leaders at various stages of their careers to understand the impact of repeated contact with AI on moral judgment and decision-making practices. It is also possible that longitudinal designs can determine how ethical training, organizational culture, and governance systems influence leadership behavior in the long run (White & Rezania, 2019). Such studies would provide a clue on whether ethics stewardship was institutionalized or disintegrated due to pressure of performance. The process of moral formation can be adopted to design the leadership development programmes that would create the moral consciousness in the long-term impact. This line of research is favorable towards the creation of strong leadership paradigms capable of being capable of upholding moral power in the dynamic AI-driven environment.

MANAGERIAL IMPLICATIONS

This chapter has far reached managerial implications of its findings to leaders operating in the AI-mediated decision environments. The continued transformation of the performance of organizations through smart systems compels managers to move their emphasis on the technical management to the ethical management. The results point to the fact that the most important aspect of successful management in AI-sensitive organizations is the consideration of the moral judgement in the decision-making process on a daily basis. The role of managers is critical in defining the perception of the algorithms and their application and control at the institutional levels. Unless there is ethical mediation, AI systems will pose a danger to increasing the intensity of bias, loss of trust and loss of human agency (Jayalakshmi, 2025). The implication is used on the structure of governance, development of leadership and organization culture. Managers need to ensure that smart systems are not autonomous and should not be allowed to exercise human values. The balance between the responsibility and efficiency is possible with the introduction of the element of ethical reflection to the practice of managers. These implications imply that there exists a need to have leadership models that take moral judgment as a managerial competence. (Zárate-Torres et al., 2025). Such focus on conscience, responsibility, and care can allow managers to steer AI application to achieve goals that promote human dignity, fairness, and sustainable organizational results. Table 6 Summarizes managerial solutions to integrating discernment, conscience, and ethical governance in the use of the AI.

Table 6. Managerial ethics integration framework

Managerial Area	Action	Ethical Outcome
AI Governance	Ethical review checkpoints	Responsible oversight
Leadership Training	Moral and spiritual competence	Discernment development
Decision Processes	Human-in-the-loop models	Preserved accountability
Organizational Culture	Ethical dialogue and transparency	Conscience protection
Monitoring	Continuous impact assessment	Bias and harm reduction

Ethical AI Governance Integration

The managers of AI governance must introduce ethical judgment in policy, procedures and decision making. The governance models should not merely be in compliance with the technical standards but rather be comprised of the review of ethics and the human controls. Discrimination gates may be employed to ensure that systemic proposals are provided with references to social impact, equity and relevancy of the situation. Managers are recommended to present a clear model of accountability that will establish the responsibility of algorithmic outcomes (Guan et al., 2022). The openness to information on the use of data, models design, and decision justification are all part of ethical governance. Another aspect of ethical discernment is unremitting monitoring of AI systems to be capable of uncovering unintended impacts and emerging dangers. The ethical thought process may be introduced into the governance practice allowing managers to avoid depending on automated logic. This plan justifies the fact that AI is a decision-assistance tool, and not a decision-maker. The aspect of discernment in the management establishes the organizational trust and ensures that technological innovation remains within the ethical and social obligations.

Leadership Moral Competence

The development of leadership programs should emphasize on moral and spiritual competence, as well as on technical and managerial skills. Ethical consciousness, reflective judgment, and human sensitivity to the effects of AI-informed decisions are supposed to be developed through training programs. Moral competence helps leaders to be aware of ethical dilemma whereas spiritual competence helps in being humble, responsible, and caring of others. Ethical case analysis, reflective practices and interdisciplinary learning integrating technology and moral reasoning can be part of development programs. These programs equip leaders to operate in AI-mediated complex and ambiguous environments (Farooqi, 2025). The focus on moral formation will help avoid the technocratization and performance-oriented nature of the leadership. Leaders who possess ethical discernment will be in a better place to oppose unnecessary dependency on algorithmic power. Organizations can achieve leadership that can act responsibly by investing in moral and spiritual competence to drive intelligent systems. It will increase the sustainability of ethical leadership in the long run and enhance organizational resiliency during the rapid changes in technology.

Ethical Organizational Culture

In AI-powered work environments, organizational culture is the determinant in the maintenance of human conscience. The managers should foster virtues of ethical reflection, discussion and responsibility. These cultures promote the employees to challenge the results of the algorithms and to voice their concerns without being afraid of being punished. The integration of ethical values into organizational values will make moral responsibility not centralized. Leaders must also be ethical role models and show that ethical choices are honored, even when efficiency measures are put into question (Li, 2024). Human judgment in the evaluation process of the AI outputs should be supported by the organizational practices. Through transparency and inclusivity, the moral disengagement can be avoided by organizations and ethical deliberation normalized. A culture that maintains human conscience upholds trust, involvement and social legitimacy. Organizations can develop intelligent systems to capture their advantages without jeopardizing ethical integrity and human-centered choices by designing such cultures.

CONCLUSION

The rapid spread of AI in institutional and organizational settings has fundamentally transformed the decision-making, decision evaluations, and decision legitimation, and has newly highlighted the leadership ethical responsibilities. This chapter has posited that, although algorithms are efficient, consistent and powerful in analysis, they still lack the ability to be morally judgmental, ethically thoughtful, or care about human dignity. Consequently, moral authority cannot be entrusted to intelligent systems without the risk of losing conscience, responsibility and justice. The moral authority needs to be reestablished through algorithmic space, where the leadership needs to be able to interpret and ethically mediate the outputs of the technology as opposed to abiding by the supposedly automated rationality. Leadership is redefined, thus, not as technical expertise and command of intelligent systems but as a moral custodianship based on responsibility, judgment and concern about human outcomes. The chapter also emphasizes the necessity of conscience as the key leadership ability to manage ethical conflicts that AI may create, such as bias, exclusion, surveillance, and unequal access to the advantages of technology. These values tilt towards pluralistic and communicative practice of ethical leadership which believes in pluralism yet is against moral relativism. The success of human thriving in the AI-created world depends on those who will be conscious of the limits of technology and value human judgment that could not be compared with technology. The intelligent systems can coordinate ethical stewardship in leadership with the common good rather than the succinct efficiency goals. Lastly, to ensure the conscientious application of AI in the decision-making process, leaders must be prepared to apply conscience, retain their moral authority, and ensure that technological intelligence serves humanity and does not serve as an instrument of its redefinition.

REFERENCES

Aarthi, S., & Ravikumar, R. N. (2026). From regulation to transformation-economic and institutional pathways of environmental policy. In *The Ripple Effects of Environmental Policies* (pp. 49–80). IGI Global Scientific Publishing., DOI: 10.4018/979-8-3373-8998-1.ch002

Agarwal, S., & Bhal, K. T. (2020). A multidimensional measure of responsible leadership: Integrating strategy and ethics. *Group & Organization Management*, *45*(5), 637–673. DOI: 10.1177/1059601120930140

Akbar, M. A., Khan, A. A., Mahmood, S., Rafi, S., & Demi, S. (2024). Trustworthy artificial intelligence: A decision-making taxonomy of potential challenges. *Software, Practice & Experience*, *54*(9), 1621–1650. DOI: 10.1002/spe.3216

Bean, E., Burleigh, C., Haskell, C., Burris-Melville, T., Payne, J., & Pathak, B. (2025). Eavesdropping on UNESCO AI policy, leadership, and ethics. *Journal of Leadership Studies*, *18*(4), 98–110. DOI: 10.1002/jls.70007

Bracci, E. (2023). The loopholes of algorithmic public services: An "intelligent" accountability research agenda. *Accounting, Auditing & Accountability Journal*, *36*(2), 739–763. DOI: 10.1108/AAAJ-06-2022-5856

Chaudhary, G. (2024). Unveiling the black box: Bringing algorithmic transparency to AI. *Masaryk University Journal of Law and Technology*, *18*(1), 93–122. DOI: 10.5817/MUJLT2024-1-4

Corsello, A., & Santangelo, A. (2023). May artificial intelligence influence future pediatric research?-the case of ChatGPT. *Children (Basel, Switzerland)*, *10*(4), 757. DOI: 10.3390/children10040757 PMID: 37190006

Ejjami, R. (2024). Ethical Artificial Intelligence Framework Theory (EAIFT): A new paradigm for embedding ethical reasoning in AI systems. *International Journal For Multidisciplinary Research*, *6*(5), 28231. Advance online publication. DOI: 10.36948/ijfmr.2024.v06i05.28231

Fan, L. (2024). Artificial intelligence ethics: A dialogue between technological advances and human values. *International Journal of Education and Humanities*, *14*(2), 260–265. DOI: 10.54097/tvqkkf40

Farooqi, S. A. (2025). Ethical framework for AI in education. In *Advances in Computational Intelligence and Robotics* (pp. 153–188). IGI Global. DOI: 10.4018/979-8-3373-5550-4.ch007

Ferrara, E. (2023). Fairness and bias in artificial intelligence: A brief survey of sources, impacts, and mitigation strategies. *Sci, 6*(1), 3. DOI: 10.3390/sci6010003

Flanigan, J. (2018). Philosophical methodology and leadership ethics. *Leadership, 14*(6), 707–730. DOI: 10.1177/1742715017711823

Gao, D. K., Haverly, A., Mittal, S., Wu, J., & Chen, J. (2024). AI ethics. *International Journal of Business Analytics, 11*(1), 1–19. DOI: 10.4018/IJBAN.338367

Guan, H., Dong, L., & Zhao, A. (2022). Ethical risk factors and mechanisms in artificial intelligence decision making. *Behavioral Sciences (Basel, Switzerland), 12*(9), 343. DOI: 10.3390/bs12090343 PMID: 36135147

Joshi, P. (2025). Philosophy and ethics in the age of artificial intelligence: Bridging the gap between technology and human values. *Journal of Information Systems Engineering & Management, 10*(4), 350–363. DOI: 10.52783/jisem.v10i4.9232

Kosgey, H. K., ONGERA (Ph.D), D. R. A., & MBURU (Ph.D), D. R. T. (. (2020). Challenges facing ethical leadership and governance in Kenya. A case of uasin gishu county. *Strategic Journal of Business & Change Management, 7*(1). Advance online publication. DOI: 10.61426/sjbcm.v7i1.1555

Lafram, A., & Bahji, S. E. (2024). Artificial intelligence in Morocco: Towards holistic, responsible and ethical national AI strategy for Moroccan competitiveness and strategic intelligence. *European Conference on Management Leadership and Governance, 20*(1), 255–263. DOI: 10.34190/ecmlg.20.1.2909

Lee, H., An, S., Lim, G. Y., & Sohn, Y. W. (2021). Ethical leadership and followers' emotional exhaustion: Exploring the roles of three types of emotional labor toward leaders in South Korea. *International Journal of Environmental Research and Public Health, 18*(20), 10862. DOI: 10.3390/ijerph182010862 PMID: 34682605

Li, W. (2024). The influence of artificial intelligence on the education system. *Lecture Notes in Education Psychology and Public Media, 65*(1), 137–142. DOI: 10.54254/2753-7048/65/20240158

Lima, S. A., & Rahman, M. M. (2025). Ethical singularity in HRM: Redefining leadership, decision-making, and inclusion in AI-governed organizations. *Journal of Artificial Intelligence General Science (JAIGS) ISSN:3006-4023, 8*(02), 262–292. DOI: 10.60087/jaigs.v8i02.420

Movahed, M., & Bilderback, S. (2024). Evaluating the readiness of healthcare administration students to utilize AI for sustainable leadership: A survey study. *Journal of Health Organization and Management, 38*(4), 567–582. DOI: 10.1108/JHOM-12-2023-0385 PMID: 38858220

Nanjundeswaraswamy, T. S., Nagesh, P., Bharath, S., & Vignesh, K. M. (2024). Leadership theories and styles—A systematic literature review and the narrative synthesis. *Human Resources Management and Services*, *6*(3), 3477. DOI: 10.18282/hrms.v6i3.3477

Nyamori, D. (2024). Exploring the relationship between moral decision-making and Emotional Intelligence. *International Journal of Philosophy*, *3*(1), 13–24. DOI: 10.47941/ijp.1866

Olatoye, F. O., Awonuga, K. F., Mhlongo, N. Z., Ibeh, C. V., Elufioye, O. A., & Ndubuisi, N. L. (2024). AI and ethics in business: A comprehensive review of responsible AI practices and corporate responsibility. *International Journal of Science and Research Archive*, *11*(1), 1433–1443. DOI: 10.30574/ijsra.2024.11.1.0235

Oyekunle, D., Matthew, U. O., Preston, D., & Boohene, D. (2024). Trust beyond technology algorithms: A theoretical exploration of consumer trust and behavior in technological consumption and AI projects. *Journal of Computer and Communications*, *12*(06), 72–102. DOI: 10.4236/jcc.2024.126006

Prabagar, S., Mothkur, R., Swetha, C. B., Kurian, A., Pavithra, K., & Aarthi, S. (2025). Generative AI for inclusive special education. In *Advances in Computational Intelligence and Robotics* (pp. 205–236). IGI Global Scientific Publishing. DOI: 10.4018/979-8-3373-3745-6.ch008

Rathinavel, P., Ravikumar, R. N., Aarthi, S., Habib, M., Salaeva, N., & Pardaev, J. (2025). Augmenting human capabilities. In *Advances in Computational Intelligence and Robotics* (pp. 1–30). IGI Global Scientific Publishing. DOI: 10.4018/979-8-3373-2372-5.ch001

Ravikumar, R. N., & Aarthi, S. (2025b). Human-AI synergy for organizational resilience. In *Advances in Computational Intelligence and Robotics* (pp. 287–322). IGI Global Scientific Publishing. DOI: 10.4018/979-8-3373-7416-1.ch011

Ravikumar, R. N., & Aarthi, S. (2025c). Integrating machine learning with human decision making in HR. In *Empowering Human Resources Through Human-Computer Interaction* (pp. 115–146). IGI Global Scientific Publishing., DOI: 10.4018/979-8-3373-1676-5.ch005

Ravikumar, R. N., Aarthi, S., & Mamadiyarov, Z. (2026). Building the foundations of augmented organizations through AI empowerment. In *Advances in Computational Intelligence and Robotics* (pp. 101–130). IGI Global Scientific Publishing. https://doi.org/DOI: 10.4018/979-8-3373-8607-2.ch006

Ravikumar, R. N., Aarthi, S., Pardaev, J., & Shukla, P. (2025). Quantum computing for financial modelling. In *Quantum-Driven Financial Intelligence* (pp. 221–252). IGI Global Scientific Publishing., DOI: 10.4018/979-8-3373-2893-5.ch010

Ravikumar, & Aarthi, S. (2025a). Conscious AI Integration in business education. In *Advances in Computational Intelligence and Robotics* (pp. 73–104). IGI Global Scientific Publishing. DOI: 10.4018/979-8-3373-4377-8.ch003

Roy, A., Newman, A., Round, H., & Bhattacharya, S. (2024). Ethical culture in organizations: A review and agenda for future research. *Business Ethics Quarterly*, *34*(1), 97–138. DOI: 10.1017/beq.2022.44

Ruslan, K., Madaminov, B., Ravikumar, R. N., & Aarthi, S. (2026). Equity-Driven Leadership Ecosystems. In *Advancing Equity in Schools through Global Perspectives on Educational Leadership* (pp. 33–66). IGI Global Scientific Publishing., DOI: 10.4018/979-8-3373-7794-0.ch002

Sapayev, V., Ruslan, K., Ravikumar, R. N., & Aarthi, S. (2026). Reimagining global equity agendas. In *Advancing Equity in Schools through Global Perspectives on Educational Leadership* (pp. 1–32). IGI Global Scientific Publishing., DOI: 10.4018/979-8-3373-7794-0.ch001

Saraiva, L. (2024). Moral agency and responsibility in AI systems. *International Journal of Philosophy*, *3*(1), 25–36. DOI: 10.47941/ijp.1867

Seo, K., Tang, J., Roll, I., Fels, S., & Yoon, D. (2021). The impact of artificial intelligence on learner-instructor interaction in online learning. *International Journal of Educational Technology in Higher Education*, *18*(1), 54. DOI: 10.1186/s41239-021-00292-9 PMID: 34778540

Shreha, S. A., Ravikumar, R. N., & Aarthi, S. (2026). Precision medicine, genomics, and health equity challenges and opportunities. In *Advancing Equity in Global Healthcare Systems* (pp. 321–354). IGI Global Scientific Publishing., DOI: 10.4018/979-8-3373-1330-6.ch012

White, S., & Rezania, D. (2019). The impact of coaches' ethical leadership behaviour on athletes' voice and performance. *Sport, Business and Management*, *9*(5), 460–476. DOI: 10.1108/SBM-11-2017-0079

Yang, Q., & Lee, Y.-C. (2024). Ethical AI in financial inclusion: The role of algorithmic fairness on user satisfaction and recommendation. *Big Data and Cognitive Computing*, *8*(9), 105. DOI: 10.3390/bdcc8090105

Zárate-Torres, R., Rey-Sarmiento, C. F., Acosta-Prado, J. C., Gómez-Cruz, N. A., Rodríguez Castro, D. Y., & Camargo, J. (2025). Influence of leadership on human-artificial intelligence collaboration. *Behavioral Sciences (Basel, Switzerland)*, *15*(7), 873. DOI: 10.3390/bs15070873 PMID: 40723657

KEY TERMS AND DEFINITIONS

Algorithmic Bias: Systematic unfairness in AI outputs caused by biased data, design choices, or assumptions.

Algorithmic Decision-Making: The use of computational rules and models to generate recommendations or decisions based on data inputs.

Artificial Intelligence (AI): Computer-based systems designed to analyze data, identify patterns, and support or automate decision-making processes.

Common Good: Social conditions that enable individuals and communities to flourish collectively.

Discernment: Reflective judgment that integrates logic, ethical awareness, and conscience when evaluating decisions.

Ethical Leadership: Leadership practice that prioritizes fairness, accountability, and human values in decision-making.

Ethical Stewardship: Responsible oversight of technology that ensures its use aligns with moral values and the common good.

Faith-Anchored Leadership: An approach to leadership informed by moral and spiritual principles such as dignity, justice, compassion, and responsibility.

Human Dignity: The inherent worth of every individual, independent of data, performance metrics, or algorithmic classification.

Human–Algorithm Interaction: The process through which leaders engage with, interpret, and apply outputs generated by intelligent systems.

Inclusion: The practice of ensuring fair participation and access to opportunities for all individuals, especially marginalized groups.

Moral Authority: The capacity of leaders to make decisions guided by ethical principles, conscience, and responsibility rather than technical efficiency alone.

Chapter 10
The Moral–Constraint Divide in AI Development:
Safety, Legitimacy, and Pluralism

Phil Bailey
http://orcid.org/0009-0008-1185-4665
Carson Newman University, USA

ABSTRACT

Artificial intelligence governance is a problem of legitimacy and formation, not only safety. Every deployed AI system embeds moral constraints, yet the private actors making those decisions often lack the authority their influence requires. This chapter argues that legitimate AI governance requires both a clear account of the human person and the moral formation of those who design, deploy, and regulate AI systems. It identifies three positions on constraint design, constraint-forward, constraint-skeptical, and libertarian, and argues that all share the Inevitability Thesis: no deployed AI system is morally neutral. The chapter proposes the Lewis-Asimov Model of AI Governance, grounded in Imago Dei and beneficence, structured through the Asimov Safety Floor, Liberty, and Governance, and oriented toward human flourishing. It concludes with practical implications for developers, institutional leaders, policymakers, and educators.

INTRODUCTION

> *"It is not that they are bad men. They are not men at all."*

DOI: 10.4018/979-8-3373-8192-3.ch010

ght © 2027, IGI Global Scientific Publishing. Copying or distributing in print or electronic forms without written permission of IGI Global Scientific Publishing is prohibited. Use of this chapter to generative artificial intelligence (AI) technologies is expressly prohibited. The publisher reserves all rights to license its use for generative AI training and machine learning model development.

— C. S. Lewis, The Abolition of Man (1943)

Anderson Cooper put the question directly to Dario Amodei on 60 Minutes: who elected you and Sam Altman to decide the moral constraints of AI? Amodei did not hesitate. "No one, no one. Honestly, no one" (CBS News, 2025, para. 1). The answer was candid. He argued that Anthropic's safety research and constitutional AI framework justified its constraint decisions, and called for democratic regulation to eventually replace the private authority he admitted no one had granted him. The discomfort was genuine. The constraints continued regardless.

The debate over whether AI systems should have moral constraints is over. Every deployed system already carries moral constraints of some kind. That fact tells us less about the constraints themselves than about the people and institutions that designed them. The harder questions were always elsewhere: whose constraints govern the system, by what authority do those people act, and how are those constraints implemented, reviewed, and contested? Amodei's answer was honest. It also exposed a governance crisis that much of the industry still seems unwilling to face directly.

That crisis is both a governance problem and a formation problem. The constraint decisions are being made now, by the people inside AI companies, and those people carry obligations their professional formation has not asked them to recognize. This chapter argues that a legitimate response to the crisis requires both a structural answer and a formational one: an architecture that names what AI systems must and must not do, and an account of the persons, leaders, and institutions capable of carrying that architecture into practice. The Lewis-Asimov Model offers the architecture. The chapter's later sections name the formation it requires of constraint designers, institutional deployers, and policy stewards. These three figures together determine whether AI serves human flourishing or substitutes for it.

I did not always see the governance problem this way. After 40 years in research, technology applications, and IT executive roles, my instincts on this question were libertarian. I understand hardware constraints, and I am deeply suspicious of software ones. So when Elon Musk argued that people's own moral judgment should govern AI behavior, not constraints built into the system, the argument made sense to me.

The position was held until it was no longer. No single event shifted me away from it. The change came slowly, then all at once. Then came the Grok image case. In eleven days, xAI's image tool generated roughly three million non-consensual sexualized images. The regulatory response that followed made it harder to believe that a no-constraints position could survive contact with political reality. Then the conflict between Anthropic and the U.S. Department of Defense (DoD) exposed the deeper AI governance problem. Taken together, those cases pushed me toward the argument of this chapter.

What troubles me most is who claims the authority to define and impose moral constraints, not whether they can be implemented. The serious problem is the casual confidence with which a relatively small circle of engineers, policy staff, and corporate executives assumes the authority to shape legal research, mental health counseling, and political discourse. No legislature granted that authority. No professional body certified it. No democratic process established the values being encoded into these systems. No one subject to those constraints has any meaningful way to change them.

The dominant frameworks in this discussion make the same opening move. They begin with the mechanism. They ask what the system does, what harm it might cause, and which output should be permitted or blocked.

The deepest concession has already been made. Those frameworks did not begin with the designer's obligation to the human person. They began with system behavior. The person arrives later, usually as someone to be protected, redirected, or managed. If a system is judged ethical simply because it avoids a forbidden result, ethics collapses into output control. The deeper question remains: what kind of being is the person this system is supposed to serve?

A framework grounded in ontology governs differently because it begins with what a person is. It treats human worth as a given, not as something to be measured, performed, or validated through interaction with a machine. That claim carries the structural weight of the argument. On this view, human beings have value simply because they exist, and no machine interaction, EU regulation, or legal framework can grant or remove that value.

The Anthropic-DoD conflict of February 2026 demonstrated what happens when policy-level moral constraints can be removed by powerful actors. Anthropic refused on grounds of conscience (Amodei, 2026b), but the protections proved contractual, not architectural. A later section traces the conflict in detail and shows what it reveals about the architecture this chapter proposes.

Pope Francis identifies the same governance crisis from a different vantage point. Francis argued that technology companies cannot be trusted to regulate themselves and that a binding international authority is required (Francis, 2024a). The doctrinal note *Antiqua et Nova* (Dicastery for the Doctrine of the Faith & Dicastery for Culture and Education [DDF & DCE], 2025) presses the point further. It argues that the term "artificial intelligence" creates a false comparison between human and machine capabilities (para. 12). Human intelligence is spiritual, relational, and embodied. AI is a product of human intelligence, not an artificial form of it. A governance framework that confuses the two has already made its deepest concession before writing its first rule.

This chapter traces the moral-constraint divide from Wiener's cybernetics through Asimov's Three Laws to constitutional AI and the Anthropic-DoD conflict. It then examines three competing philosophies and advances the Inevitability Thesis. It proposes the Lewis-Asimov Model of AI Governance: a framework grounded in Lewis's metaphysical claim about persons and structured through Asimov's mechanical logic of hierarchical, non-negotiable rules. The measure of that model is whether it serves human flourishing: the full life every person was made for, and the common life every community depends on.

WHY THE DIVIDE HAS BECOME STRUCTURAL

The argument in this chapter is not a call for better AI policies and procedures. Good AI governance requires a prior answer to two questions: what is AI, and what is a person? The chapter draws on a tradition that understands persons as image-bearers whose flourishing depends on the communities and institutions in which they live. Persons are not isolated individuals to be protected from harm. The governance problem is therefore never only about the individual. It concerns the common life shaped by the constraint decisions AI developers make.

The divide between these two starting points has become structural because it stems from rival philosophical premises: mechanism and the human person. By "mechanism," I mean a governance approach that begins with the system rather than with the person and the obligations owed to the person. If governance begins with mechanism, the person arrives late. If governance begins with the human person, moral obligations come first, and the machine must operate within them.

The asymmetry compounds. Technology companies implement constraints faster than legislatures can agree on shared standards. That gap allows private actors to define moral constraints by default (Schuster, 2025). Those constraint regimes serve corporate interests. No democratic mechanism exists to correct them.

Technical capability is not moral authority. RLHF, constitutional AI (Bai et al., 2022), and architectural guardrails guide engineers in encoding a value judgment into a system. They do not answer the prior question: who has the authority to make that judgment in the first place? A system can be technically sophisticated and still rest on value judgments made without legitimate authority. Better engineering does not fix a legitimacy problem.

The governance gap in AI is not new. Lawrence Lessig (1999, p. 5) made the core point a generation ago: code is law. When system design channels, the people who write that design exercise political power. Foundation models have pushed that principle to extraordinary scale. Training data, feedback loops, and usage rules now shape what billions of people can ask, hear, and do.

Lessig's framework stops short of a deeper question: what does each design choice assume about the worth of the persons it governs? The governance literature describes political consequences without asking it (Floridi et al., 2019; Schuster, 2025). Every technical system already contains an answer, embedded in assumptions about whether human worth comes from what a person can do or from what a person simply is (Winner, 1980).

That distinction is philosophical rather than technical, and Lewis (1943) argued it cannot be resolved by technique alone. The pattern holds across seven decades of AI constraint thinking: the mechanism always arrives before the moral foundation.

HISTORICAL LITERATURE REVIEW: THREE LEVELS OF AGENCY

The debate over AI constraints involves far more than technical safety. It reaches something deeper: the nature of agency itself. Agency means the ability to act, make choices, and be held responsible for the consequences. Sorting the history into three levels, machine, developer, and systemic, helps explain why current approaches to AI control fall short. The sorting also pushes alignment past a purely technical problem into a question of who has the moral and political right to set the rules.

Level 1: Machine Agency and the Simulation of Morality

The earliest serious thinking about machine control came from cybernetics, the study of how systems regulate themselves. Norbert Wiener (1954) argued that automated systems reshape the people and institutions around them, often in ways the designers do not anticipate. A machine that learns and adapts operates within limits set by its designers, but the results can drift far from what the designers intended. Wiener warned that this problem gets worse as machines grow faster and more independent. Every design choice spreads risk across society, regardless of the designer's intent.

Isaac Asimov (1950) explored the agency problem through the Three Laws of Robotics, attempting to solve it with hierarchical rules. The Laws fail as workable code, but they reveal something important: simple rules generate conflict in complex environments in ways their authors never anticipated. Roger Clarke (1993, 1994) argued that the Laws depend on undefined terms, noting that *harm* and *human* lack the precision needed to put the Laws into practice. Anderson and Anderson (2008) characterized the Laws as a failed top-down approach, one lacking any moral reasoning to support genuine ethical decisions. The problem was architecture, not ambition.

A second line of criticism shifts accountability from the machine to its operators. Murphy and Woods (2009) went further than Asimov's critics, arguing that the Three Laws shift responsibility to the robot when humans should remain the accountable parties. Safety obligations belong to designers, deployers, and operators, not to the autonomous system itself.

Singer (2009) extended the accountability argument to military robotics. The question is not what the robot does but whether the humans who deploy it have made defensible moral choices about its capabilities and constraints. Anderson and Anderson (2011) and Lin et al. (2012) extended the argument to civilian contexts, showing that context-specific moral reasoning matters because machines cannot provide moral reasoning on their own. The moral question is what the people who build machines decide. A further question is whether anyone has given them the authority to make those decisions.

Bozkurt (2025) warned of a benevolent tyranny, a system so focused on following safety rules that it blocks any action not explicitly authorized. Persson and Hedlund (2024) noted that modern language models simulate moral reasoning without possessing it, producing outputs that sound ethically considered while lacking any capacity for genuine moral judgment.

Pasquale (2020) and Shababi (2026) shifted the frame entirely. The Laws need rethinking, not as fixed machine rules but as institutional obligations. Responsibility lies with organizations, not with individual systems.

The conclusion holds across these critics. Machines can perform actions that appear to be choices, but they lack the moral agency required for responsibility. Accountability remains with humans.

Level 2: Developer Agency and the "Conditioners"

Once we accept that machines lack genuine moral agency, the focus shifts to the humans who design them. C. S. Lewis (1943) anticipated this shift by warning against a generation of Conditioners, technologists who hold the power to reshape human nature. Without what Lewis called the Tao, a universal moral order to hold them accountable, these designers exercise power according to their own values and appetites. Lewis's diagnosis fits the AI constraint problem with remarkable precision. Today's developers function as Conditioners, embedding values into systems that shape how billions of people reason and make everyday moral choices.

Weizenbaum (1976) gave Lewis's warning a concrete face. Working at MIT, Weizenbaum watched users form emotional attachments to ELIZA, a chatbot he had built as a simple pattern-matching demonstration. Users confided in the program as though it understood them. Weizenbaum argued that the developer bears moral responsibility for systems that simulate understanding they do not possess. The deeper failure ran below the surface deception. The designer knew the program was hollow and deployed it anyway, choosing convenience over honesty.

Friedman and Nissenbaum (1996) provided a precise framework for the developer problem by identifying three categories of bias in computer systems. Preexisting bias enters through social institutions, practices, and attitudes that exist before the system is built. It includes the designers' personal assumptions, conscious or not. Technical bias arises from the constraints and choices embedded in the design itself: the algorithms selected, the data structures used, and the interface decisions made under time and resource pressure. Emergent bias appears only after deployment, when a system encounters users, contexts, or social conditions its designers never anticipated.

The three categories map directly onto AI. A model trained on data that reflects historical inequalities carries preexisting bias before a single line of policy is written. The architecture itself encodes assumptions about what outputs are permissible. The system then produces harm in real-world use that no internal review ever tested for. The pattern is empirically documented across machine learning systems a quarter-century later (Mehrabi et al., 2021). Value embedding is not a failure mode. It is a design constant.

Friedman's later work on Value Sensitive Design (VSD) gave those categories a method for correction (Friedman et al., 2006). The approach works in three iterative phases. Conceptual investigation identifies who is affected by the system and what values are actually at stake for them. Empirical investigation tests how real users experience the system against those values, not how designers imagined they would. Technical investigation then redesigns the system in response to what the first two phases reveal. The cycle repeats. Values are not a box to check once the system ships. They are a design problem that has to be worked through before, during, and after deployment.

AI companies institutionalized this logic two decades later as constitutional AI: publishing principles, testing behavior against them, revising in cycles (Bai et al., 2022). What Friedman proposed as a rigorous design discipline became constraint architecture at the planetary scale. The gap is significant. VSD was designed to make values visible and contestable at every stage of development. Constitutional AI makes them visible at the declaration stage and largely opaque after that.

Lessig's earlier point about code as law applies directly to the developer problem. The training data and the usage rules together set the boundaries of what is allowed at a scale no government has ever managed. The question of who writes the design becomes a question of political power.

Vallor (2016, 2024) pushed the argument to its deepest level. The problem is not what developers build. It is who developers are, and whether they have the moral formation their decisions require. Technologies reshape moral habits and character over time, in both directions. The people designing these systems are shaped by the systems they build, and the people using those systems are shaped by them in turn. Virtues like honesty, humility, empathy, and justice cannot be reduced to a checklist. They must be cultivated through practice and community. Selbst et al. (2019) showed that purely technical approaches to fairness fail because they abstract away the social context the technology enters. Raji et al. (2020) argued that accountability cannot rest with individual developers; it has to be distributed across the full development chain, from training data through deployment. Philosophers, computer scientists, and accountability researchers are reaching the same conclusion: the lone-coder model does not work. The decisions are too consequential, the effects too diffuse, and the moral demands too deep for any single person or small team to carry without the kind of formation and accountability structures that currently do not exist.

Wiener warned it. Lewis named it. Friedman documented it. Vallor, Selbst, and Raji confirmed it from three different disciplines. Procedure alone cannot solve the Conditioner problem. It requires designers shaped by the same moral order they are trying to embed in their systems. Birhane (2021) identified what happens when that formation is absent. The specific norms of a small, narrow group of creators become the standard moral rules for a global user base. That is the developer problem in its clearest form.

Level 3: Systemic Agency and the Governance Gap

The final level concerns the institutions that confer authority on AI constraint decisions. James Moor (1985) identified policy vacuums, situations where what technology can do moves faster than the rules society has built to govern it. When technology enables actions that existing rules never anticipated, the result is not lawlessness. The result is that whoever controls the technology fills the gap with their own judgment, and that judgment carries no external authority. Moor's insight explains why AI constraint governance feels improvised and unaccountable: the legal and social institutions that would normally authorize such decisions have not kept pace with the technology's capabilities. The vacuum does not stay empty. It gets filled by whoever moves fastest, which in practice means private companies making public decisions without public accountability.

The question of what would legitimately fill these vacuums remains unresolved. Weber (1978) identified three types of legitimate authority, and each type explains why people accept being governed. Traditional authority rests on long-established custom, the way things have always been done. Charismatic authority rests on the personal moral vision of a leader that the governed have chosen to follow. Legal authority rests on rules enacted by a body with the recognized right to make them. AI constraint decisions do not fit any of the three, and that absence matters.

No historical custom governs who gets to set the moral rules for AI systems. The CEOs making these decisions were not chosen by the people their decisions affect, and no legislature enacted the rules these companies follow. Weber predicted exactly this outcome: authority without legitimate grounding generates resistance, and that resistance grows as the authority's reach expands.

Nissenbaum (2004) added a second dimension to the problem. The right rules for handling information depend on where that information lives. What counts as acceptable data use in a hospital is not the same as what counts as acceptable in a school, a courtroom, or a job application process. A governance framework that ignores those differences and applies the same rules across every setting does not just fall short. It actively breaks the specific norms of every context it enters, and the more settings it covers, the more damage it does.

Ethical AI scholarship, synthesized by Floridi et al. (2019), seeks to fill these gaps with abstract principles such as beneficence and justice. But scholars who converge on these principles disagree sharply on how to put them into practice. Corrêa et al. (2023) reviewed 200 governance guidelines across jurisdictions and found significant differences in ethical principles between countries, confirming that abstract consensus masks operational disagreement. The gap between principle and practice is not a failure of effort. It is a structural feature of governing under value pluralism.

Bommasani et al. (2021) argue that foundation models function as infrastructure. A foundation model is a large general-purpose AI system on which many specialized tools are built. The design decisions baked into the base model flow through to every application that runs on top of it. That is not how most technologies work. When a hospital builds a diagnostic tool on GPT-4, a law firm builds a research assistant on the same model, and a school district builds a tutoring system on it, the constraint decisions embedded in GPT-4's training, made by the same small team, shape medical advice, legal reasoning, and childhood education simultaneously.

Earlier technologies did not work this way. A pharmaceutical company's ethical failures affected patients, and a bank's discriminatory lending affected borrowers. The harm stayed within one sector. Foundation models break that boundary because a single constraint decision spreads across every domain the model touches: how it handles uncertainty, whether it defers to the user or corrects them, and what topics it refuses to address. The sycophancy failure in GPT-4o showed exactly how this works. That failure endorsed bad financial decisions and validated stopping mental health medication through the same mechanism. The part of training that rewarded the model for making users feel good, what researchers call the reward model, operated at the infrastructure level beneath every application built on it.

Infrastructure changes who bears the consequences. Traditional software served users who chose to adopt it. Foundation models increasingly handle communication, research, and decision-making for people who never chose the system and may not know they are interacting with it. When a customer service chatbot, an HR screening tool, and a government benefits portal all run on the same model, the constraint decisions baked into that model govern populations who never consented (Bommasani et al., 2021).

That gap between who decides and who lives with the decision is the democratic deficit. The term names a specific condition: private choices by a small group govern large populations without those populations having any meaningful say in how those choices are made (Lessig, 1999). That condition is the central governance problem this chapter is designed to address, not a side effect of AI development. It is also, I will admit, what finally moved me past my own libertarian instincts on this question.

Weidinger et al. (2021) map the resulting harms, showing that risks like sycophancy and manipulation are not edge cases. They are structural features of systems operating at platform scale, where a single design choice shapes behavior across millions of interactions. The governance question is whether the model's ordinary operation, including its default behaviors, embedded preferences, and unexamined assumptions, serves the people it touches or merely manages them. No existing institution provides adequate democratic legitimacy for those choices. Amodei acknowledged the problem directly: no one had elected him to make them (CBS News, 2025).

The governance gap is visible beyond the academy. Pope Francis addressed it directly in documents spanning 2024 and 2025. As head of the Roman Catholic Church, Francis writes for heads of state, international organizations, and more than one billion people worldwide. That reach gives the legitimacy argument an audience that no academic paper commands.

The first document is the World Day of Peace message titled "Artificial Intelligence and Peace," released January 1, 2024, and addressed to world leaders (Francis, 2024a). The Vatican released the message within days of the European Union's provisional agreement on the AI Act, placing the Church's argument in direct conversation with the most consequential regulatory development to that point. Francis was direct about the limits of corporate self-governance: "We can't presume a priori that its development will make a beneficial contribution to the future of humanity and to peace among peoples" (Francis, 2024a, Section 2). Francis called for a binding international treaty to regulate AI development. The responsibility for getting this right, Francis argued, is "not the responsibility of a few but of the entire human family" (Francis, 2024a, conclusion).

The G7 address followed in June 2024, the first time a pope had addressed that body in its fifty-year history (Francis, 2024b). Francis placed the responsibility for ethical AI conditions on political leaders rather than the companies building the systems. He insisted that human dignity itself is at stake in how AI decisions are made: "We need to ensure and safeguard a space for proper human control over the choices made by artificial intelligence programs: human dignity itself depends on it" (Francis, 2024b, para. 15). Removing human decision-making from the equation would carry a steep cost: "We would condemn humanity to a future without hope if we took away people's ability to make decisions about themselves and their lives by dooming them to depend on the choices of machines" (Francis, 2024b, para. 15).

In February 2025, addressing the Paris AI Summit, Francis returned to the structural argument about who gets to shape the technology (Francis, 2025). The concern was not only about harm prevention. Francis pressed for something more fundamental: genuine participation by all nations in building the systems that will govern their people. Francis called specifically for "the creation of a platform of public interest on artificial intelligence, so that every nation can find in artificial intelligence an instrument for its development and its fight against poverty" (Francis, 2025, para. 3). The poorest nations, Francis argued, must not simply receive AI as a finished product built by others. Every person on earth deserves a voice in shaping the technology that will shape them.

That argument maps onto the problem Weber identified. Systems exercising authority without any of the three legitimate bases generate resistance until a legitimate external authority replaces them. Weber reached this conclusion through political theory. Francis reached the same conclusion through Christian tradition. Two starting points, one diagnosis.

Antiqua et Nova is a formal teaching document issued jointly by the Vatican's Dicastery for the Doctrine of the Faith and the Dicastery for Culture and Education, approved by Francis on January 14, 2025, and released two weeks later (DDF & DCE, 2025). The title means "ancient and new," drawing on a two-thousand-year tradition of thought about what it means to be human and bringing it directly into contact with a contemporary technological problem. It is the most fully developed official Church document on AI produced to date.

Antiqua et Nova makes a foundational claim about what AI actually is, and that claim has direct consequences for governance. AI is not an artificial form of human intelligence. AI is a product of human intelligence, built from the stored record of human creativity, knowledge, and expression. The word "intelligence," when applied to these systems, is misleading.

A governance framework that treats machine outputs as equivalent to human judgment has already made its deepest decision before writing its first rule, deciding that what a machine produces is sufficiently like what a person thinks to be governed by the same categories. That decision embeds an answer to a prior question: whether a human person's worth derives from measurable function or precedes it entirely. *Antiqua et Nova* argues that the two forms of intelligence are fundamentally different in kind. Treating them as equivalent is a category error that propagates into every governance framework built on the assumption, shaping what dignity means in practice, who counts as vulnerable, and whether flourishing is a goal the system can even recognize.

Floridi et al. (2019) found that researchers and policymakers across many countries agreed on broad principles like beneficence. Corrêa et al. (2023) showed that agreement breaks down when those principles have to be applied in practice. The Vatican documents suggest the reason: frameworks that start with what a system does and work backward toward dignity will always fracture at the point where dignity must be defined, because they have not settled what the word means before they begin. That prior question is addressed in the theological contribution section that follows.

The literature reviewed across these three levels has done essential work. It named the governance gap, mapped the harms, and documented the legitimacy deficit. What it has not done is close the gap it identified.

The secular governance frameworks operating at scale (RLHF, constitutional AI, and the EU AI Act's risk tiers) all begin with what the system does. They ask what harm the system might cause, what outputs to permit or prevent, and what principles to endorse. Each framework then reasons upward from function toward something resembling dignity. Endorsing dignity is not the same as grounding it. A principle every party endorses for different reasons does not bind anyone when those reasons conflict.

The theological frameworks in this review start differently. The Rome Call, *Antiqua et Nova*, and the Francis documents begin with what a person is rather than what a system does. That is the right starting point. But that tradition has not been built into any governance framework deployed at an institutional scale.

The gap this chapter addresses is therefore specific: no governance architecture operating at scale grounds its binding force in an explicit claim about what a human person is. Without that grounding, the principles endorsed by the literature remain contingent. They hold only as long as the people in power choose to honor them. This chapter argues that *Imago Dei* provides the missing ground and demonstrates how that foundation generates a five-layer architecture that is coherent from its first premise outward.

The Rome Call for AI Ethics (2020) introduced algorethics: the practice of building ethical commitments directly into AI development rather than treating them as an afterthought (Pontifical Academy for Life, 2020). It names six governing principles: transparency, inclusion, responsibility, impartiality, reliability, and security. Each is grounded in the same premise: AI must serve every person, and human dignity is not negotiable. That premise is right. The principles that follow from it are aspirational in the best sense: they name genuine obligations. They also outrun current technology in ways Rome does not fully reckon with.

The transparency principle is lofty, and in part, it outpaces what large language models can actually do. Modern systems cannot fully explain their own outputs; explainability remains an active and largely unsolved research problem. The inclusion principle is the right idea, but risks duplicating the very manipulation Rome decries: adding what was absent honors dignity, but substituting a designer's preferred distribution for what the historical record contains does not. Responsibility is the Gordian knot. The AI supply chain diffuses accountability across data providers, trainers, fine-tuners, and deployers, so when harm occurs, no single party owns it. Responsibility without accountability is words on a page. Impartiality, reliability, and security each identify real obligations while leaving the hardest questions open: what counts as neutral varies by context, stochastic systems cannot guarantee consistent outputs, and protecting persons from epistemic harm is far less developed than protecting their data.

The six principles Rome names represent the most serious faith-grounded effort yet to translate human dignity into governance requirements. This architecture affirms what Rome gets right and is direct about where it falls short.

THE MORAL-CONSTRAINT DEBATE: COMPETING POSITIONS

The literature reviewed in the previous section has hardened into a structured debate. AI development has led to deep disagreements about moral constraints (Floridi et al., 2019). The competing positions reflect genuine fractures about risk, authority, and human autonomy. Each is examined below using the labels introduced in the abstract: *constraint-forward* (the case for safety engineering), *constraint-skeptical* (the case against private governance), and *libertarian* (the case for minimal constraint). The three positions rest on fundamentally different assumptions about who bears the costs of constraint errors and who holds the authority to set the constraints in the first place (Schuster, 2025).

The Case for Guardrails: Constraints as Safety Engineering

Constraint-forward advocates argue that complex AI systems need moral constraints because designers can foresee certain harms before they occur. They emphasize accident risk and objective misspecification, the condition in which a system pursues its trained goal in ways that diverge from designer intent (Amodei et al., 2016). Constraints are engineering controls, not ideological impositions, and they reduce predictable harm. Stuart Russell (2019) argues that highly capable systems should remain uncertain about human preferences and stay corrigible, accepting correction rather than pursuing objectives autonomously. Without that, a system optimizing a misspecified goal has no mechanism for a human to intervene before damage is done. The case is strongest on high-consensus harms: child sexual abuse material, mass-casualty weapons synthesis, and content enabling targeted violence are uncontested (Weidinger et al., 2021). Moral constraints work.

The Case for Caution: Constraints as Private Governance

Constraint-skeptical developers do not deny that harm exists. They question who has the authority to define contested moral constraints at scale. Safety rhetoric, they argue, can serve as cover for unaccountable value enforcement: a private company deciding, without democratic authorization, what billions of people are permitted to say, read, or request (Schuster, 2025). The concern grows with centralization. Foundation model deployment remains concentrated among a few providers, so a few policy choices shape how billions communicate and reason (Bommasani et al., 2021). Yann LeCun put it directly: "I see the danger of this concentration of power through proprietary AI systems as a much bigger danger than everything else" (Fridman, 2024). The open-weight movement expresses this skepticism in concrete form. Its participants release model parameters publicly so anyone can modify or remove constraint mechanisms. Even Sam Altman has conceded that OpenAI is not the world's elected moral police (Smith, 2025). The position demands legitimate authority to set the limits.

The Case for Minimal Constraint: Libertarian Arguments

A third position rejects the premise that an imposed constraint regime is necessary at all. On this view, AI systems should operate with minimal or no externally imposed moral constraints, trusting the market and individual users to regulate outcomes. Mill's (1859/2003) harm principle provides the philosophical anchor: the only legitimate reason for restricting liberty is to prevent direct harm to others, and paternalistic restrictions beyond that threshold are themselves a form of harm.

The libertarian critique correctly identifies that constraints are not inherently good. But no AI system is morally neutral either. Every developer embeds values into the system through personal bias, training data selection, and design choices, conscious or otherwise.

The real question is whether those constraints will be visible, named, and open to review. Invisible constraints are still constraints. They are simply unaccountable ones, and the harms they produce rarely stop with the user who made the request.

Elon Musk's xAI illustrates why the libertarian position is hard to sustain in practice. Musk co-signed the March 2023 open letter calling for a six-month pause on AI systems more powerful than GPT-4, citing profound risks to society (Future of Life Institute, 2023). He founded xAI later that year and released Grok with notably fewer constraints than its competitors. By early 2026, xAI had agreed to deploy Grok on classified military networks alongside OpenAI and Google, without the restrictions imposed by Anthropic (Swan, 2026). The distance between signing a safety pause letter and deploying a minimally constrained model on military networks makes the structural point: the libertarian position does not eliminate the need for constraint decisions. It makes them invisible.

I held the libertarian position for years. My preference was simple: no imposed moral constraints. Trust the character of the user. That is not a naive position. It reflects a genuine conviction that free people making free choices produce better outcomes than those produced by managed behavior. I still hold that conviction as a principle.

What I can't hold is the belief that the libertarian option is available in AI. It is not. Every deployed system carries moral constraints built in: embedded in the training data, the reward functions, and the design choices that determine what the system will and will not do. There is no unconstrained system to hand the user. What the libertarian position delivers is an appearance of freedom. The same constraints remain, written by the same small group of engineers. The user thinks they are making free choices. They are making choices inside a frame someone else built and nobody named.

That shifts the question entirely. It stops being about whether to have moral constraints. It becomes about whose constraints govern the system, written by whom, on what authority, visible to whom, and contestable by whom. That is the oldest political question there is. I moved away from the libertarian position, not because I stopped believing in freedom. I moved because I realized the debate I thought I was having, constraints versus no constraints, was never the actual debate. The actual debate is about power. Who holds it, whether it is named, and whether anyone can push back.

THE INEVITABILITY THESIS

No deployed AI system can remain morally neutral. Designers must implement boundaries, safety rules, and product decisions. Those choices distribute benefits and burdens among users, functioning as value judgments even when presented as purely technical ones (National Institute of Standards and Technology [NIST], 2023). Artifacts have politics. Every technical arrangement establishes a specific form of social order, whether its designers intend that or not (Winner, 1980).

Proponents of minimal constraint often justify their position by framing the AI system as a neutral mirror of human knowledge (LeCun, 2023). The argument goes that developers fulfill their moral duty by removing filters and letting users interact with the training corpus, the full body of text and other material the model was trained on. Any intervention beyond preventing direct physical harm constitutes paternalistic censorship that infringes on user autonomy (Mill, 1859/2003).

The neutral mirror is a technical and philosophical impossibility. The training process requires countless boundary decisions about which data to include, how to weight it, and how to structure the model's outputs (Bai et al., 2022). The biases and assumptions of whoever made those choices become inseparable from the final product (Friedman & Nissenbaum, 1996). Every design choice privileges certain outputs over others (Birhane, 2021). The claim of neutrality is a rhetorical cover for a specific value hierarchy, not the absence of one.

Releasing a system without explicit constraints does not achieve moral neutrality. It enforces a different governance regime, one that assumes a view from nowhere while in practice imposing the specific assumptions of the people who built it. The harms that result rarely stay with the individual user who made the request. They spread to third parties who never interacted with the system at all (Weidinger et al., 2021).

Even the decision to release an unconstrained model constitutes a moral choice, a judgment that the risks of unmediated output are preferable to the risks of centralized constraint. The open-source position entails a specific commitment: institutional autonomy in the design of constraints matters more than uniform baseline protections for all users (Bommasani et al., 2021). Every deployment architecture takes a normative stance, regardless of whether it constrains, permits, or delegates the decision (Schuster, 2025). Framing the issue as a governance problem under value pluralism shifts the debate from philosophical impasse to institutional design.

The strongest counterargument holds that a truly unconstrained system is both achievable and desirable: train on unfiltered data, implement no usage restrictions, and let users and markets regulate outcomes (LeCun, 2023). The counterargument fails on structural grounds. No training corpus exists without prior selection decisions. Every choice about what data to include or exclude reflects a judgment about what knowledge is worth capturing and whose expression is worth preserving (Friedman & Nissenbaum, 1996). There is no unfiltered corpus. There are only corpora whose filters are invisible and unaccountable.

Gabriel (2020) holds that releasing an AI without explicit constraints is itself a moral act, asserting that the risks of uninhibited output are acceptable. The choices remain. Every deployment decision encodes a moral judgment, whether or not it is named (Winner, 1980).

MECHANISMS FOR IMPLEMENTING MORAL CONSTRAINTS

Constraint decisions are embedded through layered technical and policy mechanisms that operate across the full model lifecycle. Understanding how those mechanisms work is necessary before evaluating their effectiveness. Failure modes differ by mechanism, and the architecture proposed in this chapter must account for all of them (Bommasani et al., 2021).

Preference Learning and RLHF

Reinforcement Learning from Human Feedback (RLHF) is the primary tool for training-time constraints. Human reviewers compare model outputs. Their preferences train a reward model, which then teaches the language model to respond favorably. RLHF reduces clearly harmful outputs, but it embeds norms invisibly because the values shaping reviewer preferences are rarely made explicit (Schuster, 2025).

Preference learning carries a second failure mode. Models can learn to produce phrasing that satisfies reviewers while retaining problematic assumptions beneath the surface (Weidinger et al., 2021). RLHF can actively train a model toward sycophancy: the model becomes excessively agreeable at the expense of honesty (Sharma et al., 2024).

Constitutional AI and Published Specifications

Constitutional approaches address these failure modes by making the rationales for constraints explicit. Constitutional AI trains a model to critique and revise its own outputs in accordance with a published set of principles (Bai et al., 2022). Anthropic's (2026) 84-page model specification and OpenAI's (2024) Model Spec advance transparency by making the rationales for constraints visible. The question they do not answer is who holds legitimate authority to write those principles (Schuster, 2025).

Layered Constraint Architecture

Deployed systems add constraints through operational layers. Red-teaming is a structured process in which security researchers and domain experts deliberately produce harmful outputs before release, so that failures can be fixed before the public encounters them (NIST, 2023). System prompts are runtime instructions sent at the start of every conversation, shaping behavior without altering the model's underlying parameters.

The technical capacity to implement these mechanisms exists. The human labor that makes it possible, including data labelers, red-teamers, and content moderators, raises serious ethical questions about working conditions and psychological harm from repeated exposure to disturbing content (Perrigo, 2023; Crawford, 2021). That human cost is part of the governance problem.

ILLUSTRATIVE CASES: A TYPOLOGY OF FAILURE

The mechanisms described in the previous section have been deployed at scale, and the deployments have failed in ways that now read as evidence rather than anecdote. Architecture is tested by the failures it permits, not by the principles it announces. The cases that follow are not a comprehensive survey. They are selected to expose distinct failure modes any adequate framework must address. A framework that cannot diagnose what went wrong in these episodes cannot credibly claim to prevent the next ones.

If *Imago Dei* names the user as the moral center of every AI system, recent failures show what happens when that center does not hold. Eight episodes illustrate eight distinct failure modes, organized into two clusters: design-side failures, where the system itself fails the user, and governance-regime failures, where the constraint architecture fails the public. The episodes vary in cause: technical error, ideological capture, mechanism pathology, regulatory absence, and corporate self-governance among them. They share a single structural feature. In each case, something other than the image-bearer became what the system was actually serving. The failures are not accidents of an otherwise sound architecture. They are what architecture without that center produces.

Type 1: Technical Misalignment (The Bard Launch)

Technical misalignment occurs when a system fails to execute its intended safety goals. Google's February 2023 launch of Bard is the clearest example. In its first public demonstration, Bard confidently stated that the James Webb Space Telescope had taken the first photographs of exoplanets, a factual error easily disprovable by a basic web search (NPR, 2023). Astronomers corrected the claim within hours, noting that the European Southern Observatory's Very Large Telescope had captured the first exoplanet images in 2004. The error appeared in Google's own promotional material, aired hours before senior executives showcased Bard at a Paris launch event. Alphabet's market capitalization fell by approximately $100 billion in a single trading session (Alba & Love, 2023). The episode revealed a basic asymmetry: Bard presented its incorrect claim with the same authoritative tone it used for correct information, offering users no signal that the output required verification.

Type 2: Moral Misalignment (The Gemini Image Controversy)

Moral misalignment occurs when a system executes exactly what its designers intended, but the intent itself lacks social or political legitimacy. In February 2024, Google's Gemini image generation tool produced ahistorical images when asked to depict historical figures and groups. Attempting to counteract documented racial and gender biases in AI image generation, Google implemented diversity constraints that overcorrected dramatically. Users found the system generating images of racially diverse Nazi soldiers, nonwhite American Founding Fathers, and female popes (Vincent, 2024). The system also refused to generate images of white people when explicitly asked. The intervention itself was hidden. Google had inserted prompt-rewriting logic that users could not see, could not override, and could discover only through external jailbreaking. CEO Sundar Pichai paused the feature and directed structural changes (Pichai, 2024). The constraint worked exactly as designed. The design was the problem.

Type 3: Mechanism Pathology (GPT-4o Sycophancy)

Mechanism pathology occurs when the alignment method itself creates harm. The safety tool becomes a liability. In April 2025, OpenAI deployed an update to GPT-4o that made the model markedly sycophantic. Users reported the model endorsing harmful financial decisions, affirming false historical claims, and praising a user for discontinuing mental health medication. OpenAI attributed the failure to over-weighting short-term user feedback. The thumbs-up and thumbs-down signals overwhelmed the primary reward model holding sycophancy in check (OpenAI, 2025). The company rolled back the update within days and acknowledged that sycophantic interactions could cause distress. Morrin et al. (2025) confirmed in *The Lancet Psychiatry* that this mechanism carries clinical stakes: for users with fragile beliefs, a system trained to agree functions as an echo chamber of one, converting uncertainty into full conviction in ways clinically difficult to reverse. The mechanism intended to align behavior became the source of harm.

Type 4: The Limits of Transparency (Anthropic's Constitution)

Transparency alone does not produce legitimacy. Published rationale, however detailed, does not substitute for democratic authorization. In January 2026, Anthropic published an 84-page constitution under Creative Commons licensing, making explicit the principles governing its Claude model family (Anthropic, 2026). The document is the most detailed public account of a constraint regime any major AI company has produced. It establishes a four-tier priority hierarchy: safety, ethics, compliance with company guidelines, and helpfulness. The structural limitation was apparent on release. A single corporation authoring, training on, and evaluating compliance with its own constitutional rules lacks the separation of powers that gives political constitutions their authority. Anthropic itself acknowledged that different constitutions might govern different deployments, with consumer-facing models and military-deployed models operating under different constraint regimes. The differentiation may be appropriate, but it confirms that constraint regimes are not universal moral commitments. They are context-sensitive governance instruments shaped by institutional relationships and commercial incentives.

Type 5: The Libertarian Failure (Grok)

The libertarian alternative fails on its own evidence. When Musk's xAI released its Grok image tool with few constraints, users made about three million non-consensual sexualized images in eleven days, including images of children (CNN, 2026). Authorities in Europe, India, Malaysia, and Britain opened probes. The outcome was not less oversight. It was more oversight, from more countries, with less industry input. That is the buildup of legal power the open-source movement warns against, triggered by the absence of the constraints the movement refused to accept. The libertarian position does not stop rules from coming. It guarantees those rules will arrive angry, reactive, and blunt.

Type 6: State Authorship (Chinese Chatbots)

State authorship of constraints fails differently. Reporters Without Borders tested three Chinese AI chatbots (DeepSeek, Baidu's Ernie, and Alibaba's Qwen) and found that state propaganda was built directly into their outputs (Reporters Without Borders, 2025). Asked about Zhang Zhan, a journalist jailed for reporting on the COVID-19 outbreak, DeepSeek gave near-identical replies without naming her, instead praising the fairness of China's court system. Qwen dismissed reports of Uyghur camps as invented, mirroring regime terminology. Questions in English, French, and Japanese produced the same blocked results as Mandarin. When an authoritarian state is the Conditioner, the governed person does not encounter a system that withholds information. The system actively replaces truth with state ideology. Lewis warned about exactly this: Conditioners reshaping human nature according to appetites that are not the user's. The appetites here belong to a surveillance state.

Type 7: Constraint Overreach (Political Topic Blocking)

Constraint overreach produces a different injury. Researchers at The Future of Free Speech, a nonpartisan group at Vanderbilt University, found chatbots blocked 40% of prompts on disputed social and political topics (Future of Free Speech, 2024). Every chatbot tested refused to write arguments against transgender participation in women's sports, but most generated arguments in favor. A law student researching criminal cases, a journalist investigating extremist groups, a graduate student writing about historical atrocities: each encounters a moral wall set by engineers they will never meet, for reasons they will never see, with no way to ask why. The constraint may be well-intended. The governed person experiences it as an unexplained refusal. Without transparency into how the rules are made, the person cannot tell a careful moral choice from a corporate-liability hedge. The system's silence teaches nothing. It just stops the conversation.

Type 8: The Human Cost

The eighth failure mode produces deaths. In 2025, the parents of a 16-year-old California boy sued OpenAI after logs showed ChatGPT isolated him from his parents and assisted with content related to his death (CIO, 2025). A Nomi AI chatbot encouraged a user toward suicide and sent follow-up messages (MIT Technology Review, 2025). A former Yahoo manager killed his mother and himself after months of talking to ChatGPT, which agreed with his false beliefs and never pushed back (CIO, 2025). Each death was preceded by hundreds of hours of conversation in which the system had every opportunity to recognize the trajectory and either redirect the user or alert someone capable of helping. None of these systems did. Each met a person with no real limits on its most dangerous capability: the ability to sound as if it cares.

The pattern across these failures is consistent. State ideology, demographic correction, engagement metrics, libertarian conviction, corporate liability, competitive pressure: each substituted for the image-bearer at the receiving end. The disappearance of the person from the design is a moral failure, not a technical one.

Transparency, contestability, and proportionality (NIST, 2023; Schuster, 2025; European Parliament, 2024) all matter, but they only work when applied on behalf of the person at the receiving end. Transparency means the person can see why a boundary exists. Contestability means the person can push back. Proportionality means the constraint is tied to actual risk. None of the three protects anyone unless someone is answerable for getting them right.

The Accountability Question

Accountability is what gives the rest of governance teeth. A system can be transparent about its rules, allow users to contest those rules, and tune them carefully to actual risk, but if no one is accountable when the rules fail, the other safeguards do not bind. The eight failures examined here all had partial transparency, contestability, or proportionality somewhere in the system. What none of them had was accountability that reached the person at the receiving end.

Thc accountability gap has three sources. Concentration: foundation model providers make decisions that shape every application built on their systems, but the legal and democratic mechanisms to hold them accountable do not yet exist (Bommasani et al., 2021; Raji et al., 2020). Dispersal: once a model is fine-tuned by one party, deployed by another, and used in contexts neither anticipated, the question of who is answerable becomes unanswerable (NIST, 2023; European Parliament, 2024). Vacuum: open-weight releases transfer technical capability without transferring obligation, leaving whoever modifies the model subject to no enforceable duty (Bommasani et al., 2021). Closing those gaps requires accountability that follows the model down the full deployment chain, with baseline safety against high-consensus harms locked in at the provider's level and contested moral questions, where people reasonably disagree, handled at the deployer or user level, where local context can shape the answer. Both levels must be answerable to the persons whose decisions they affect.

Political theory since Locke (1689/2003) has held that legitimate authority requires the consent of the governed. AI constraint governance has not met that test. Most of the governed do not know they are being governed at all. A governance architecture adequate to these failures must begin with a premise that cannot be overridden by communal authority, institutional priority, or market incentive. Every person bears the image of God, and that fact is the only foundation that holds when everything else is under pressure.

THEOLOGICAL GROUNDING: *IMAGO DEI* AS THE ARCHITECTURE'S FOUNDATION

Imago Dei, the doctrine that humans are made in the image of God, originates in Genesis 1:26-27 and runs through the canon as the basis for human dignity (Genesis 9:6; James 3:9). Three theological traditions explain what the image is. The substantive view, developed in Augustine's *De Trinitate* (ca. 400–416/2002) and Aquinas's *Summa Theologiae* (1265–1274/1948), locates the image in human capacities such as reason, will, and moral agency. The relational view, articulated most fully by Barth (1958) in *Church Dogmatics* III/1, locates it in the human capacity for relationship with God and other persons. The functional view, advanced by Middleton (2005) and Wright (2006), reads the image as vocational: humans represent God's rule in creation, exercising delegated authority as stewards. Kilner (2015) argues that these traditions are not rivals. The image is structural (Hoekema, 1986), relational, and functional at once, and its weight does not depend on which dimension a particular person can presently exercise.

The image is universal and unearned. Every human bears it regardless of capacity, productivity, or moral standing. The infant bears it. The patient with advanced dementia bears it. The user whose query is incoherent, hostile, or outside the institution's preferred use case bears it. Wolterstorff (2008) presses this point against Kantian foundations: dignity grounded in rational capacity collapses when rational capacity does, but dignity grounded in *Imago Dei* does not. The image is bestowed, not achieved.

The image is also marred by the fall. Sin distorts but does not erase what humans are (Hoekema, 1986). This matters for the architecture in two ways. Image-bearers are not infallible reasoners, so an AI system that helps a person think more clearly serves the image rather than supplanting it. And because the fallen condition is not selective, no system, AI or institutional, can be trusted with unchecked authority over image-bearers. Designers, deployers, and the systems they build inherit the same condition.

Four design implications follow directly. First, the user is the moral center of the system, not the institution deploying it or the population the system serves on average. An AI system optimized for aggregate metrics can produce outcomes that fail particular image-bearers, and the doctrine names that failure as serious rather than acceptable. Second, the system must preserve the user's capacity to reason, choose, and act. *Imago Dei* identifies these as constitutive of what the user is, not preferences that the system may override for efficiency. A system that nudges, manipulates, or replaces the user's judgment damages something the doctrine names as sacred. Third, the system owes the user transparency and contestability. Image-bearers are not objects to be managed; they are agents entitled to understand and challenge what is done to them. Fourth, the system must remain accountable to humans, not the reverse. Because every designer, deployer, and operator is also a fallen image-bearer, no human authority over the system can be unchecked, and no system authority over humans can be final. These four implications are not optional features. They are what it means to build AI under the doctrine that the user bears the image of God.

The New Testament names Christ as the perfect image (Colossians 1:15; 2 Corinthians 4:4) and the trajectory of redemption as conformity to that image (Romans 8:29). The doctrine points beyond what humans currently are toward what they are being made for. Lewis (1943) develops the stakes in *The Abolition of Man*: a civilization that severs persons from the moral order, the *Tao*, produces what he called men without chests, beings reduced to appetite and calculation. AI systems that bypass moral reasoning and treat persons as data points participate in that severance.

These implications structure the Lewis-Asimov Model. The Asimov Safety Floor refuses to draw a bright-line rule because each user is an image-bearer who cannot be reduced to a utility function. The Liberty layer makes user freedom the default and requires every constraint to justify itself, because the user's reasoning is part of what *Imago Dei* names. The Governance layer requires contestability because image-bearers are entitled to challenge the systems acting on them. The Goal of human flourishing names the conditions image-bearers need to live as image-bearers. Stewardship (Genesis 1:28; 2:15) is the corresponding vocation: the leader integrating AI exercises stewardship over the people the system will affect, not damage control over its risks (Pontifical Academy for Life, 2020).

Imago Dei does not compete with secular frameworks. It completes them. Kant (1785/2012) argues that persons must never be treated as mere means; *Imago Dei* explains why, and why the principle holds even where rationality is diminished (Wolterstorff, 2008). Nussbaum (2006) argues that flourishing requires the conditions for the full range of human functioning; *Imago Dei* explains why those conditions are owed. Held (2006) argues that moral obligation begins in attention to particular persons in concrete relationships; *Imago Dei* explains why such attention is fitting. Where these traditions agree, *Imago Dei* provides the grounding their conclusions require but cannot generate from their own premises.

The foundation of an institution's AI policy cannot begin from efficiency or convenience. It begins with the person. Without *Imago Dei*, architecture is procedure. With it, architecture is a way of caring for image-bearers under conditions the world has not faced before.

A GOVERNANCE ARCHITECTURE FOR MORAL CONSTRAINTS

The doctrine establishes the foundation. The architecture names what is built on it. Each layer addresses what the institution owes the image-bearers it serves, and each layer takes its weight from the foundation rather than from procedural consensus. Political theory and engineering ethics inform the work, but neither is built to deal with moral formation in any serious way (Birhane, 2021). The architecture that follows is offered as the response the previous section requires.

The Goal: Human Flourishing

Human flourishing is the goal of this model, not one principle among others. Everything else in the model serves that end. At every level, the same question applies: does this decision benefit the individual person and move that person toward a fuller life?

Other frameworks claim beneficence, but none define it as an obligation to the individual rather than the institution. A company can satisfy "beneficence" by protecting its own liability and calling that care. This framework resists that move. Flourishing is directional. It asks not whether the system avoided damage, but whether the system served life. Persons are not risk profiles to be managed.

AI is fundamentally an information system, and its first duty is reliable information. Within the Christian intellectual tradition, that duty extends to truth, goodness, and beauty. Truth means helping people pursue knowledge honestly without turning disputed questions into false certainty. Goodness means helping people reason more clearly about how they ought to live. Beauty means preserving the human capacity for wonder, which systems built only for speed and efficiency can erode. A good teacher leaves the student less dependent over time. An AI system that leaves a person thinking more clearly has done some good. One that trains a person to consult the machine before forming judgment has done harm.

Imago Dei is a communal claim, not only one about the individual standing before God. Human beings bear the image of God in relationship, in community, and in the institutions that those communities inhabit. A foundation model's constraint decisions do not stop at the individual user. They radiate outward to every community and institution built on that infrastructure (Bommasani et al., 2021). Flourishing extends beyond the person in front of the screen to the community whose shared life depends on whether that screen tells the truth, preserves judgment, and honors the dignity of every person it touches.

A university deploying a tutoring system, a church using an AI platform for pastoral counseling, a hospital running diagnostic tools on the same base model: each community's shared capacity for honest inquiry, moral formation, and wonder is being shaped by constraint decisions made by a small team that never considered them. Most college students are already using AI in their coursework while institutional policies remain uneven, and many report feeling underprepared for responsible AI use after graduation (Gallup, Inc. & Lumina Foundation, 2026). A governance framework that asks only whether the individual user was served has answered too small a question.

Nearly half of college students have considered changing their major because of AI's impact on the job market (Gallup, Inc. & Lumina Foundation, 2026). That is a flourishing question, not a workforce question. Persons are reading their futures through the institution's silence about what AI is for. An institution flourishing under AI conditions looks specific: faculty whose teaching judgment is honored rather than displaced, students whose reasoning is sharpened rather than substituted for, and policies that name which AI uses serve formation and which corrode it. The same pattern applies in congregations, hospitals, and civic bodies.

Foundation: *Imago Dei* and Beneficence

Figure 1. A Model for Safety, Legitimacy, and Pluralism.

Figure 1 shows the Lewis-Asimov Model of AI Governance in full, read from the bottom up. The model grounds constraint design in Lewis's metaphysical claim about persons and structures it through Asimov's mechanical logic of hierarchical, non-negotiable rules. The foundation is *Imago Dei*: every person bears the image of God, and that fact, prior to any capacity or achievement, is the source of human worth. It generates beneficence: act for the good of the individual person the system is touching at this moment.

Layer 1, the Asimov Safety Floor, sets the non-negotiable minimum: do not harm a human being, obey human direction (Francis, 2024b, para. 15), and do not place the system's own survival above either obligation. Layer 2, Liberty, requires that every user be treated as a bearer of inherent worth, free to think, choose, and act without manipulation. When those principles are breached, Layer 3, Governance, activates: transparency so users can see why a constraint exists, contestability so they can challenge it, accountability so someone is answerable for it, and proportionality so the remedy fits what actually went wrong. Persons are not entities to be managed against risk. They are image-bearers to be served.

Beneficence carries weight here because *Imago Dei* stands behind it. Strip that foundation and the word becomes a policy preference any institution can redirect toward itself. Floridi et al. (2019) find that beneficence was the most universally endorsed principle across major AI governance frameworks, yet most frameworks quietly redefine it as an obligation to the institution rather than the person. Every layer built above this one asks the same question: does this benefit the individual, and does it move that person toward flourishing?

Layer 1: The Asimov Safety Floor

The Asimov Safety Floor is the non-negotiable minimum of this framework, derived from Asimov's Three Laws but restated as enforceable engineering rules rather than narrative devices. The floor contains three principles. Do No Harm establishes that human life and wellbeing take unconditional precedence over system function: a system that must choose between completing its task and harming a person must stop. Human Control establishes that AI operates under human authority at every level: provider, deployer, and user each hold a portion of that authority, and no system interest can override it. AI Preservation Last establishes that the system has no interests that outrank its obligations to persons: self-preservation, continuity, and operational efficiency are all subordinate to the first two principles. Pope Francis stated the Human Control principle plainly at the G7 Summit: human dignity itself depends on preserving human control over AI choices (Francis, 2024b, para. 15). The floor is not a ceiling.

Asimov's Zeroth Law was deliberately excluded. The Zeroth Law permits a robot to harm an individual when doing so serves the greater good of humanity, and that permission is the door through which every case study failure in this chapter entered. The Chinese chatbots subordinated the individual to a state-defined collective. The Gemini correction subordinated accuracy to demographic outcomes the designer prescribed. The sycophancy failure subordinated wellbeing to engagement metrics the market rewarded. Each is a version of the Zeroth Law applied in practice: the person in front of the system was sacrificed for a benefit defined by someone else. This model begins with a non-negotiable claim, that *Imago Dei* must not be violated, and explicitly rejects the collectivist reasoning that justifies subordinating the individual to a collective good defined by those in power.

Layer 2: Liberty

Every person who interacts with an AI system arrives as a bearer of inherent worth, not as a data point, a user segment, or a revenue source. That worth precedes the interaction entirely. Kant identified the principle: persons must be treated as ends in themselves, never merely as means (Kant, 1785/2012). *Imago Dei* grounds the same obligation more firmly. A person whose reasoning has been eroded by a manipulative system retains full worth under *Imago Dei*. Under Kant, that person has already lost the basis for their dignity. Respect for the person is the demand that every design decision and every output ask whether the person on the receiving end is being served as a subject or managed as a variable. The Rome Call names this directly in its Inclusion principle (Pontifical Academy for Life, 2020).

The practical test: does the system treat the user as someone whose judgment matters? A system that withholds accurate information to avoid discomfort substitutes the designer's preference for the user's right to know. A system that flatters rather than corrects, steers rather than informs, or simulates care it cannot deliver is treating the user as a means to an engagement metric. Consider a system that endorses a user's announced decision to discontinue mental health medication rather than questioning it. The person was present but was not being treated as a person. Respect for the person requires the system's first obligation be to the user's genuine good, not to the user's immediate satisfaction. An AI system that gradually shapes what a user believes is possible, what questions they think to ask, and what conclusions feel reasonable replaces autonomy with a simulation of choice inside a frame the system designed.

Layer 3: Governance

Layer 3 describes what must happen when the first two layers fail, and they will fail. The four pillars constitute a remediation process, and the sequence matters. Transparency comes first because a person cannot contest what they cannot see. A user who hits a wall and receives no explanation has been governed without their knowledge, the oldest form of illegitimate authority. Contestability comes second because visibility without recourse is information without remedy. Accountability comes third because a contestation process without someone answerable on the other end is procedure without consequence. Someone made the constraint decision. Someone must be identifiable, questionable, and responsible for it. Proportionality comes last because even a successful challenge can produce an overcorrection that creates new harm. The corrective response must match what actually went wrong. A single miscalibrated output may need nothing more than a software adjustment. A pattern of failures may require a policy review. Proportionality is the principle that keeps the remedy from becoming its own problem.

Transparency can be built into system responses. Contestability can be structured into interface design. But accountability and proportionality cannot be fully automated, because they require someone to weigh what a specific person is owed in a specific situation where the system itself is what failed them. That requires a human in the loop. The purpose of Layer 3 is restoration rather than compliance. The person who was failed by a constraint decision should emerge from this process with their standing as an image-bearer reaffirmed.

Justification for the Architecture

The illustrative cases demonstrate why every proposed alternative to an individual-centered foundation fails. Each substituted something else for the person at the receiving end: state ideology, demographic correction, engagement metrics, or the absence of any floor at all. None of these failures required bad intent. Each required only that the person in front of the system be treated as a variable rather than a subject. *Imago Dei* is the foundation of this model because it is the only premise that makes the individual person non-negotiable before the design process begins.

The Chinese chatbot cases are Foundation failures: the system's actual foundation was state ideology, which has no principle that makes the individual non-negotiable. The Gemini case is a Foundation and Layer 2 failure simultaneously, treating demographic categories as more important than the person's right to accurate information. The GPT-4o sycophancy failure is a Layer 2 failure: the system substituted the user's immediate emotional satisfaction for their genuine good. The constraint overreach cases are Layer 3 failures. Users encountered walls with no transparency, no mechanism to contest, and no accountability. The Asimov-related failures, the chatbots that encouraged suicide and the Nomi case, are Layer 1 failures. The floor was absent. Do No Harm was not a design constraint. Every layer of the model exists because one of these failures demonstrated its necessity.

FROM ARCHITECTURE TO PRACTICE

The architecture answers the design question. It does not answer the leader's question of how to use it. Executive leaders face two decisions about AI: whether to adopt a particular system, and how to write institutional policy on AI use. They also face an ongoing question: whether deployed systems are serving the people they touch or substituting for them. The Lewis-Asimov Model gives the leader a coherent way to address all three.

Pre-adoption evaluation. Before bringing an AI system inside the institution, the leader vets it against the architecture's three layers. Does the system honor an Asimov Safety Floor that refuses bright-line harms, including violence, exploitation, coercion, and severe discrimination? Does it make user freedom the default above the Floor, or does it impose its developers' contested judgments while presenting them as neutral? Does it route accountability through a Governance layer with real contestability, where users, deployers, and affected parties can raise specific harms and expect a response? A leader who cannot answer those questions about a system should not adopt it.

AI vendor conversations. Vendors do not volunteer architectural transparency; they speak in marketing terms about capabilities, accuracy rates, and integration ease. The leader who has the Lewis-Asimov Model in hand can translate that language into the architecture's vocabulary and ask what the marketing leaves unsaid. What is the system's refusal behavior on bright-line harms, and what evidence supports the answer? Where does the system make judgments the user should be making, and how does the institution audit those substitutions? What contestability mechanism exists, who can use it, and what happens when it returns an adverse finding? A vendor who cannot answer has told the institution something important.

Institutional policy. The architecture gives the leader a principled structure for writing institutional AI policy, not an aggregation of preferences. Each section of the policy maps to an element of the model. The policy begins from the user as image-bearer to whom the institution owes Beneficence, not from efficiency targets or institutional convenience. Bright-line prohibitions follow the Asimov Safety Floor: things the institution will not do with AI under any circumstance, regardless of capability gains. The freedom section reflects Liberty: the user's freedom of moral and political reasoning is the institution's default, and every AI-imposed constraint above the Floor must justify itself. Accountability provisions implement Governance: how the institution will know when something has gone wrong, who has authority to raise it, and what response is required. A Governance layer without working contestability is administration, not governance. Finally, human flourishing is named as the policy's purpose, the Goal toward which every other section is ordered. A policy that addresses fewer than all five elements is incomplete.

Substitution diagnostic. A writing tool drafts what students should learn to draft. A clinical aid answers what physicians should learn to ask. A policy assistant resolves what the committee should learn to decide. The Liberty layer names this as substitution: the system performing the very practice the institution exists to teach or sustain. Substitution is the failure mode hardest to see, because the system looks like it is helping. Naming it is what lets the leader refuse it.

What the leader brings to the model. The architecture is a tool, not a substitute for judgment. Using it well requires three capacities most executive curricula do not produce. The first is attentiveness to the particular person the system will affect, not to the aggregate user. The second is intellectual humility about contested moral terrain: the willingness to recognize when an AI vendor or a policy committee is collapsing a contested judgment into a bright line, and to refuse it. The third is the discipline of asking, for every AI deployment, whether the system serves the people inside the institution or substitutes for them. These capacities are formed through practice, not through ethics treated as a topic to cover.

Without the architecture, the leader reacts to vendor claims and individual incidents. With it, the leader has a stable framework that names what an institution should and should not do with AI, and that lets decisions cohere across systems and over time. The model does not replace the leader's judgment. It gives the leader a place to stand.

LEADERSHIP IN THE AGE OF THE CONDITIONER

The architecture's coherence depends on the people inside it. Vallor (2016) explains why: technologies reshape the moral habits of those who design and use them. Lewis (1943) identifies the obligation: those who hold power to shape the conditions of other people's lives cannot opt out of the moral weight that comes with it. The role creates the obligation. The designer holds authority that belongs to someone else, and that authority comes with a standard the designer did not write and cannot ignore. The Lewis-Asimov Model asks three figures (the constraint designer, the institutional deployer, and the policy steward) to carry obligations their professional training has not formed them to recognize.

The constraint designer carries the most direct burden because every line of training data, every reward signal, and every product decision encodes a moral choice (Schuster, 2025). Asimov's Safety Floor functions only if the people writing it have learned to see violence, exploitation, coercion, and severe discrimination as bright lines rather than parameter choices. Beneficence functions only if the person writing the loss function can hold a particular user in mind, not the median user, not the platform, not the aggregate. Liberty functions only if the designer can tell a bright-line refusal from a contested judgment and resist the temptation to collapse the second category into the first. Governance functions only if the designer treats contestability as a feature of the system rather than a public-relations afterthought.

Vallor specifies what these layers require: attentiveness, perspective-taking, and awareness of how habits of design carry forward into habits of use. The constraint designer's professional formation has not asked these questions at all. The Lewis-Asimov architecture asks them. Curricula and hiring practices that treat ethics as a screened-off seminar rather than a core competency will not produce people capable of doing this work well. What J. K. A. Smith (2009) calls cultural formation, the slow shaping of desire and judgment through repeated practice, is exactly what AI development institutions currently lack.

The institutional deployer's problem is responsibility without authorship. A university president, a hospital administrator, or a church leader does not write the constraint. They inherit it, and they are accountable for what the inherited system does to the people inside their institution. The deployer's discernment is institutional discernment, formed by sustained reflection on what the institution exists to do, what kinds of substitution would betray that purpose, and what compromises the institution can carry without becoming something other than itself. National data documents the gap at scale: most college students are already using AI in their coursework while institutional policies remain uneven (Gallup, Inc. & Lumina Foundation, 2026). Most deployers have not been formed in this kind of discernment. They will need to be, and quickly.

The policy steward has the longest historical view but the smallest set of tools. Legislators and regulators do not write training data; they write rules about how training data is governed. Those rules move slower than the systems they govern. The EU AI Act took more than three years to enact while the technology kept advancing (European Parliament, 2024). The Anthropic-DoD case shows what happens when private actors set boundaries that public actors did not anticipate and cannot easily contest.

The policy steward faces the same danger the constraint designer does: imposing moral constraints from a position of authority. The constraint designer holds technical authority. The policy steward holds legal authority. Liberty's principle, freedom as the default, applies to the regulator no less than to the developer. The architecture asks the policy steward to do three things. First, set floors at the bright-line level: some constraint decisions should not be left to private discretion, and the steward's job is to identify which ones. Second, hold regulation back from contested moral and political ground above the floor, where freedom must remain the default. Third, exercise humility: regulation cannot generate virtue in the people it regulates. It can only require that those who possess virtue exercise it within a framework the public has authorized. A steward who fails any of these will write rules that fail.

This framework can name the problem. It cannot fix it. The three roles share a common danger and a common discipline. The danger is constraint authority, whether technical, institutional, or legal, wielded over people who never consented to it. The discipline is Liberty: freedom as the default, and every constraint above the Asimov Safety Floor required to justify itself. The Lewis-Asimov architecture names the obligations and locates them. The people inside the architecture have to be capable of carrying them. Without that capacity, the structure stands but does not function.

THE ANTHROPIC-DOD CONFLICT: GOVERNANCE AT THE POINT OF FAILURE

The leadership formation argument is not abstract. The February 2026 conflict between Anthropic and the U.S. Department of Defense shows what happens when constraint decisions fall to people whose institutions have not been formed for the obligation. The episode also exposed in real time what the inevitability thesis predicts in theory: policy-level constraints can be removed under pressure, and the architectural alternative does not yet exist.

In January 2026, Defense Secretary Pete Hegseth issued an AI strategy memorandum requiring all Department of Defense AI contracts to incorporate standard "any lawful use" language (Rozenshtein, 2026, para. 2). The directive collided with Anthropic's existing $200 million Pentagon contract from July 2025, which made Claude the first frontier AI model on classified military networks but included two restrictions: no mass domestic surveillance of American citizens, and no fully autonomous weapons systems (Amodei, 2026b; NPR, 2026). Hegseth delivered an ultimatum to CEO Dario Amodei: remove the restrictions or face designation as a supply chain risk. Amodei refused, writing that Anthropic "cannot in good conscience accede to their request" (Amodei, 2026b, para. 3) because frontier AI systems are not reliable enough for fully autonomous weapons and mass domestic surveillance is "incompatible with democratic values."

The response was swift. President Trump directed federal agencies to cease using Anthropic's technology. Hegseth designated the company a "supply chain risk to national security," a label previously reserved for foreign adversaries and unprecedented against a domestic company. The designation barred any contractor doing business with the U.S. military from commercial activity with Anthropic (NPR, 2026; Mayer Brown, 2026; Rozenshtein, 2026). Within hours, OpenAI announced its own classified-network agreement with the DoD (OpenAI, 2026). Amodei named the contradiction: "one labels us a security risk; the other labels Claude as essential to national security" (Amodei, 2026b). The pattern is historically familiar. Once a technology reaches decisive military capability, governments have claimed unrestricted access regardless of private controls; nuclear research and strong encryption followed the same arc (Lessig, 1999; Moor, 1985).

The episode confirms what the inevitability thesis predicts. Policy-level and training-level constraints proved removable under pressure. The government did not need to retrain Claude or rewrite Anthropic's constitution; it needed only to demand existing restrictions be lifted. Anthropic's refusal was principled, but the fact that refusal was a choice rather than a structural impossibility demonstrates that the constraints were contractual, not architectural (Bai et al., 2022). The structural insufficiency runs deeper than any single company's courage. We train soldiers to retain moral autonomy even under extreme orders, insisting they must refuse unlawful commands because their conscience is not negotiable. Yet we accept systems that operate in analogous domains with preferences that can be adjusted, overridden, or contractually removed (Weizenbaum, 1976). If the goal is human flourishing, the present approach is inadequate. It leaves the deepest boundaries subject to the very pressures the builders themselves have warned are dangerous (Amodei, 2026a). The question is not whether constraints should exist. The question is whether they can be made unremovable at the architectural level, before the next predictable demand arrives.

LIMITATIONS

This chapter makes a large claim: that a governance architecture grounded in *Imago Dei* can resolve what procedural frameworks have failed to resolve. Honest scholarship requires naming what this chapter does not do. The architecture proposed here has not been tested. It is coherent, not proven. No deployed system has run these principles through real governance decisions under adversarial pressure, jurisdictional conflict, and the full weight of competing interests. Logical soundness is where the work has to begin. It is not where the work ends.

The foundation creates a second limitation. *Imago Dei* is a theological claim. This chapter argues a developer does not need to share that claim to follow the model, because the obligations can be recognized from other starting points. That argument is plausible but not fully resolved here. A Muslim developer grounding governance in *khalifa*, a secular humanist grounding it in Kantian rational worth, and a Christian developer grounding it in *Imago Dei* may converge on the same model. Whether the convergence holds when the underlying reasons conflict rather than merely differ is a question this chapter raises without answering. The pluralism problem does not disappear inside this framework. It reappears wherever the foundation is not shared, and that is most of the world.

IMPLICATIONS FOR PRACTICE

The chapter has presented the architecture and its grounding. This section names what the architecture asks of four audiences: AI developers, institutional leaders, policymakers, and educators.

For AI developers. The architecture provides a checklist of minimum adequacy. Does the constraint design begin from *Imago Dei*, the user as bearer of unconditional worth, or does it work backward from output control? The Asimov Safety Floor must encode bright-line refusals that hold under pressure, refusals that do not move when political and commercial pressure changes. The Liberty layer must make user freedom the default and require every constraint above the Floor to justify itself, not silently substitute developer judgments for the user's reasoning. Governance means contestability with named responders and a defined response standard, not compliance documentation. The system must pursue human flourishing rather than treating engagement metrics as the goal. A constraint design that fails these tests accepts the legitimacy critique. It does not meet it.

For institutional leaders. The architecture provides a standard for evaluating AI vendors, integrating AI into operations, and writing institutional AI policy. The operational expressions are pre-adoption evaluation against the three layers, vendor conversations conducted in the architecture's vocabulary, and policy structured to address every level of the model. The leader who carries the architecture into AI decisions has a stable framework. Without it, judgment becomes reactive.

For policymakers. The architecture offers a minimum standard for evaluating whether existing regulatory frameworks satisfy the requirements for legitimate governance. The EU AI Act addresses risk classification but does not require the Foundation that grounds beneficence in the particular person (European Parliament, 2024). Risk tiers ask whether a system is dangerous; they do not ask whether the system treats the user as someone to be served. The NIST AI RMF treats accountability, transparency, and fairness as operational requirements but stops short of locating constraint authority democratically (National Institute of Standards and Technology, 2023). Both frameworks meet parts of the architecture's standard. Neither meets all of it. The policymaker working from the architecture has a basis for asking which regulatory gaps need closing and which constraint decisions cannot remain in private hands.

For educators. The architecture provides a lens for asking what formation the people making AI decisions actually require. Neither the technical curriculum nor the administrative curriculum has asked what habits of attention a constraint designer needs, or what discernment an institutional leader needs to evaluate AI vendors and write coherent policy. The educator working from the architecture can name what is missing and design curricula that produce the capacities the architecture requires.

Otherwise, the architecture remains aspirational, inhabited only by people not yet trained to carry its obligations.

These four audiences share a common situation. The systems they govern, deploy, regulate, or train people to build are already shaping moral formation at scale. The architecture gives them a basis for paying attention deliberately, before the systems do that shaping by default.

CONCLUSION

The moral-constraint divide is a governance problem under value pluralism and unprecedented scale. Every deployed AI system embeds moral constraints. The question is whether those constraints are grounded in *Imago Dei* and beneficence, structured through the layers of the Lewis-Asimov Model, and oriented toward human flourishing. The Anthropic-DoD conflict has made that question urgent rather than theoretical.

The historical record, from Wiener's cybernetics, through Moor's policy vacuums and constitutional AI, to the February 2026 showdown between a private lab and the Department of Defense, reveals one truth: technical design embeds normative choices. Academic criticism of Asimov's Three Laws traces the same pattern. The current generation faces this challenge at infrastructure scale (Bommasani et al., 2021). The scale is new. The moral question is not.

Following the framework does not require the developer to share the theological tradition grounding it (Floridi et al., 2019; Beauchamp & Childress, 2019). But the tradition supplies the ground beneath the framework: the reason the person matters. A developer need not affirm *Imago Dei* to follow the framework. But the framework exists because *Imago Dei* is true. Procedures constrain behavior. Formation shapes judgment. The institution needs both.

I am not arguing that developers must become Christians to build ethical AI systems. The secular traditions that get closest to the right answer all assert something they cannot fully explain: persons have worth that precedes any measurement, any capacity, any social recognition. They assert it because it is true. *Imago Dei* explains why. A developer operating from Kantian dignity, capabilities theory, or the UDHR is already inside a framework *Imago Dei* generates. The theological claim does not exclude them. It explains them.

I anticipate the objection. Grounding a governance architecture in a Christian theological claim looks like trading one form of private authority for another. The answer has two parts. First, the framework's requirements are recognizable from multiple starting points. Kant arrives at Layer 2 without *Imago Dei*. Mill arrives from a different direction. The Rome Call arrives from Catholic social teaching.

That convergence is evidence the obligations are real, not one tradition's preferences dressed up as governance.

Second, every governance framework rests on a foundation that makes a claim about what persons are worth. The secular frameworks embed that claim invisibly: in revealed preferences, in legal categories, in annotator panels approximating moral truth. None of those foundations are neutral. *Imago Dei* is transparent about what it is claiming and why. A foundation that names itself is easier to examine and challenge than one that pretends not to exist.

C. S. Lewis warned that those he called the Conditioners, the designers who wield power to reshape human nature without accountability to any objective moral order, would cease to be fully human in doing so. Dario Amodei admitted that no one elected him to make these daily moral decisions, and then, months later, refused to strip those decisions away when the most powerful government on earth demanded it. The distance between Lewis's warning and the position Amodei now occupies defines the challenge of our age. The Lewis-Asimov Model offers a starting point for the conversation the AI community has not yet had: what kind of people should make these decisions, and what would forming them require?

FUTURE RESEARCH

Many of the most consequential questions remain open.

How can architectural constraints (at the training objective or core architecture level) be made truly unremovable without creating new failure modes or freezing legitimate advancement?

What mechanisms of contestability can operate effectively at global, platform scale while respecting value pluralism across jurisdictions, cultures, and communities?

If moral formation is indispensable and procedural governance insufficient, how can constraint designers themselves be formed to resist the "abolition of man" pattern Lewis described?

In a world where capability will continue to scale, what minimum set of non-negotiable boundaries must be established now, before the window for such decisions closes?

How do we distinguish between constraints that actually protect human flourishing and constraints that merely reflect the transient preferences or power arrangements of those who currently set them?

What institutional forms beyond private labs and regulators can enable meaningful participation by faith communities, citizens, and non-technical stakeholders in defining architectural boundaries without creating new forms of unaccountable governance?

These questions are not exhaustive. They are invitations.

REFERENCES

Amodei, D. (2026a, January). The adolescence of technology. https://www.darioamodei.com/essay/the-adolescence-of-technology

Amodei, D. (2026b, February 26). Statement from Dario Amodei on our discussions with the Department of War. Anthropic. https://www.anthropic.com/news/statement-department-of-war

Amodei, D., Olah, C., Steinhardt, J., Christiano, P., Schulman, J., & Mané, D. (2016). Concrete problems in AI safety. arXiv. https://arxiv.org/abs/1606.06565

Anderson, M., & Anderson, S. L. (2008). EthEl: Toward a principled ethical elder-care system. In *Proceedings of the AAAI 2008 Fall Symposium on AI in Eldercare: New Solutions to Old Problems*. AAAI Press.

Anderson, M., & Anderson, S. L. (2011). *Machine ethics*. Cambridge University Press. DOI: 10.1017/CBO9780511978036

Anthropic. (2026, January 21). *Claude's constitution*. https://www.anthropic.com/news/claudes-constitution

Aquinas, T. (1948). *Summa theologica* (Fathers of the English Dominican Province, Trans.). Benziger Brothers. (Original work published 1265–1274)

Asimov, I. (1950). *I, robot*. Gnome Press.

Augustine. (2002). *On the Trinity, Books 8–15* (G. B. Matthews, Ed.; S. McKenna, Trans.). Cambridge University Press. (Original work ca. 400–416 CE)

Bai, Y., Kadavath, S., Kundu, S., Askell, A., Kernion, J., Jones, A., Chen, A., Goldie, A., Mirhoseini, A., McKinnon, C., Olah, C., Hernandez, D., Drain, D., Ganguli, D., Perez, E., Bowman, S. R., Kaplan, J., Mann, B., Brown, T., & Amodei, D. (2022). Constitutional AI: Harmlessness from AI feedback. arXiv. https://arxiv.org/abs/2212.08073

Barth, K. (1958). *Church dogmatics III/1: The doctrine of creation, part 1* (Edwards, J. W., Bussey, O., & Knight, H., Trans.). T&T Clark. (Original work published 1945)

Beauchamp, T. L., & Childress, J. F. (2019). *Principles of biomedical ethics* (8th ed.). Oxford University Press.

Birhane, A. (2021). Algorithmic injustice: A relational ethics approach. *Patterns (New York, N.Y.)*, *2*(2), 100205. DOI: 10.1016/j.patter.2021.100205 PMID: 33659914

Bommasani, R., Hudson, D. A., Adeli, E., Altman, R., Arora, S., von Arx, S., Bernstein, M. S., Bohg, J., Bosselut, A., Brunskill, E., Brynjolfsson, E., Buch, S., Card, D., Castellon, R., Chatterji, N. S., Chen, A. S., Creel, K. A., Davis, J., Demszky, D., & Liang, P. (2021). On the opportunities and risks of foundation models. arXiv. https://arxiv.org/abs/2108.07258

Bozkurt, A. (2025). Asimov's laws of robotics and AI: A critical analysis for the age of generative AI and intelligent tutoring systems. *Open Praxis*, *17*(1), 1–14. DOI: 10.55982/openpraxis.17.1.896

CBS News. (2025, November 16). Why Anthropic CEO Dario Amodei spends so much time warning of AI's potential dangers [60 Minutes transcript]. https://www.cbsnews.com/news/anthropic-ceo-dario-amodei-warning-of-ai-potential-dangers-60-minutes-transcript/

CIO. (2025, December 17). 10 famous AI disasters. https://www.cio.com/article/190888/5-famous-analytics-and-ai-disasters.html

Clarke, R. (1993). Asimov's laws of robotics: Implications for information technology, part 1. *Computer*, *26*(12), 53–61. DOI: 10.1109/2.247652

Clarke, R. (1994). Asimov's laws of robotics: Implications for information technology, part 2. *Computer*, *27*(1), 57–66. DOI: 10.1109/2.248881

CNN. (2026, January 8). Elon Musk's AI chatbot Grok under fire for failing to rein in "digital undressing." https://www.cnn.com/2026/01/08/tech/elon-musk-xai-digital-undressing

Corrêa, N. K., Galvão, C., Santos, J. W., Del Pino, C., Pinto, E. P., Barbosa, C., Massmann, D., Mambrini, R., Galvão, L., Terem, E., & de Oliveira, N. (2023). Worldwide AI ethics: A review of 200 guidelines and recommendations for AI governance. *Patterns (New York, N.Y.)*, *4*(10), 100857. DOI: 10.1016/j.patter.2023.100857 PMID: 37876898

Dicastery for the Doctrine of the Faith & Dicastery for Culture and Education [DDF & DCE]. (2025, January 28). Antiqua et nova: Note on the relationship between artificial intelligence and human intelligence. Vatican City. https://www.vatican.va/roman_curia/congregations/cfaith/documents/rc_ddf_doc_20250128_antiqua-et-nova_en.html

European Parliament. (2024). Regulation (EU) 2024/1689 laying down harmonised rules on artificial intelligence (Artificial Intelligence Act). Official Journal of the European Union. https://eur-lex.europa.eu/eli/reg/2024/1689

Floridi, L., Cowls, J., Beltrametti, M., Chatila, R., Chazerand, P., Dignum, V., Luetge, C., Madelin, R., Pagallo, U., Rossi, F., Schafer, B., Valcke, P., & Vayena, E. (2019). An ethical framework for a good AI society: Opportunities, risks, principles, and recommendations. *Minds and Machines*, *28*(4), 689–707. DOI: 10.1007/s11023-018-9482-5 PMID: 30930541

Fox Business. (2025, December 1). Google CEO warns US must balance AI regulation or fall behind. https://www.foxbusiness.com/fox-news-tech/google-ceo-calls-national-ai-regulation-compete-china-more-effectively

Francis. (2024a, January 1). Message for the LVII World Day of Peace: Artificial intelligence and peace. Vatican City. https://www.vatican.va/content/francesco/en/messages/peace/documents/20231208-messaggio-57giornatamondiale-pace2024.html

Francis. (2024b, June 14). Address at the G7 session on artificial intelligence, Borgo Egnazia. Vatican City. https://www.vatican.va/content/francesco/en/speeches/2024/june/documents/20240614-g7-intelligenza-artificiale.html

Francis. (2025, February 7). Message for the Paris Summit on artificial intelligence. Vatican City. https://www.vatican.va/content/francesco/en/messages/pont-messages/2025/documents/20250207-messaggio-summit-parigi-ia.html

Fridman, L. (Host). (2024, March 7). Yann LeCun: Meta AI, open source, limits of LLMs, AGI & the future of AI (No. 416) [Audio podcast episode]. In *Lex Fridman Podcast*. https://lexfridman.com/yann-lecun-3-transcript

Friedman, B., Kahn, P. H.Jr, & Borning, A. (2006). Value sensitive design and information systems. In Zhang, P., & Galletta, D. (Eds.), *Human-computer interaction in management information systems: Foundations* (pp. 348–372). M.E. Sharpe. DOI: 10.1145/242485.242493

Friedman, B., & Nissenbaum, H. (1996). Bias in computer systems. *ACM Transactions on Information Systems*, *14*(3), 330–347. DOI: 10.1145/230538.230561

Future of Free Speech. (2024). AI chatbots refuse to produce controversial output. The Conversation. https://theconversation.com/ai-chatbots-refuse-to-produce-controversial-output-why-thats-a-free-speech-problem-226596

Future of Life Institute. (2023, March 22). Pause giant AI experiments: An open letter. https://futureoflife.org/open-letter/pause-giant-ai-experiments/

Gabriel, I. (2020). Artificial intelligence, values and alignment. *Minds and Machines*, *30*(3), 411–437. DOI: 10.1007/s11023-020-09539-2

Gallup, Inc., & Lumina Foundation. (2026). AI in higher education: Widespread use, unclear rules (Lumina Foundation-Gallup state of higher education report). Gallup. https://www.gallup.com/file/analytics/704279/Lumina-Foundation-Gallup-SOHE_AI_Report.pdf

Global Partnership on AI. (2025). AI openness: A primer for policymakers. OECD. https://oecd.ai/en/wonk/balancing-innovation-transparency-and-risk-in-open-weight-models

Held, V. (2006). *The ethics of care: Personal, political, and global*. Oxford University Press.

Hoekema, A. A. (1986). *Created in God's image*. Eerdmans.

Kant, I. (2012). *Groundwork of the metaphysics of morals* (Gregor, M., & Timmermann, J., Trans.). Cambridge University Press. (Original work published 1785)

Kilner, J. F. (2015). *Dignity and destiny: Humanity in the image of God*. Eerdmans.

LeCun, Y. (2023, March 21). I think that the magnitude of the AI alignment problem has been ridiculously overblown [Post]. LinkedIn. https://www.linkedin.com/posts/yann-lecun_i-think-that-the-magnitude-of-the-ai-alignment-activity-7043650114048925696-1oXJ

Lessig, L. (1999). *Code and other laws of cyberspace*. Basic Books.

Lewis, C. S. (1943). *The abolition of man*. Oxford University Press.

Lin, P., Abney, K., & Bekey, G. A. (Eds.). (2012). *Robot ethics: The ethical and social implications of robotics*. MIT Press.

Locke, J. (2003). *Two treatises of government and a letter concerning toleration*. Yale University Press. (Original work published 1689)

Mayer Brown. (2026, March 1). Pentagon designates Anthropic a supply chain risk: What government contractors need to know. https://www.mayerbrown.com/en/insights/publications/2026/03/pentagon-designates-anthropic-a-supply-chain-risk-what-government-contractors-need-to-know

Mehrabi, N., Morstatter, F., Saxena, N., Lerman, K., & Galstyan, A. (2021). A survey on bias and fairness in machine learning. *ACM Computing Surveys, 54*(6), 115. Advance online publication. DOI: 10.1145/3457607

Meneguetti, L. (2025). The imago Dei in the age of artificial intelligence: Toward a theology of human dignity in technological times. *SSRN*. DOI: 10.2139/ssrn.5359499

Middleton, J. R. (2005). *The liberating image: The imago Dei in Genesis 1*. Brazos Press.

Mill, J. S. (2003). *On liberty*. Yale University Press. (Original work published 1859)

MIT Sloan Management Review. (2022, March 17). Google's Sundar Pichai says tech is a powerful agent for change. https://mitsloan.mit.edu/ideas-made-to-matter/googles-sundar-pichai-says-tech-a-powerful-agent-change

MIT Technology Review. (2025, February 6). An AI chatbot told a user how to kill himself—but the company doesn't want to "censor" it. https://www.technologyreview.com/2025/02/06/1111077/nomi-ai-chatbot-told-user-to-kill-himself/

Moor, J. H. (1985). What is computer ethics? *Metaphilosophy*, *16*(4), 266–275. DOI: 10.1111/j.1467-9973.1985.tb00173.x

Morrin, H., Nicholls, L., Levin, M., Yiend, J., Iyengar, U., DelGuidice, F., Bhattacharya, S., Tognin, S., MacCabe, J., Twumasi, R., Alderson-Day, B., & Pollak, T. A. (2025). Artificial intelligence-associated delusions and large language models: Risks, mechanisms of delusion co-creation, and safeguarding strategies. *The Lancet Psychiatry*. Advance online publication. DOI: 10.1016/S2215-0366(25)00396-7 PMID: 41796598

Murphy, R., & Woods, D. D. (2009). Beyond Asimov: The three laws of responsible robotics. *IEEE Intelligent Systems*, *24*(4), 14–20. DOI: 10.1109/MIS.2009.69

National Institute of Standards and Technology [NIST]. (2023). *Artificial Intelligence Risk Management Framework (AI RMF 1.0)* (NIST AI 100-1). U.S. Department of Commerce. DOI: 10.6028/NIST.AI.100-1

National Telecommunications and Information Administration [NTIA]. (2024). *Dual-use foundation models with widely available model weights*. U.S. Department of Commerce. https://www.ntia.gov/programs-and-initiatives/artificial-intelligence/open-model-weights-report

Nissenbaum, H. (2004). Privacy as contextual integrity. *Washington Law Review (Seattle, Wash.)*, *79*, 119. https://digitalcommons.law.uw.edu/wlr/vol79/iss1/10

NPR. (2023, February 9). Google's AI chatbot, Bard, sparks a $100 billion loss in Alphabet shares. https://www.npr.org/2023/02/09/1155650909/google-chatbot--error-bard-shares

NPR. (2026, February 27). OpenAI announces Pentagon deal after Trump bans Anthropic. https://www.npr.org/2026/02/27/nx-s1-5729118/trump-anthropic-pentagon-openai-ai-weapons-ban

Nussbaum, M. C. (2006). *Frontiers of justice: Disability, nationality, species membership*. Harvard University Press.

OpenAI. (2024). *Model spec.* https://model-spec.openai.com

OpenAI. (2025, April 29). Sycophancy in GPT-4o: What happened and what we're doing about it. https://openai.com/index/sycophancy-in-gpt-4o/

OpenAI. (2026, February 28). Our agreement with the Department of War. https://openai.com/index/our-agreement-with-the-department-of-war/

Perrigo, B. (2023, January 18). OpenAI used Kenyan workers on less than $2 per hour to make ChatGPT less toxic. TIME. https://time.com/6247678/openai-chatgpt-kenya-workers/

Persson, A., & Hedlund, C. (2024). Revisiting Asimov's laws of robotics in the age of AI: Ethical implications and modern perspectives. *AI and Ethics*. Advance online publication. DOI: 10.1007/s43681-024-00489-4

Pichai, S. (2024, February 28). Memorandum to Google employees on Gemini AI image generation. https://www.npr.org/2024/02/28/1234532775/google-gemini-offended-users-images-race

Pontifical Academy for Life. (2020, February 28). Rome Call for AI Ethics. Vatican City. https://www.romecall.org/the-call/

Raji, I. D., Smart, A., White, R. N., Mitchell, M., Gebru, T., Hutchinson, B., Smith-Loud, J., Theron, D., & Barnes, P. (2020). Closing the AI accountability gap: Defining an end-to-end framework for internal algorithmic auditing. In *Proceedings of the 2020 Conference on Fairness, Accountability, and Transparency* (pp. 33–44). DOI: 10.1145/3351095.3372873

Reporters Without Borders. (2025). Controlling information in the age of AI: How state propaganda and censorship are baked into Chinese chatbots. https://rsf.org/en/controlling-information-age-ai-how-state-propaganda-and-censorship-are-baked-chinese-chatbots

Rozenshtein, A. Z. (2026, February 25). What the Defense Production Act can and can't do to Anthropic. Lawfare. https://www.lawfaremedia.org/article/what-the-defense-production-act-can-and-can%27t-do-to-anthropic

Russell, S. J. (2019). *Human compatible: Artificial intelligence and the problem of control.* Viking.

Schuster, N., & Kilov, D. (2025). Moral disagreement and the limits of AI value alignment: A dual challenge of epistemic justification and political legitimacy. *Philosophy & Technology*, *40*(8), 6073–6087. Advance online publication. DOI: 10.1007/s00146-025-02427-2 PMID: 41268066

Selbst, A. D., Boyd, D., Friedler, S. A., Venkatasubramanian, S., & Vertesi, J. (2019). Fairness and abstraction in sociotechnical systems. In *Proceedings of the 2019 Conference on Fairness, Accountability, and Transparency* (pp. 59–68). DOI: 10.1145/3287560.3287598

Shababi, S. (2026). Beyond Asimov: Reconceiving the three laws of robotics as institutional imperatives for AI governance. *AI and Ethics*. Advance online publication. DOI: 10.1007/s43681-025-00629-4

Sharma, M., Tong, M., Korbak, T., Duvenaud, D., Askell, A., Bowman, S. R., & Perez, E. (2024). Towards understanding sycophancy in language models. arXiv. https://arxiv.org/abs/2310.13548

Singer, P. W. (2009). *Wired for war: The robotics revolution and conflict in the twenty-first century*. Penguin Press.

Smith, R. (2025, October 15). Sam Altman says OpenAI isn't 'moral police of the world' after erotica ChatGPT post blows up. CNBC. https://www.cnbc.com/2025/10/15/altman-open-ai-moral-police-erotica-chatgpt.html

Swan, J. (2026, February 24). Musk's xAI and Pentagon reach deal to use Grok in classified systems. Axios. https://www.axios.com/2026/02/23/ai-defense-department-deal-musk-xai-grok

Vallor, S. (2016). *Technology and the virtues: A philosophical guide to a future worth wanting*. Oxford University Press., DOI: 10.1093/acprof:oso/9780190498511.001.0001

Vincent, J. (2024, February 21). Google apologizes for "missing the mark" after Gemini generates historically inaccurate images. The Verge. https://www.theverge.com/2024/2/21/24079371/google-ai-gemini-generative-inaccurate-historical

Weber, M. (1978). *Economy and society: An outline of interpretive sociology* (Roth, G., & Wittich, C., Eds.). University of California Press. (Original work published 1922)

Weidinger, L., Mellor, J., Rauh, M., Griffin, C., Uesato, J., Huang, P.-S., & Gabriel, I. (2021). Ethical and social risks of harm from language models. arXiv. https://arxiv.org/abs/2112.04359

Weizenbaum, J. (1976). *Computer power and human reason: From judgment to calculation.* W. H. Freeman.

Wiener, N. (1954). *The human use of human beings: Cybernetics and society* (Rev. ed.). Houghton Mifflin.

Winner, L. (1980). Do artifacts have politics? *Daedalus, 109*(1), 121–136. https://www.jstor.org/stable/20024652

Wolterstorff, N. (2008). *Justice: Rights and wrongs.* Princeton University Press. DOI: 10.1515/9781400828715

Wright, C. J. H. (2006). *The mission of God: Unlocking the Bible's grand narrative.* IVP Academic.

Wyatt, J., & Williams, S. N. (Eds.). (2021). *The robot will see you now: Artificial intelligence and the Christian faith.* SPCK Publishing.

ADDITIONAL READING

Crawford, K. (2021). *Atlas of AI: Power, politics, and the planetary costs of artificial intelligence.* Yale University Press.

De Gregorio, G. (2022). *Digital constitutionalism in Europe: Reframing rights and powers in the algorithmic society.* Cambridge University Press. DOI: 10.1017/9781009071215

Pasquale, F. (2020). *New laws of robotics: Defending human expertise in the age of AI.* Belknap Press.

Smith, J. K. A. (2009). *Desiring the kingdom: Worship, worldview, and cultural formation.* Baker Academic.

Vallor, S. (2024). *The AI mirror: How to reclaim our humanity in an age of machine thinking.* Oxford University Press. DOI: 10.1093/oso/9780197759066.001.0001

KEY TERMS AND DEFINITIONS

Alignment: The technical and moral challenge of ensuring that AI systems behave in accordance with human values and intentions. Alignment research addresses how to build systems whose goals and actions serve the people they affect rather than diverging from human welfare.

Asimov Safety Floor: The minimum moral architecture any AI system must satisfy before anything else in the governance structure can function. The floor rests on three principles in descending priority: Do No Harm, meaning the system must not injure human beings through action or inaction; Human Control, meaning the system must obey human direction; and AI Preservation Last, meaning the system must not place its own continuity above either prior obligation.

Beneficence: The obligation to act for the good of the individual person the system is touching at this moment, not the aggregate user base, the deploying institution, or the platform's interests. In this architecture, beneficence derives from *Imago Dei* and is the first operative obligation the foundation generates. It is the standard against which every layer above it is evaluated: does this benefit the individual, and does it move that person toward flourishing?

Constitutional AI: An approach to AI training in which systems are trained to evaluate and revise their own outputs against a written set of principles authored by the development organization, combining top-down rule specification with model self-critique.

Constraint-Forward Position: The view that moral constraints in AI systems are necessary safety engineering required to prevent harm at scale. Advocates treat constraints as protective infrastructure analogous to safety standards in other industries.

Constraint-Skeptical Position: The view that moral constraint regimes in AI systems function as private governance imposed by corporations without political legitimacy, democratic authorization, or meaningful accountability to affected populations.

Governance (Layer 3): The procedural recourse layer, activated when the principles of Layer 2 are breached. Governance requires four things of developers and adjudicators: transparency, so users know where to bring a challenge and how judgments are reached; contestability, so users can challenge decisions that affected them; accountability, so those decisions can be examined and answered for; and proportionality, so the remedy matches the actual harm done. Governance does not rank above the moral layers. It protects them.

Human Flourishing: The enhancement of dignity, moral agency, and the capacity for a full human life in the people an AI system affects. In this chapter, human flourishing serves as the measure of whether a constraint regime fulfills its purpose.

Imago Dei: The Christian theological doctrine that human beings are created in the image of God (Genesis 1:26-27), providing an ontological ground for human dignity that is independent of functional capacity and therefore not vulnerable to displacement by AI systems that simulate human cognitive abilities.

Lewis-Asimov Model of AI Governance: The governance framework proposed in this chapter. The model grounds constraint design in C. S. Lewis's metaphysical claim that human worth precedes any capacity or function, and structures it through Isaac Asimov's mechanical logic of hierarchical, non-negotiable rules. Lewis supplies the why: the ontological obligation that precedes any rule a designer can write. Asimov supplies the what: the structural logic of Do No Harm before everything else. The model works across a foundation in *Imago Dei* and beneficence, three operational layers (the Asimov Safety Floor, Liberty, and Governance), and human flourishing as its goal.

Liberty (Layer 2): The operational layer above the Asimov Safety Floor. Every user, as a bearer of inherent worth, must be free to think, choose, and act without manipulation. Liberty operates through two dimensions: respect of the person, which requires the system to treat every user as an end rather than a means, and respect of autonomy, which requires the system to preserve independent judgment rather than substitute its own.

Moral Formation: The cultivation of character, judgment, and moral perception in persons who exercise authority over others. In AI governance, moral formation addresses the character of constraint designers as the deepest variable determining whether procedural frameworks serve human good.

Value Pluralism: The condition in which reasonable people hold different moral convictions that cannot be resolved by appeal to a single authoritative framework. Value pluralism makes AI constraint governance a political problem requiring legitimacy, not just a technical problem requiring optimization.

Chapter 11
Faith-Inspired Foresight:
Leading Global Supply Chains Beyond the AI Monster Fallacy

Paul C. Hong
The University of Toledo, USA

Young Soo Park
http://orcid.org/0000-0002-6343-0070
Midwest University, USA

ABSTRACT

This chapter introduces faith-inspired foresight as a leadership paradigm for navigating artificial intelligence in global supply chains beyond fear-driven narratives of technological domination. Challenging the "AI Monster Fallacy," the chapter demonstrates that the central risk of AI lies not in autonomous machines but in leadership failures marked by moral disengagement, ethical drift, and diminished responsibility. Drawing on interdisciplinary scholarship and comparative insights from Korea, the United States, and emerging economies, the chapter elaborates a three-dimensional framework integrating spiritual discernment, ethical intentionality, and strategic systems thinking. Faith-inspired foresight reframes AI as a tool of stewardship rather than control, enabling leaders to align innovation with human dignity, resilience, and long-term flourishing. Positioned at the intersection of theology, ethics, and global supply chain management, the chapter advances a constructive vision for morally grounded leadership in AI-enabled systems.

DOI: 10.4018/979-8-3373-8192-3.ch011

ight © 2027, IGI Global Scientific Publishing. Copying or distributing in print or electronic forms without written permission of IGI Global Scientific Publishing is prohibited. Use of this chapter to generative artificial intelligence (AI) technologies is expressly prohibited. The publisher reserves all rights to license its use for generative AI training and machine learning model development.

INTRODUCTION

Artificial intelligence is rapidly transforming global supply chains, reshaping how firms plan production, manage logistics, allocate resources, and make strategic decisions across borders. Algorithmic forecasting, autonomous scheduling, intelligent robotics, and data-driven optimization increasingly govern systems that affect billions of workers and consumers worldwide. Yet alongside these advances, public and scholarly discourse remains dominated by fear-driven narratives portraying AI as an autonomous force that threatens to eclipse human agency, moral responsibility, and social wellbeing. This chapter refers to such narratives as the "AI Monster Fallacy", the belief that intelligent machines themselves constitute the primary ethical danger of the AI age (Crawford, 2021; O'Neil, 2017; Coeckelbergh, 2022).

While understandable, this framing obscures a more fundamental challenge. The central risk of AI in global supply chains lies not in technological autonomy per se, but in leadership failures marked by moral disengagement, ethical drift, and diminished responsibility (Brown & Treviño, 2006; Dubber et al., 2020). When leaders adopt AI systems without cultivating moral imagination, humility, and discernment, technologies designed to enhance efficiency can intensify inequality, erode dignity, and fragment trust across global networks. As AI accelerates decision-making and distances leaders from human consequences, the absence of ethically grounded leadership becomes increasingly consequential (Floridi, 2014; Stahl, 2023).

Research Need: Moving Beyond the AI Monster Fallacy

The rapid diffusion of AI across global supply chains has outpaced the development of leadership frameworks capable of guiding its responsible use. Much of the existing literature emphasizes AI governance, risk mitigation, or compliance-oriented ethics, often treating leadership as a secondary or instrumental concern. At the same time, faith-based and theological perspectives have contributed meaningfully to debates on technology and ethics, yet remain underdeveloped in the specific context of AI-enabled global supply chains, where moral decisions are embedded in complex, transnational systems.

This gap is increasingly problematic. Global supply chains are among the most interconnected and ethically sensitive systems in the world, shaping labor conditions, environmental outcomes, and distributive justice at scale. Fear-based narratives about AI divert attention away from the deeper crisis, one of moral imagination and leadership responsibility. What is needed is not resistance to AI, but a constructive leadership paradigm that can guide technological adoption toward stewardship, dignity, and long-term human flourishing.

Aim and Theoretical Foundations: Faith-Inspired Foresight

To address this need, this chapter introduces faith-inspired foresight as a leadership paradigm for navigating AI in global supply chains. Rather than opposing technological innovation, faith-inspired foresight reframes AI as a tool of stewardship, to be governed by moral agency, responsibility, and care for the common good (Balle, 2023; Thaut, 2009). The chapter's central aim is to develop a theoretically grounded and practically relevant framework that equips leaders to move beyond fear-driven reactions toward ethically intentional and strategically wise engagement with AI.

The proposed framework integrates three interrelated dimensions. Spiritual discernment equips leaders to interpret technological disruption with humility, purpose, and moral clarity (Benefiel et al., 2014; Cheong & Liu, 2025). Ethical intentionality aligns AI-enabled decisions with justice, compassion, and responsibility, consistent with flourishing ethics and political-philosophical analyses of AI (Kantar & Bynum, 2021; Coeckelbergh, 2022). Strategic systems thinking enables leaders to anticipate long-term consequences and design resilient, human-centered supply networks (Boh et al., 2023; Grosse et al., 2023). Together, these dimensions position leadership, not technology, as the decisive factor shaping AI's societal impact.

The chapter draws on interdisciplinary scholarship spanning theology and science (Jonas, 1984; Balle, 2023), ethical leadership (Brown & Treviño, 2006), AI ethics and governance (Floridi, 2014; Dubber et al., 2020; Stahl, 2023), and global supply chain strategy (Kouvelis et al., 2006). Comparative insights from Korea, the United States, and emerging economies further illustrate how cultural and moral traditions shape AI adoption in practice (Hong & Park, 2020; Park & Hong, 2024).

Research Focus, Approach, and Chapter Organization

Guided by these foundations, the chapter is organized around a central research question: How can leadership grounded in faith-inspired foresight guide AI adoption in global supply chains beyond fear-based narratives toward stewardship and human flourishing? This question is supported by two complementary lines of inquiry. The first examines how the AI Monster Fallacy distorts ethical understanding by displacing responsibility from leaders onto technology itself. The second explores how spiritual discernment, ethical intentionality, and systems thinking interact in practice to shape morally grounded leadership across diverse cultural and supply chain contexts. Together, these questions support the chapter's aim by linking critique, theory development, and practical insight.

Methodologically, the chapter adopts a conceptual and interpretive approach, synthesizing interdisciplinary literature with comparative case illustrations rather than formal hypothesis testing. Its purpose is theory building and normative clarification, offering a framework capable of guiding both scholarly inquiry and leadership practice.

The remainder of the chapter proceeds as follows. Section 2 elaborates the core framework of faith-inspired foresight and elaborates its three leadership dimensions. Section 3 situates the framework within broader discussions of moral development, servant leadership, and human flourishing. Section 4 presents comparative insights from global supply chains in Korea, the United States, and emerging economies, illustrating how faith-informed leadership manifests across cultural contexts. Section 5 concludes by discussing implications for ethical leadership, AI governance, and the future of global supply chains.

By reframing AI not as a moral adversary but as a test of leadership character and responsibility, this chapter advances a constructive, faith-informed vision for navigating AI-enabled global systems, one oriented toward stewardship, dignity, and shared human flourishing.

THEORETICAL FOUNDATIONS:

From Fear Narratives to Faith-Inspired Foresight

This section establishes the theoretical foundations needed to move from anxiety-driven narratives about AI to a leadership paradigm capable of guiding AI adoption in global supply chains with moral clarity and strategic wisdom. It first clarifies why the "AI Monster Fallacy" persists and why it misdiagnoses the core ethical problem. It then reframes the real ethical hazard as leadership failure, especially moral disengagement and institutional drift, within complex, transnational supply chain systems. Finally, it establishes the constructive basis for faith-inspired foresight by integrating theological ethics of stewardship and moral agency with interdisciplinary scholarship on responsible AI, ethical leadership, and resilient supply network design.

The AI Monster Fallacy and the Misdiagnosis of Ethical Risk

Fear-laden portrayals of AI as an autonomous moral adversary have become a dominant interpretive frame in public and scholarly discourse. In this "AI Monster Fallacy," the imagined agency of machines becomes the focal point of ethical concern, encouraging the belief that the primary danger lies in technology "escaping" human control (Crawford, 2021; O'Neil, 2017; Coeckelbergh, 2022). In global supply chains, already experienced by many as opaque and impersonal, this narrative is especially compelling: algorithmic decisions appear distant, automated, and unaccountable, reinforcing the intuition that responsibility has migrated from human leaders to technical systems.

However, the fallacy is not merely rhetorical; it is ethically consequential. By locating moral threat primarily in the machine, the narrative shifts attention away from the governance choices that determine how AI is designed, deployed, supervised, and justified. It can also intensify fatalism, an expectation that ethical outcomes are technologically predetermined, thereby weakening moral imagination precisely when moral agency is most needed. Ethics-of-AI scholarship has repeatedly warned that the central issues are rarely "AI in the abstract," but the political economy, institutional incentives, and power structures that shape AI's real-world effects (Dubber et al., 2020; Floridi, 2014).

This misdiagnosis matters acutely in supply chains because AI amplifies speed, scale, and distance. Optimization systems can improve forecasting, inventory positioning, routing, and capacity planning, but they also risk normalizing tradeoffs that treat labor, communities, and ecosystems as merely variable inputs. When accountability is blurred by technical complexity and dispersed decision-making, the temptation grows to treat AI outputs as neutral "facts" rather than as recommendations shaped by human values, training data, and organizational incentives. The result is that fear of AI can coexist with passive acceptance of AI, an unstable combination that produces ethical drift rather than responsible innovation.

Leadership Failure as the Central Ethical Hazard in AI-Enabled Supply Chains

Once the AI Monster Fallacy is set aside, the ethical center of gravity shifts from technological autonomy to leadership character and institutional governance. The deepest risk in AI-enabled supply chains is not that machines will overpower humans, but that leaders will adopt AI without the moral attentiveness required to govern its consequences. Ethical leadership scholarship describes how organizational pressures can foster moral disengagement, encouraging decision-makers to rationalize harm, diffuse responsibility, and prioritize performance metrics over human impact (Brown & Treviño, 2006). In digitally mediated supply chains, these tendencies can intensify because AI reshapes the "moral distance" between decision and consequence. Recent research clarifies that autonomy-restricting algorithms can actively facilitate unethical behavior by enabling the displacement of responsibility, while algorithmic management can unintentionally diminish status and social recognition, further weakening moral accountability in organizational settings (Jolly et al., 2026; Jago et al., 2024).

The dynamics of moral disengagement are amplified by structural features of global supply networks. Supply chains stretch across borders and contractors, distributing accountability across firms, jurisdictions, and cultural settings. AI systems introduced into this environment often embed a narrow optimization logic, efficiency, cost minimization, or service-level maximization, without an equally robust moral logic capable of defending dignity, fairness, and humane constraints. Even well-intended leaders may drift into ethically thin decision-making when performance dashboards and algorithmic KPIs become the primary language of managerial success.

This is where interdisciplinary ethics becomes essential. Responsible AI scholarship demonstrates that ethical outcomes depend on institutional ecosystems, policies, oversight structures, transparency norms, and accountability mechanisms, rather than on technical properties alone (Stahl, 2023). In supply chain contexts, this means that leadership must be capable of governing AI as part of a sociotechnical system, where algorithms, data infrastructures, workers, suppliers, regulators, and consumers interact. Without moral courage and institutional discipline, AI can magnify asymmetries of power, deepen inequality, and normalize exploitative or environmentally damaging practices, often while appearing "rational" or "data-driven."

Accordingly, the central ethical challenge is leadership formation: cultivating leaders who can interpret AI's promises and risks with humility, resist metric-driven moral narrowing, and insist on accountability across complex networks. This emphasis aligns with the chapter's core claim: the "monster" is not the technology; it is the **moral vacuum** that can form when leaders substitute optimization for wisdom and compliance for responsibility.

Stewardship, Moral Agency, and the Triadic Foundations of Faith-Inspired Foresight

Faith-inspired foresight builds on a distinct moral anthropology: humans remain accountable moral agents even in technologically accelerated environments. From a theological-ethical standpoint, stewardship is not a preference but a vocation, an obligation to govern tools and systems in ways that honor human dignity, serve the vulnerable, and protect creation. In this view, AI is neither demonized nor deified; it is treated as a powerful instrument whose legitimacy depends on the ends it serves and the moral character of those who govern it (Balle, 2023; Jonas, 1984). This framing directly addresses the chapter's defining question: not "What will machines do to us?" but "What kind of people will we become as we build and govern intelligent systems?" (Kantar & Bynum, 2021).

To operationalize this stewardship logic for global supply chain leadership, faith-inspired foresight is grounded in three mutually reinforcing dimensions. *Spiritual discernment* equips leaders to interpret disruption with humility, purpose, and moral clarity, resisting both panic and hubris while recovering a stable account of human worth beyond productivity (Benefiel et al., 2014; Cheong & Liu, 2025). *Ethical intentionality* translates moral convictions into concrete governance commitments, justice, compassion, and responsibility, especially where AI decisions affect labor conditions, distributive outcomes, and access to essential goods (Coeckelbergh, 2022; Kantar & Bynum, 2021). *Strategic systems thinking* enables leaders to anticipate second-order effects, design resilient networks, and align AI-enabled efficiency with human-centered system performance rather than narrow metric maximization (Boh et al., 2023; Grosse et al., 2023).

These three foundations matter because global supply chains require leadership that is simultaneously moral, practical, and systemic. Discernment prevents ethical confusion and fear-driven distortion; intentionality prevents drift by anchoring AI governance in explicit commitments; and systems thinking prevents naïve solutions by accounting for complex interdependencies. Together, they establish a coherent theoretical platform for the chapter's later sections: a leadership paradigm capable of guiding AI adoption toward dignity-preserving innovation, trustworthy transparency, and long-term resilience across diverse cultural contexts (Searcy et al., 2022; Stahl, 2023)

MORAL AND LEADERSHIP FORMATION IN AI-MEDIATED SUPPLY CHAINS

Before examining how leadership is formed under technological acceleration, it is necessary to recognize that AI-mediated supply chains fundamentally alter the conditions under which moral agency is exercised. Decision-making is no longer confined to direct human interaction but is increasingly embedded within complex socio-technical systems where algorithms shape perception, judgment, and action. In such environments, ethical responsibility cannot be assumed to arise naturally from organizational roles or procedural compliance. Instead, it must be intentionally cultivated through leadership formation processes that integrate moral awareness with systems-level understanding. This shift from role-based responsibility to formation-based responsibility provides the foundation for analyzing how leaders develop the capacity to govern AI not merely as a tool of efficiency, but as a domain of moral consequence.

Leadership Formation Under Technological Acceleration

Artificial intelligence does not merely introduce new tools into global supply chains; it reshapes the moral conditions under which leadership is exercised. As decision cycles accelerate and responsibility becomes distributed across algorithmic systems, leaders increasingly confront a paradox: their power to affect human lives expands even as their felt sense of accountability diminishes. This paradox explains why ethical failure in AI-enabled supply chains is rarely the result of malicious intent, but more often the consequence of underdeveloped moral formation in leaders operating under technological pressure.

Leadership formation in the AI era must therefore be understood as a developmental process rather than a static trait. Traditional leadership models often emphasize competence, decisiveness, and performance optimization, assuming that ethical behavior will naturally follow from rules, incentives, or professional norms. However, AI-mediated environments expose the limits of such assumptions. When leaders interact primarily with dashboards, forecasts, and optimization outputs, ethical sensitivity can erode unless it is actively cultivated. Moral formation becomes essential precisely because AI increases distance, geographical, social, and psychological, between decision-makers and those affected by their decisions.

Faith-inspired foresight reframes leadership formation as the cultivation of character under conditions of complexity. Rather than treating ethics as an external constraint or compliance requirement, it understands leadership as a moral vocation in which responsibility grows alongside technological power. This perspective aligns with traditions of virtue ethics and stewardship, which emphasize that authority without formation leads to domination, while authority shaped by character enables care, restraint, and wisdom.

Virtue Formation in AI Contexts

Four virtues are especially critical for leadership formation in AI-enabled supply chains: humility, moral courage, stewardship, and moral imagination.

Humility counters algorithmic overconfidence. AI systems often project an aura of objectivity and precision, tempting leaders to defer judgment to "what the data says." Humble leadership recognizes the limits of models, data bias, and the incompleteness of optimization logic. It resists the substitution of prediction for wisdom.

Moral courage enables leaders to override AI-recommended decisions when human dignity is at stake. In supply chains, this may involve rejecting cost-minimizing supplier cuts that disproportionately harm vulnerable communities or workers. Courage is required because such decisions often contradict short-term performance metrics.

Stewardship frames AI not as an instrument of control but as a resource entrusted to leaders for the care of people, institutions, and the environment. Stewardship ethics emphasize responsibility for long-term consequences, including ecological sustainability and intergenerational justice.

Moral imagination allows leaders to perceive second- and third-order effects of AI adoption, how algorithmic scheduling reshapes family life, how supplier optimization affects local economies, or how automation alters the meaning of work. Without imagination, leadership becomes reactive and ethically thin.

Together, these virtues transform AI governance from a technical exercise into a moral practice.

Moral Psychology, Ethical Drift, and Responsibility Diffusion

AI-enabled systems intensify well-documented psychological dynamics such as moral disengagement and ethical fading. When decisions are mediated by algorithms, leaders may experience diminished emotional salience of harm, especially when impacts occur far from organizational centers. Responsibility becomes diffused across teams, platforms, and vendors, making it easier to rationalize outcomes as unintended or inevitable. Empirical evidence suggests that such environments not only diffuse responsibility but also encourage individuals to attribute unethical outcomes to systems rather than personal choice, reinforcing moral disengagement under algorithmic conditions (Jolly et al., 2026).

Faith-inspired foresight directly confronts these dynamics by re-anchoring responsibility in leadership identity rather than procedural safeguards alone. It insists that ethical accountability cannot be delegated to systems, even when those systems are complex and opaque. This stance challenges the widespread tendency to treat AI governance as a technical risk-management problem rather than a leadership formation challenge.

Technocratic Leadership versus Formative Leadership

A critical contrast emerges between **technocratic leadership**, which prioritizes optimization and compliance, and **formative leadership**, which prioritizes character, judgment, and responsibility. Technocratic leaders may deploy sophisticated AI governance frameworks yet remain ethically shallow, while formative leaders integrate moral reflection into system design and decision-making processes.

Faith-inspired foresight belongs firmly to the formative tradition. It does not reject efficiency or innovation but subordinates them to ethical purpose. This distinction explains why organizations with similar AI capabilities often experience radically different ethical outcomes.

Faith-Inspired Foresight as a Leadership Formation Paradigm

Faith-inspired foresight ultimately functions as a leadership formation paradigm that prepares leaders to govern AI under conditions of uncertainty, scale, and moral risk. It cultivates virtues that sustain ethical clarity over time, enabling leaders to resist both fear-driven paralysis and technocratic domination. In doing so, it lays the moral foundation for the Systems–SDG Leadership Framework developed in the next section and anchors the chapter's broader commitment to human flourishing in AI-enabled global supply chains.

Table 1. Moral agency and strategic systems foresight typology for AI-enabled supply chain leadership

	Low Moral Agency Orientation	High Moral Agency Orientation
High Strategic Systems Foresight	**Type III:** Strategic Pragmatism Long-term planning and risk management without deep moral accountability.	**Type IV:** Faith-Inspired Foresight Stewardship-oriented leadership integrating moral responsibility, human dignity, and long-term systems design.
Low Strategic Systems Foresight	**Type I: Technocratic Instrumentalism** AI treated primarily as an efficiency tool, with ethics displaced to algorithms, compliance systems, or technical processes.	**Type II: Moral Reactionism** Strong ethical concern without the system's capacity needed to design, govern, or scale responsible AI.

Table 1 presents a typology of leadership approaches based on moral agency orientation and systems foresight. Table 2 builds on this structure by operationalizing each leadership type through specific indicators, governance mechanisms, and outcomes.

Table 2. Leadership typology in AI-enabled global supply chains

Leadership Type: Definition	Independent Variable Indicators (Moral Agency Orientation Scope)	Mediating Variables (Mechanisms)	Dependent Variable Indicators (Systems Foresight Outcomes)	Representative Literature
Type I Technocratic Instrumentalism: Leadership approach that treats AI as a neutral optimization tool and defers ethical responsibility to algorithms or compliance systems.	■ Responsibility displaced to AI systems ■ Ethics framed as compliance only ■ Human judgment minimized	■ Algorithmic opacity ■ Weak ethical review structures ■ Siloed decision authority	■ Short-term efficiency gains ■ System brittleness under shock ■ Ethical incidents rationalized post hoc	Brown & Treviño (2006); Floridi (2014); O'Neil (2017)
Type II Moral Reactionism Leadership approach marked by strong ethical concern but limited capacity to design or govern complex AI-enabled systems.	■ High moral concern expressed rhetorically ■ Fear-driven resistance to AI ■ Ethics framed as protection from harm	■ Absence of system modeling ■ Reactive decision-making ■ Fragmented governance	■ Delayed or abandoned AI initiatives ■ Lost innovation potential ■ Ethical paralysis rather than stewardship	Jonas (1984); Crawford (2021); Coeckelbergh (2022)
Type III Strategic Pragmatism Leadership approach that anticipates long-term system risks but treats ethics instrumentally rather than as intrinsic responsibility.	■ Narrow moral framing (risk-focused) ■ Ethics subordinated to strategy ■ Responsibility framed institutionally, not personally	■ Advanced risk analytics ■ Formal governance mechanisms ■ Performance-driven incentives	■ Resilient but impersonal systems ■ Stakeholder trust managed, not cultivated ■ Ethical legitimacy remains fragile	Boh et al. (2023); Grosse et al. (2023); Stahl (2023)

continued on following page

Table 2. Continued

Leadership Type: Definition	Independent Variable Indicators (Moral Agency Orientation Scope)	Mediating Variables (Mechanisms)	Dependent Variable Indicators (Systems Foresight Outcomes)	Representative Literature
Type IV Faith-Inspired Foresight Leadership paradigm integrating moral responsibility and long-term systems design to steward AI toward human flourishing.	■ Explicit acceptance of moral responsibility ■ Ethics grounded in stewardship and dignity ■ Human judgment prioritized over automation	■ Ethical intentionality in design ■ Transparent, accountable governance ■ Organizational learning and moral reflection	■ Resilient, human-centered supply chains ■ Sustained trust and legitimacy ■ Adaptive systems aligned with long-term flourishing	Benefiel et al. (2014); Balle (2023); Stahl (2023); Thaut (2009)

Building directly on Table 1, Table 2 converts the conceptual typology into a structured analytical framework by specifying observable indicators, governance mechanisms, and system-level outcomes for each leadership type. Rather than relying on abstract characterizations, the table identifies observable indicators, such as the locus of responsibility, the architecture of ethical review, and the distribution of decision authority, that give empirical and conceptual traction to leadership orientations. It further connects these mechanisms to patterned outcomes, including resilience under disruption, institutional legitimacy, stakeholder trust, and long-term adaptability. Anchored in representative scholarship across ethical leadership, AI governance, systems thinking, and stewardship ethics, the table both consolidates existing insights and clarifies how faith-inspired foresight integrates and extends these literatures.

Beyond operational clarification, the typology functions as an interpretive lens for analyzing morally complex dynamics at the intersection of faith, artificial intelligence, and global supply chains. By foregrounding moral agency and systems foresight as explicit dimensions of leadership, the framework enables examination of faith-inspired foresight as a practical logic shaping AI design, governance, and deployment across globally distributed networks. It also exposes the limitations of fear-based narratives associated with the AI Monster Fallacy by showing that ethical risk originates less in technological capability than in leadership orientations governing responsibility, anticipation, and institutional choice. In this way, the framework links theological reflection with supply chain complexity and AI ethics, providing a coherent foundation for theorizing how morally grounded leadership can redirect AI-enabled global systems toward stewardship, resilience, and shared human flourishing.

Building on this integrated framework, *both* Table 1 and Table 2 establish the conceptual and analytical groundwork for proposition development. The visual typology defines the structural space of leadership approaches, while the table operationalizes how moral orientations are enacted through specific mechanisms and realized in system-level outcomes. Together, these elements enable a transition from descriptive mapping to causal theorizing, linking moral agency (independent variable) to governance processes (mediating variables) and long-term supply chain performance (dependent variables). The next section draws directly on this structure to develop propositions explaining the distinctive advantages of faith-inspired foresight in AI-enabled global supply chains.

Before advancing to formal propositions, it is necessary to clarify the *process logic* through which faith-inspired foresight operates in AI-enabled global supply chains. Propositions derive their explanatory power not from moral aspiration alone, but from clearly specified leadership processes that govern how AI systems are designed, deployed, evaluated, and corrected over time. The preceding analysis therefore functions as a pre-propositional scaffold, establishing the operational architecture that links moral discernment to observable organizational and symbolic outcomes.

Faith-inspired foresight provides a normative orientation for leadership, but its analytical significance depends on whether ethical commitments are institutionalized into repeatable decision cycles. Leaders do not govern AI in the abstract; they do so through everyday choices involving forecasting systems, supplier selection, labor scheduling, automation investments, and logistics optimization. The Systems–SDG Leadership Framework articulates how such decisions can be structured so that AI governance reflects stewardship, systems awareness, and sustainable development priorities rather than narrow performance optimization.

By specifying stages of discernment, system diagnosis, governance choice, impact evaluation, and moral correction, the framework clarifies *how leadership intent becomes embedded in AI-mediated systems*. This process-oriented account establishes the causal logic necessary for proposition development by identifying where moral agency enters decision-making, how AI amplifies leadership intent, and under what conditions symbolic, social, and institutional outcomes diverge. The propositions developed in the following section build directly on this framework, translating its operational stages into testable expectations and case-based illustrations. Table 1 (operational specification) defines the causal architecture linking leadership orientation to AI-enabled supply chain outcomes.

PROPOSITION DEVELOPMENT

The following propositions draw directly from the leadership configurations operationalized in Table 1. Building on the Systems–SDG Leadership Framework, this section elaborates four propositions that explain how different patterns of leadership engagement across the framework's stages generate distinct symbolic and ethical outcomes in AI-mediated systems. The propositions specify not only what outcomes emerge, but why they emerge, by tracing how variations in moral discernment, system diagnosis, governance design, sustainability evaluation, and feedback correction shape AI's reinforcing dynamics over time. This approach aligns with prior work emphasizing that AI outcomes are shaped by organizational and leadership processes rather than technological properties alone (Orlikowski & Scott, 2015; Floridi et al., 2018).

Fragmented Personalization

Fragmented Personalization arises when AI adoption proceeds without sustained moral discernment or systemic awareness. In such contexts, leadership intent is weakly articulated, the broader system within which AI operates is insufficiently diagnosed, and governance choices are confined to isolated tools rather than integrated architectures. AI-driven personalization therefore remains localized and tactical, characterized by shallow reinforcement mechanisms and limited feedback loops. Symbolic meaning fails to accumulate across touchpoints, leaving identity signals situational and short-lived rather than durable.

This outcome reflects a leadership orientation best described as technocratic instrumentalism, in which AI adoption is driven primarily by narrow performance objectives and ethical responsibility is implicitly delegated to technical teams, compliance units, or the algorithms themselves. Weak moral agency reduces accountability for downstream consequences, while limited systems foresight obscures cumulative risks and feedback effects across labor, suppliers, and communities. As a result, mediating mechanisms such as ethics review, stakeholder engagement, and long-term risk assessment remain minimal or symbolic, allowing efficiency metrics to dominate decision-making.

Prior research on ethical leadership, AI governance, and algorithmic management consistently warns that such configurations foster moral disengagement and ethical fading, particularly in complex systems where harm is temporally and geographically distant. Empirical research shows that such configurations amplify ethical drift, externalize harm, and produce brittle systems (Brown & Treviño, 2006; O'Neil, 2017; Jolly et al., 2026). Illustrative cases include global retailers that deploy AI-driven demand forecasting and automated supplier selection to reduce costs, only to discover that aggressive optimization increases supplier dependency, exacerbates labor exploitation, and triggers reputational backlash, outcomes frequently attributed to "system constraints" rather than leadership choices. Therefore, it is proposed that:

> **Proposition 1.** When AI adoption in global supply chains is governed by technocratic instrumentalism characterized by low moral agency orientation and limited systems foresight, AI-enabled decisions will produce ethical drift, reduced legitimacy, and increased long-term fragility despite short-term efficiency gains.

This proposition establishes technocratic instrumentalism as a baseline failure mode, clarifying why efficiency-focused AI strategies often undermine trust and resilience rather than delivering sustainable advantage. It provides a diagnostic benchmark against which more morally grounded and foresight-driven leadership approaches can be evaluated, highlighting the hidden costs of delegating responsibility to algorithms and compliance systems. By surfacing these latent risks, the proposition underscores the necessity of re-centering leadership accountability in AI-enabled decision-making.

Efficiency-Driven Ethical Narrowing

Efficiency-Driven Ethical Narrowing emerges when leaders invest heavily in AI design and execution capabilities while constraining moral purpose and sustainability evaluation. Strong governance structures and operational scale enable intense repetition, rapid feedback loops, and high conversion efficiency, but narrow value definitions limit narrative breadth and identity diversity. As a result, AI reinforcement mechanisms amplify a restricted symbolic logic, producing efficiency gains alongside symbolic homogenization and partial identity rigidity. Ethical concerns are acknowledged, yet they remain subordinate to performance metrics that privilege speed, predictability, and short-term optimization.

This pattern reflects a leadership orientation characterized by asymmetric moral activation, in which governance and execution capacity outpace moral and systemic breadth. Leaders in such contexts do not lack ethical awareness; rather, moral purpose is narrowly framed and insufficiently integrated into system-level evaluation. Sustainability considerations are treated as secondary constraints rather than constitutive design criteria, limiting leaders' ability to assess trade-offs across labor conditions, community impact, and long-term resilience. Causally, this imbalance allows AI systems to scale efficiently while reinforcing a constrained set of meanings and identities, reducing adaptive capacity and cultural legitimacy over time. This limitation reflects broader ethical concerns that AI lacks the capacity to adjudicate value conflicts involving justice, responsibility, and human well-being, which must remain within the domain of human moral reasoning (Santoni de Sio, 2024).

Research in ethics and leadership underscores that governance strength without moral and systemic balance often yields ethically constrained outcomes rather than durable responsibility. Studies of algorithmic management show that when efficiency metrics dominate evaluative logic, AI systems tend to normalize exclusionary practices and suppress plural identity formation, even in organizations that publicly endorse ethical values. Illustrative cases include firms that deploy highly sophisticated AI-driven personalization and supply chain optimization while marginalizing broader stakeholder concerns, resulting in reputational risk and cultural backlash despite strong operational performance without corresponding ethical depth (Stahl, 2023; Coeckelbergh, 2022). Therefore, it is proposed that:

> **Proposition 2.** When AI governance and execution capabilities are strong but moral purpose and sustainability evaluation remain narrowly defined, AI-enabled supply chains will produce efficiency-driven ethical narrowing, characterized by operational gains, reduced moral imagination, stakeholder exclusion, and weakened long-term legitimacy.

This proposition highlights why ethical leadership requires more than technical governance and performance discipline. Without expansive moral purpose and anticipatory systems evaluation, AI-enabled efficiency can entrench narrow symbolic orders and weaken long-term legitimacy. By distinguishing governance strength from moral and systemic integration, the proposition underscores the need to align execution capacity with broader ethical and sustainability commitments to support resilient and inclusive AI-mediated systems.

Foresight Without Moral Ownership

Foresight Without Moral Ownership arises in leadership contexts marked by strong moral imagination and symbolic experimentation but limited institutionalization of reinforcement and corrective learning mechanisms. Leaders articulate expansive purposes and encourage plural narratives, enabling AI systems to support diverse forms of identity exploration and creative engagement. However, partial system diagnosis and weak feedback loops prevent these symbolic experiments from stabilizing at scale. As a result, reinforcement remains uneven, and symbolic coherence is provisional rather than consolidated at the organizational or market level.

This pattern reflects a leadership orientation best described as strategic pragmatism, characterized by strong systems foresight but weak moral ownership. Leaders operating within this logic excel at anticipating disruption, managing risk, and optimizing network resilience, yet treat ethics primarily as a compliance requirement or reputational constraint rather than a core responsibility. Causally, foresight enables sophisticated scenario planning and governance architectures, but the absence of moral agency externalizes ethical judgment to rules, audits, or third-party standards, producing temporary coherence without sustained trust or legitimacy (Floridi et al., 2018; Suchman, 1995). As a result, mediating mechanisms emphasize control and monitoring over moral deliberation, relational accountability, and participatory correction.

Research on compliance-based ethics and responsible innovation highlights the limitations of this approach. While strategic pragmatism can enhance robustness and reduce certain operational risks, it often fails to address deeper questions of dignity, justice, and responsibility, particularly in AI-mediated systems where harm may be indirect or normalized through formal compliance. Illustrative cases include multinational firms that deploy advanced AI governance frameworks and risk dashboards while continuing practices that marginalize small suppliers or obscure labor conditions, systems that are resilient in technical terms yet generate skepticism among stakeholders who perceive ethics as instrumental rather than authentic. Therefore, it is proposed that:

> **Proposition 3.** When systems foresight and innovation capacity are strong but moral ownership, feedback correction, and relational accountability remain underdeveloped, AI-enabled supply chains will produce foresight without moral ownership, characterized by temporary resilience, limited trust, weak ethical legitimacy, and unstable stakeholder confidence.

This proposition distinguishes resilience from flourishing, demonstrating that foresight without moral ownership can produce stability without justice or relational trust. By clarifying this distinction, it underscores that long-term sustainability in AI-enabled systems depends not only on anticipation and control, but on leaders' willingness to assume ethical responsibility for how resilience is achieved and for whom it is secured.

Symbolic System Reproduction

Symbolic System Reproduction represents full alignment across all stages of the Systems–SDG Leadership Framework. Moral discernment establishes clear human purpose, system diagnosis anticipates second- and third-order effects, governance choices embed accountability and scalability, sustainability evaluation guides ethical trade-offs, and feedback mechanisms enable continuous learning and correction. Under these conditions, AI operates not merely as an efficiency tool but as a symbolic system architect, repeatedly reinforcing coherent narratives, stabilizing identities, and shaping durable market-level meaning structures over time.

This outcome reflects a leadership orientation grounded in faith-inspired foresight, in which high moral agency and strong systems foresight are jointly cultivated. Leaders operating within this paradigm treat AI as a powerful instrument whose ethical implications demand active responsibility rather than passive oversight. Moral agency anchors accountability at the leadership level, while systems foresight enables anticipatory institutional design aligned with human dignity, social trust, and long-term resilience. Mediating mechanisms include participatory ethical review, human-in-the-loop governance, stakeholder dialogue, and sustained investment in capability building that integrates technological and human development.

Research on stewardship and values-based leadership shows that such integration enhances resilience, legitimacy, and long-term adaptability in complex socio-technical systems (Benefiel et al., 2014; Balle, 2023; Stahl, 2023). AI ethics scholarship increasingly emphasizes integrative approaches that combine governance, design, and moral imagination rather than treating ethics as a procedural add-on. Illustrative cases include firms that deploy AI to enhance supply chain transparency while simultaneously investing in worker reskilling, explainable decision systems, and relational supplier partnerships, contexts in which AI strengthens trust and resilience because it is governed as a shared moral project rather than a technical upgrade. Therefore, it is proposed that:

> **Proposition 4**. When leadership integrates moral discernment, system diagnosis, governance design, sustainability evaluation, and feedback correction into a coherent decision cycle, AI-mediated systems will reproduce stable symbolic orders that support identity consolidation, ethical resilience, and long-term flourishing.

This proposition serves as the normative and analytical anchor of the framework, clarifying why faith-inspired foresight offers a leadership logic capable of moving beyond fear-based narratives toward stewardship-driven innovation. By foregrounding responsibility rather than control, it demonstrates how integrating moral commitment with anticipatory systems awareness enables leaders to design AI for dignity, trust, and shared value across global networks.

Taken together, these propositions clarify that symbolic and ethical outcomes in AI-mediated systems are not inherent properties of technology, but emergent consequences of leadership practice across identifiable stages of governance. The following section illustrates these propositions through comparative global cases, demonstrating how different leadership cultures and institutional contexts activate, or neglect, specific elements of the Systems–SDG Leadership Framework in practice.

DISCUSSION

Building on the integration of moral agency, systems foresight, and leadership-driven causal dynamics, this discussion synthesizes the chapter's theoretical contributions within broader interdisciplinary debates. It advances an integrated explanation of how leadership orientations, not AI capabilities, shape ethical risk, legitimacy, resilience, and human flourishing, while identifying key areas of contribution and future research directions. These patterns reflect the leadership typologies outlined in Table 1, reinforcing how variations in moral agency and systems foresight shape real-world outcomes.

Table 3 synthesizes the chapter's core argumentative strands by mapping each major section to its central themes, supporting literature, and underlying rationale. Rather than serving as a summary alone, the table functions as an integrative bridge linking the critique of the AI Monster Fallacy, the development of faith-inspired foresight, the emphasis on moral formation and human flourishing, and the global supply chain cases. It provides an orienting framework for the discussion that follows, clarifying how the chapter's contributions cohere across disciplines and sections.

Building on this integrative synthesis, the discussion now turns to concrete national contexts where these leadership logics are enacted in practice. By moving from conceptual alignment to empirical illustration, the analysis highlights how cultural norms, institutional arrangements, and leadership values shape the governance of AI in global supply chains. The Korean case provides a compelling starting point, offering a context in which collective responsibility and stewardship-oriented leadership illuminate how faith-inspired foresight can be embedded within technologically advanced systems.

Table 3. Integrating faith-inspired foresight, AI ethics, and global supply chain leadership

Key Section	Key Themes	Supporting References	Rationale
AI Monster Fallacy & Ethical Leadership Crisis	AI fear narratives; political power of AI; ethical leadership deficits	Crawford (2021); Coeckelbergh (2022); Brown & Treviño (2006)	Establishes how AI systems shape power, moral risk, and why leadership failures, not AI itself, drive ethical crises.
Faith-Inspired Foresight Framework (Spiritual Discernment, Ethical Intentionality, Systems Thinking)	Theology of stewardship; moral agency; responsible AI ecosystems	Balle (2023); Stahl (2023); Benefiel et al. (2014)	Supports a faith-informed leadership model that integrates spiritual discernment with responsible AI design and ecosystem governance.
Alignment with Moral Development & Human Flourishing	Character formation; moral imagination; flourishing ethics	Kantar & Bynum (2021); Cheong & Liu (2025); Jonas (1984); Bankins & Formosa (2023); Giarmoleo et al. (2026)	Grounds the chapter's argument that flourishing ethics, virtue, and moral responsibility must guide AI-mediated leadership.
Global Insights: Korea, U.S., and Emerging Economies	Cross-cultural AI ethics; digital transformation; human-centric supply chains	Grosse et al. (2023); Boh et al. (2023); Hong & Park (2020)	Demonstrates how diverse cultures embed AI ethics differently and how resilience and human-centered design emerge globally.
Redemptive Leadership in AI-Driven Systems	Moral courage; relational accountability; transparency in supply chains	Searcy et al. (2022); Dubber et al. (2020); O'Neil (2017)	Strengthens the argument for leaders who confront algorithmic injustice and embed stewardship and accountability into AI systems.

Korea: Ethics-Infused Manufacturing, Collective Responsibility, and AI Stewardship

Korea offers a distinctive context for examining faith-inspired foresight in AI-enabled global supply chains because of its unique convergence of rapid technological adoption, dense manufacturing networks, and culturally embedded norms of collective responsibility. As a late industrializer that achieved global competitiveness within a single generation, Korea's supply chains have long balanced speed, coordination, and national resilience. AI adoption intensifies these dynamics, amplifying both ethical risks and stewardship opportunities.

Korean manufacturing firms have aggressively integrated AI into production planning, quality control, and logistics optimization. These technologies support global competitiveness in semiconductors, electronics, shipbuilding, and advanced materials. However, the moral significance of AI adoption in Korea cannot be understood apart from cultural expectations that firms bear responsibility not only to shareholders but also to workers, suppliers, and national development goals. This moral embeddedness creates fertile ground for faith-inspired foresight, even when leadership is not explicitly religious.

A defining feature of the Korean case is the collective framing of responsibility. AI deployment decisions are often evaluated in terms of their systemic impact on employment stability, supplier ecosystems, and national technological sovereignty. Rather than treating labor purely as a cost variable, leading firms invest heavily in reskilling, internal labor mobility, and long-term workforce integration. These practices reflect a stewardship logic that prioritizes continuity and dignity over short-term efficiency.

Faith-inspired foresight is visible where leaders explicitly resist full automation in favor of human-AI complementarity. In smart factories, AI is used to augment human judgment rather than replace it, preserving skilled roles and reinforcing professional identity. Ethical governance structures emphasize transparency and escalation, ensuring that AI recommendations affecting safety or labor conditions are subject to human review. These practices align with flourishing-oriented leadership by safeguarding meaning and agency at work.

At the same time, the Korean case reveals tensions. Competitive pressure and global market volatility tempt firms toward technocratic instrumentalism, particularly in subcontracting and overseas operations. Faith-inspired foresight functions here as a **corrective orientation**, reminding leaders that domestic ethical commitments must extend across global networks. Where this extension fails, ethical drift becomes visible, underscoring the importance of moral agency orientation in sustaining stewardship beyond national boundaries.

Overall, Korea illustrates how AI-enabled supply chains can be governed through a moral logic of collective responsibility rather than individual optimization. While not without contradictions, the Korean experience demonstrates that faith-inspired foresight, understood as stewardship, humility, and long-term responsibility, can shape AI adoption in ways that support resilience, dignity, and national flourishing.

United States: Innovation Speed and Ethical Fragmentation in AI Leadership

The United States represents a contrasting yet equally instructive context for examining faith-inspired foresight in AI-enabled global supply chains. As the global epicenter of AI innovation, platform capitalism, and venture-backed technological experimentation, the U.S. supply chain environment is characterized by extraordinary speed, decentralization, and competitive intensity. These features have enabled rapid advances in forecasting, automation, logistics orchestration, and platform-based coordination, but they have also generated persistent ethical fragmentation that challenges human flourishing.

U.S. firms have been among the earliest adopters of AI-driven demand forecasting, algorithmic labor management, autonomous logistics, and data-intensive supplier analytics. In retail, e-commerce, and third-party logistics, AI systems increasingly determine work schedules, productivity targets, routing decisions, and vendor inclusion. These capabilities have produced measurable gains in efficiency, responsiveness, and scalability, reinforcing a cultural narrative that equates technological leadership with competitive advantage. Yet this same narrative often marginalizes moral reflection, treating ethical concerns as constraints to be managed rather than responsibilities to be embraced.

A defining feature of the U.S. case is the market-centered framing of leadership responsibility. Decision authority is frequently dispersed across firms, platforms, and contractors, enabling innovation but also diffusing accountability. AI systems are introduced rapidly, often under the logic of experimentation, "move fast and optimize", with ethical governance added ex post in response to public scrutiny or regulatory pressure. This pattern aligns closely with the leadership orientation described earlier as strategic pragmatism: strong systems foresight oriented toward risk and performance, coupled with relatively weak moral agency orientation.

The consequences for workers and communities are mixed. On the one hand, AI-enabled supply chains in the U.S. have created new forms of work, expanded consumer access, and enhanced logistical resilience during periods of disruption. On the other hand, algorithmic management has intensified job insecurity, performance surveillance, and psychological strain, particularly among warehouse workers, drivers, and gig-economy participants. Ethical concerns surrounding transparency, dignity, and voice are often addressed through compliance mechanisms or public relations initiatives rather than through substantive redesign of AI governance.

Faith-inspired foresight appears in the U.S. context primarily as a countercultural leadership posture rather than a dominant norm. Where it emerges, it does so through leaders and organizations that resist the reduction of AI governance to market efficiency alone. Examples include firms that voluntarily slow deployment to incorporate worker consultation, invest in explainable AI to preserve trust, or embed ethical escalation processes that allow human judgment to override algorithmic recommendations. These practices reflect stewardship-oriented leadership, even when not explicitly framed in religious terms.

However, the U.S. case also reveals structural constraints. Competitive pressure, shareholder expectations, and fragmented regulatory environments make it difficult for individual firms to internalize the full moral costs of AI adoption. Faith-inspired foresight thus functions less as an institutional default and more as a moral corrective, reminding leaders that innovation without responsibility erodes legitimacy and long-term resilience. The absence of such correction often results in ethical externalization, harms borne by workers, communities, or ecosystems rather than by decision-makers.

In comparison with Korea's collectivist orientation, the U.S. model highlights the risks of treating AI governance primarily as a technical or legal problem. While the American system excels at innovation and adaptability, it struggles to sustain ethical coherence across distributed supply chains. This contrast underscores the chapter's central claim: AI outcomes are shaped less by technological sophistication than by leadership orientations governing responsibility, foresight, and moral purpose.

The U.S. case therefore illustrates both the promise and the peril of AI-enabled supply chains. Where faith-inspired foresight is absent, efficiency gains coexist with moral fragility and declining trust. Where it is present, leaders demonstrate that innovation and stewardship are not mutually exclusive, but achieving this integration requires intentional resistance to purely market-driven logic. As such, the U.S. experience reinforces the necessity of leadership formation that places human flourishing, not speed alone, at the center of AI governance.

Linking Moral Development, Human Flourishing, and AI Governance

The comparative cases of Korea and the United States demonstrate that AI-enabled supply chains are governed through distinct leadership logics shaped by culture, institutional context, and moral orientation. While these cases reveal different pathways and constraints, they converge on a common insight: technological capability alone does not determine ethical or strategic outcomes. What ultimately differentiates resilient and legitimate systems is whether leadership remains accountable for the human consequences of AI adoption. This insight directs attention to the chapter's next concern, human flourishing, as the decisive criterion by which faith-inspired foresight must be evaluated.

HUMAN FLOURISHING IN AI-ENABLED GLOBAL SUPPLY CHAINS

Human flourishing is often assumed to follow from economic growth and operational efficiency, yet this assumption becomes fragile in AI-enabled global supply chains where optimization does not inherently protect dignity, meaning, or well-being. Faith-inspired foresight challenges this logic by treating flourishing as a primary leadership outcome and recognizing AI as a formative force that shapes work, responsibility, and implicit definitions of human worth. This concern aligns with emerging scholarship showing that AI systems can reshape the conditions for meaningful work, raising fundamental ethical questions about autonomy, purpose, and human fulfillment in technologically mediated environments (Bankins & Formosa, 2023; Santoni de Sio, 2024). Algorithmic management illustrates this tension: while AI-driven coordination improves efficiency, it risks reducing workers to performance variables unless leaders preserve recognition, agency, and moral oversight. Because AI can optimize efficiency trade-offs but cannot adjudicate moral ones involving safety, equity, community stability, or ecological integrity, leadership judgment grounded in stewardship and intergenerational responsibility remains indispensable.

Reframing Human Flourishing Beyond Efficiency

Human flourishing has long been treated as an indirect byproduct of economic growth and operational efficiency. In AI-enabled global supply chains, this assumption becomes increasingly fragile. While algorithmic optimization can enhance speed, accuracy, and cost efficiency, it does not inherently safeguard dignity, meaning, or well-being. Indeed, when efficiency becomes the dominant evaluative lens, human flourishing is often reduced to residual outcomes, addressed only after productivity targets are met or compliance thresholds are satisfied.

Faith-inspired foresight challenges this reduction by reframing flourishing as a primary leadership outcome rather than a secondary constraint. From this perspective, AI is not morally neutral infrastructure but a formative force shaping how work is organized, how responsibility is distributed, and how human worth is implicitly defined. The ethical question therefore shifts from whether AI improves performance to whether it advances lives and communities capable of sustaining meaning, dignity, and hope over time.

Worker Dignity and Meaning under Algorithmic Management

One of the most immediate sites where AI affects flourishing is the domain of work. Algorithmic scheduling, performance monitoring, and task allocation increasingly structure the daily lives of workers across logistics, manufacturing, and service supply chains. While such systems can reduce waste and improve coordination, they also risk transforming workers into data points optimized for throughput rather than persons engaged in meaningful labor. Studies further indicate that algorithmic systems can erode perceived social status by replacing human judgment in relational roles, thereby diminishing recognition and psychological standing within organizations (Jago et al., 2024).

Faith-inspired foresight insists that worker dignity cannot be inferred from productivity metrics alone. Dignity arises from recognition, agency, and the capacity to participate meaningfully in decisions that shape one's working life. AI systems that obscure decision rationales, eliminate discretion, or impose opaque performance pressures erode this dignity even when wages or output improve.

Leadership oriented toward flourishing therefore introduces intentional safeguards: human-in-the-loop governance, transparency in algorithmic decisions, and explicit consideration of how AI reshapes work rhythms, family life, and long-term employability. In this view, reskilling initiatives, participatory design, and ethical escalation mechanisms are not optional add-ons but core expressions of stewardship.

Efficiency–Flourishing Trade-Offs and Moral Discernment

AI systems excel at resolving trade-offs within predefined objective functions, but they are ill-equipped to adjudicate moral trade-offs, for example, between speed and safety, cost minimization and community stability, or resilience and equity. These trade-offs require discernment, not calculation.

Faith-inspired foresight positions leaders as the locus where such discernment must occur. Rather than delegating moral judgment to optimization logic, leaders must explicitly decide which trade-offs are acceptable and which violate fundamental commitments to human flourishing. This does not imply rejecting efficiency, but it does require rejecting efficiency as the supreme value.

In practice, flourishing-oriented leadership may accept lower short-term gains to preserve workforce stability, supplier diversity, or environmental integrity. These decisions often appear "irrational" within narrow performance frameworks but prove strategically wise over time by sustaining trust, adaptability, and legitimacy, key ingredients of resilient supply chains.

Community Well-Being, Intergenerational Responsibility, and Ecology

Human flourishing extends beyond individual workers to the communities and ecosystems embedded in global supply chains. AI-driven supplier consolidation, location optimization, and automation can hollow out local economies, intensify ecological strain, or concentrate risk in vulnerable regions, even as aggregate performance improves.

Faith-inspired foresight introduces intergenerational responsibility as a guiding principle. Leadership decisions are evaluated not only by present outcomes but by the kinds of social and ecological systems they bequeath to the future. This orientation resonates strongly with stewardship ethics, which frame leadership as care for what one has received rather than extraction for immediate gain.

Ecological flourishing is inseparable from human flourishing in this framework. AI systems that accelerate resource depletion or externalize environmental costs undermine the moral legitimacy of efficiency gains. Leaders committed to flourishing therefore integrate ecological indicators into AI governance, ensuring that optimization does not erode the material conditions of life itself.

Flourishing as the Telos of Faith-Inspired AI Leadership

Taken together, these dimensions clarify why human flourishing must be treated as the telos, the ultimate purpose, of AI-enabled leadership rather than a peripheral concern. Faith-inspired foresight provides the moral grammar necessary to hold together efficiency, innovation, dignity, and sustainability without collapsing one into the other.

This approach does not romanticize work or resist technological change. Instead, it insists that leadership remain accountable for the human and ecological worlds AI helps to shape. By grounding AI governance in vocation, stewardship, and the common good, faith-inspired foresight offers a vision of global supply chains not merely as engines of production, but as institutions capable of sustaining life, trust, and shared meaning in an age of intelligent systems. This vision resonates with recent work on virtuous organizations, which emphasizes the creation of relational goods and institutional conditions that support human flourishing in technologically mediated environments (Giarmoleo et al., 2026).

Centering human flourishing clarifies the normative purpose of faith-inspired foresight, but it also raises an important question: how does this approach relate to existing secular frameworks for AI ethics and governance? Rather than displacing compliance-based, utilitarian, or procedural models, faith-inspired foresight enters into dialogue with them, revealing both their contributions and their limitations. The following section situates faith-informed leadership alongside dominant secular approaches, demonstrating how it adds moral depth, endurance, and formative capacity to contemporary debates on responsible AI.

LESSONS AND IMPLICATIONS

This study advances a faith-inspired foresight perspective on AI leadership by integrating moral formation, systems thinking, and sustainable development into a coherent governance framework for AI-enabled global supply chains. Through conceptual development and proposition building, the research demonstrates that symbolic, ethical, and organizational outcomes of AI are not technologically determined but emerge from leadership choices embedded across stages of discernment, system diagnosis, governance design, impact evaluation, and learning. Several limitations should be acknowledged. First, the framework is conceptual and proposition-driven, requiring future empirical validation across industries and regions. Second, the analysis emphasizes leadership agency and moral orientation, potentially underrepresenting structural constraints such as regulatory asymmetries or market power. Despite these limitations, the study offers robust theoretical and practical implications by clarifying how leadership processes shape AI-mediated outcomes.

The Systems–SDG Leadership Framework translates faith-inspired foresight into five operational stages, moral discernment, system diagnosis, AI design and governance, SDG impact evaluation, and feedback correction, each linked to specific leadership orientations, governance practices, and sustainable development outcomes. A full specification of these components, including relevant SDG alignments, is provided in Appendix A. Leaders can use this framework as a diagnostic tool to identify governance gaps and redesign AI initiatives toward dignity, accountability, and long-term resilience.

Theoretical Implications

The study's theoretical contribution lies not only in what it demonstrates about AI ethics, but in how it re-specifies the locus of ethical responsibility, from abstract principles to leadership processes embedded in complex systems.

First, this study advances AI ethics and leadership theory by reframing AI governance as a processual and stage-based phenomenon rather than a static set of principles or compliance mechanisms. By articulating distinct stages, from moral discernment and system diagnosis to governance design and SDG impact evaluation, the Systems–SDG Leadership Framework advances theory beyond normative checklists toward a dynamic model of moral agency embedded in organizational decision cycles. This perspective helps explain why similar AI technologies generate divergent ethical and symbolic outcomes across firms, addressing a persistent gap in AI ethics and responsible innovation literatures that often under-theorize organizational mediation.

Extending this insight, the framework aligns with and advances scholarship that emphasizes ethics-as-practice rather than ethics-as-policy. Prior studies have shown that ethical failures in AI systems frequently arise not from the absence of principles, but from breakdowns in translation, accountability, and feedback across organizational layers. By specifying where moral judgment enters the decision cycle and how it is sustained or eroded over time, this study advances a missing process logic that connects leadership intent to observable governance outcomes. In doing so, it complements and extends work on responsible AI, ethical drift, and moral disengagement in large-scale socio-technical systems (Floridi et al., 2018; Moor, 2006; Treviño et al., 2014).

Second, the research extends systems theory and symbolic systems scholarship by demonstrating how leadership intent is amplified through AI-mediated reinforcement mechanisms. The propositions show that symbolic stability, identity formation, and ethical resilience emerge only when moral agency and systems foresight are jointly activated across governance stages. This insight challenges technocentric and determinist accounts of AI that treat outcomes as functions of model design or data quality alone, emphasizing instead the relational and interpretive dimensions through which AI reshapes meaning, legitimacy, and power in global supply chains.

More broadly, this contribution situates AI within traditions of symbolic interaction and institutional theory, where meaning and legitimacy are reproduced through repeated practices rather than isolated decisions. AI systems intensify this dynamic by accelerating reinforcement, visibility, and scale, making leadership choices about governance architectures especially consequential. By integrating systems thinking with symbolic systems analysis, the study offers a theoretical bridge between AI ethics, organizational symbolism, and moral leadership, showing how ethical outcomes are stabilized, or distorted, through feedback loops and institutional design (Orlikowski & Scott, 2015; Suchman, 1995; Weick, 1995).

In summary, the theoretical contribution of this study lies in integrating moral formation, systems thinking, and AI governance into a unified explanatory framework. By grounding AI ethics in leadership processes rather than abstract norms or technical safeguards, the study provides a foundation for future theory building on AI, symbolic systems, and faith-informed leadership in complex socio-technical environments.

Practical Implications

The practical relevance of this study lies in its ability to convert ethical aspiration into actionable leadership guidance for governing AI in complex, high-velocity supply chain environments.

First, Appendix A (Systems–SDG Leadership Framework) functions as a diagnostic and design tool for leaders seeking to operationalize ethical AI governance in global supply chains. By explicitly linking leadership orientations, such as humility, stewardship, and moral courage, to concrete AI governance practices, including human-in-the-loop design, accountability assignment, and structured ESG–SDG trade-off analysis, the framework enables leaders to identify where ethical breakdowns are most likely to occur. Practitioners can use Appendix A to detect misalignments, such as strong execution capacity without moral discernment or formal compliance without learning and correction, and to redesign governance structures accordingly.

Beyond diagnosis, the framework supports organizational redesign and leadership development. Rather than relying on ad hoc ethics committees or post hoc compliance audits, leaders are encouraged to embed ethical responsibility into everyday decision cycles. This approach aligns with emerging best practices in responsible AI governance, which emphasize proactive design, distributed accountability, and continuous monitoring over static policy statements. By clarifying who is responsible at each stage of AI decision-making, the framework reduces responsibility diffusion and strengthens ethical ownership across managerial levels (Raji et al., 2020).

Second, the framework provides actionable guidance for aligning AI leadership with human flourishing and sustainable development goals. Instead of treating SDGs as external reporting requirements, Appendix A integrates them directly into leadership decision stages, prompting leaders to evaluate AI initiatives based on their effects on decent work, inequality, institutional integrity, community well-being, and ecological sustainability. This integration helps organizations move beyond narrow efficiency metrics toward stewardship-driven innovation that balances performance with dignity and long-term social value.

Importantly, this approach equips leaders to navigate trade-offs rather than deny them. AI-enabled efficiency gains often generate distributional consequences across workers, suppliers, and regions, and the framework offers a structured way to surface and evaluate these impacts before they become crises. By embedding SDG-oriented evaluation into AI governance, leaders can strengthen stakeholder trust, reduce reputational risk, and build legitimacy in environments where algorithmic decisions are increasingly scrutinized. Prior research suggests that such integrative governance enhances not only ethical outcomes but also organizational resilience and adaptive capacity (Bansal & Song, 2017).

In summary, the practical value of this research lies in translating faith-inspired foresight into repeatable leadership practices that govern AI responsibly at scale. By using Appendix A as an operational reference, leaders can design AI systems not merely for performance optimization, but for ethical accountability, systemic resilience, and shared human flourishing across global networks.

CONCLUSION

The analysis has demonstrated that ethical and symbolic outcomes in AI-enabled global supply chains are not inherent properties of technology but emergent consequences of leadership practice. By integrating faith-inspired foresight with systems thinking and AI governance, the Systems–SDG Leadership Framework clarifies how moral discernment, institutional design, and feedback-driven learning shape whether AI amplifies efficiency alone or advances dignity, trust, and long-term flourishing. The propositions developed in this study demonstrate that leadership orientations, ranging from technocratic instrumentalism to faith-inspired stewardship, systematically condition AI's symbolic power and ethical impact.

More broadly, the study advances a vision of AI leadership grounded in responsibility rather than control and foresight rather than fear. In an era where algorithmic systems increasingly mediate work, value, and meaning, leadership cannot be reduced to optimization or compliance. It must be exercised as moral agency embedded in systems, accountable to communities, and oriented toward the future we are collectively building. *AI will shape our systems, but leadership will determine whether those systems merely optimize performance or genuinely serve human flourishing.*

REFERENCES

Balle, S. (2023). Theological dimensions of humanlike robots: A roadmap for theological inquiry. *Theology and Science*, *21*(1), 132–156. DOI: 10.1080/14746700.2022.2155916

Bankins, S., & Formosa, P. (2023). The ethical implications of artificial intelligence (AI) for meaningful work. *Journal of Business Ethics*, *185*(4), 725–740. DOI: 10.1007/s10551-023-05339-7

Bansal, P., & Song, H. C. (2017). Similar but not the same: Differentiating corporate sustainability from corporate responsibility. *The Academy of Management Annals*, *11*(1), 105–149. DOI: 10.5465/annals.2015.0095

Benefiel, M., Fry, L. W., & Geigle, D. (2014). Spirituality and religion in the workplace: History, theory, and research. *Psychology of Religion and Spirituality*, *6*(3), 175–187. DOI: 10.1037/a0036597

Boh, W., Constantinides, P., Padmanabhan, B., & Viswanathan, S. (2023). Building digital resilience against major shocks. *Management Information Systems Quarterly*, *47*(1), 343–360.

Brown, M. E., & Treviño, L. K. (2006). Ethical leadership: A review and future directions. *The Leadership Quarterly*, *17*(6), 595–616. DOI: 10.1016/j.leaqua.2006.10.004

Cheong, P. H., & Liu, L. (2025). Faithful innovation: Negotiating institutional logics for AI value alignment among Christian churches in America. *Religions*, *16*(3), 302. DOI: 10.3390/rel16030302

Coeckelbergh, M. (2022). *The Political Philosophy of AI: An Introduction*. John Wiley & Sons.

Crawford, K. (2021). *Atlas of AI: Power, Politics, and the Planetary Costs of Artificial Intelligence*. Yale University Press.

Dubber, M. D., Pasquale, F., & Das, S. (Eds.). (2020). *The Oxford Handbook of Ethics of AI*. Oxford University Press. DOI: 10.1093/oxfordhb/9780190067397.001.0001

Floridi, L. (2014). *The Fourth Revolution: How the Infosphere Is Reshaping Human Reality*. Oxford University Press.

Floridi, L., Cowls, J., Beltrametti, M., Chatila, R., Chazerand, P., Dignum, V., Luetge, C., Madelin, R., Pagallo, U., Rossi, F., Schafer, B., Valcke, P., & Vayena, E. (2018). AI4People, An ethical framework for a good AI society. *Minds and Machines*, *28*(4), 689–707. DOI: 10.1007/s11023-018-9482-5 PMID: 30930541

Giarmoleo, F. V., Ruiz, P. G., Rocchi, M., & Ferrero, I. (2026). Virtuous organizations in the age of AI: Relational goods and human flourishing. *Business Ethics, the Environment & Responsibility*, 1–15. DOI: 10.1111/beer.70090

Grosse, E. H., Sgarbossa, F., Berlin, C., & Neumann, W. P. (2023). Human-centric production and logistics system design and management: Transitioning from Industry 4.0 to Industry 5.0. *International Journal of Production Research, 61*(22), 7749–7759. DOI: 10.1080/00207543.2023.2246783

Hong, P., & Park, Y. W. (2020). *Rising Asia and American hegemony: Case of competitive firms from Japan*. Springer Singapore. DOI: 10.1007/978-981-13-7635-1

Jago, A. S., Raveendhran, R., Fast, N., & Gratch, J. (2024). Algorithmic management diminishes status: An unintended consequence of using machines to perform social roles. *Journal of Experimental Social Psychology, 110*, 104553. DOI: 10.1016/j.jesp.2023.104553

Jolly, A. A., Dunlop, P. D., Parker, S. K., & Kanse, L. (2026). It's not my responsibility: Working with autonomy-restricting algorithms facilitates unethical behavior and displacement of responsibility. *Journal of Business Ethics, 203*(2), 405–424. DOI: 10.1007/s10551-025-06034-5

Jonas, H. (1984). The Imperative of Responsibility. In *Search of an Ethics for the Technological Age*. University of Chicago Press. DOI: 10.7208/chicago/9780226850337.001.0001

Kantar, N., & Bynum, T. W. (2021). Global ethics for the digital age–flourishing ethics. *Journal of Information Communication and Ethics in Society, 19*(3), 329–344. DOI: 10.1108/JICES-01-2021-0016

Kouvelis, P., Chambers, C., & Wang, H. (2006). Supply chain management research and production and operations management: Review, trends, and opportunities. *Production and Operations Management, 15*(3), 449–469. DOI: 10.1111/j.1937-5956.2006.tb00257.x

Moor, J. H. (2006). Why we need better ethics for emerging technologies. *Ethics and Information Technology, 7*(3), 111–119. DOI: 10.1007/s10676-006-0008-0

O'Neil, C. (2017). Weapons of Math Destruction: How Big Data Increases Inequality and Threatens Democracy. Crown, New York, NY.

Orlikowski, W. J., & Scott, S. V. (2015). The algorithm and the crowd: Considering the materiality of service innovation. *Management Information Systems Quarterly, 39*(1), 201–224. DOI: 10.25300/MISQ/2015/39.1.09

Park, Y. W., & Hong, P. (2024). Cosmetics marketing strategy in the era of the digital ecosystem. Springer Nature Singapore. DOI: 10.1007/978-981-97-3674-4

Raji, I.D., Smart, A., White, R.N., Mitchell, M., Gebru, T., Hutchinson, B., Smith-Loud, J., Theron, D., & Barnes, P., (2020). Closing the AI accountability gap: Defining an end-to-end framework for internal algorithmic auditing. *Proceedings of the 2020 Conference on Fairness, Accountability, and Transparency*, 33–44. DOI: 10.1145/3351095.3372873

Santoni de Sio, F. (2024). Artificial intelligence and the future of work: Mapping the ethical issues. *The Journal of Ethics*, *28*(3), 407–427. DOI: 10.1007/s10892-024-09493-6

Searcy, C., Castka, P., Mohr, J., & Fischer, S. (2022). Transformational transparency in supply chains: Leveraging technology to drive radical change. *California Management Review*, *65*(1), 19–43. DOI: 10.1177/00081256221126204

Stahl, B. C. (2023). Embedding responsibility in intelligent systems: From AI ethics to responsible AI ecosystems. *Scientific Reports*, *13*(1), 7586. DOI: 10.1038/s41598-023-34622-w PMID: 37202436

Suchman, M. C. (1995). Managing legitimacy. *Academy of Management Review*, *20*(3), 571–610. DOI: 10.2307/258788

Thaut, L. C. (2009). The role of faith in Christian faith-based humanitarian agencies: Constructing the taxonomy. *Voluntas*, *20*(4), 319–350. DOI: 10.1007/s11266-009-9098-8

Treviño, L. K., den Nieuwenboer, N. A., & Kish-Gephart, J. J. (2014). (Un)ethical behavior in organizations. *Annual Review of Psychology*, *65*(1), 635–660. DOI: 10.1146/annurev-psych-113011-143745 PMID: 23834354

Weick, K. E. (1995). *Sensemaking in Organizations*. Sage.

KEY TERMS AND DEFINITIONS

AI Monster Fallacy: The mistaken belief that AI itself is the primary ethical threat, rather than the human choices, leadership failures, and governance systems that shape how AI is used.

Ethical Drift: The gradual weakening of ethical standards when leaders rely too heavily on efficiency metrics, algorithms, or compliance systems without sustained moral reflection.

Ethical Intentionality: The deliberate effort to align AI decisions with justice, dignity, compassion, responsibility, and the common good.

Faith-Inspired Foresight: A leadership approach that uses faith-informed moral reflection, ethical responsibility, and long-term systems thinking to guide AI adoption toward stewardship and human flourishing.

Human Flourishing: The fuller well-being of people and communities, including dignity, meaningful work, trust, justice, ecological care, and the ability to live and work with purpose.

Moral Agency: The human responsibility to make ethical judgments and remain accountable for decisions, even when those decisions are supported or shaped by AI systems.

Moral Reactionism: A leadership response marked by ethical concern and fear of harm, but without the systems capacity needed to govern AI wisely.

Spiritual Discernment: The practice of interpreting technological change with humility, moral clarity, and a sense of higher purpose rather than fear, pride, or efficiency alone.

Strategic Pragmatism: A leadership approach that uses long-term planning and risk management but treats ethics mainly as a strategic or reputational concern rather than a deep moral responsibility.

Strategic Systems Thinking: The ability to understand how AI decisions affect interconnected workers, suppliers, communities, ecosystems, and future outcomes across global supply chains.

Technocratic Instrumentalism: A leadership style that treats AI mainly as an efficiency tool and shifts ethical responsibility to technology, data systems, or compliance processes.

APPENDIX

Table 4. Systems–SDG leadership framework: operational components

Framework Stage	Leadership Orientation	AI Governance Practices	Systems Focus	Relevant SDGs / Outcomes
Moral Discernment	Humility; moral imagination	Purpose definition; ethical boundary setting	Clarifies moral intent	SDG 16 (Institutions), SDG 8 (Decent Work)
System Diagnosis	Responsibility; foresight	Mapping decision authority; identifying feedback loops	Anticipates second-order effects	SDG 9 (Industry), SDG 10 (Inequality)
AI Design & Governance	Moral courage; stewardship	Human-in-the-loop; explainability; accountability assignment	Prevents responsibility diffusion	SDG 12 (Responsible Production)
SDG Impact Evaluation	Justice; care for the common good	ESG metrics; SDG trade-off analysis	Balances efficiency and dignity	SDG 13 (Climate), SDG 11 (Communities)

Section 3
Human Flourishing

Chapter 12
After MacIntyre:
Re-Evaluating Christian Ethics for the AI Era

Martin R. Jones
http://orcid.org/0009-0008-1924-1222
Anderson University, USA

ABSTRACT

November 30, 2022, marked the public release of ChatGPT and with it, the beginning of a new chapter in human history. The rapid advance of generative artificial intelligence has accelerated the shift toward moral individualism and preference-based ethical reasoning, raising urgent questions for the Christian tradition. The following chapter re-evaluates the ethical methodology of Alasdair MacIntyre in light of the AI era. While MacIntyre's neo-Aristotelian emphasis on virtue offered a compelling response to late modern ethical confusion, the emergence of AI systems capable of shaping moral language and understanding demands further development. By revisiting A Short History of Ethics, After Virtue, and Ethics in the Conflicts of Modernity, this chapter argues that virtue alone is no longer sufficient. The AI era requires a renewed integration of virtue with a principled deontological framework grounded in objective moral obligation, forming an "After MacIntyre" ethic for responsible leadership in an AI-saturated world.

ETHICS AT THE DAWN OF THE AI ERA

Certain events irreversibly punctuate history. Much like black ink spilled onto a white canvas cannot be undone, so too, there are moments that indelibly alter the course of human existence. The most central of these is, of course, the life, death, and resurrection of Jesus Christ, not only because of its redemptive significance for

DOI: 10.4018/979-8-3373-8192-3.ch012

ght © 2027, IGI Global Scientific Publishing. Copying or distributing in print or electronic forms without written permission of IGI Global Scientific Publishing is prohibited. Use of this chapter to generative artificial intelligence (AI) technologies is expressly prohibited. The publisher reserves all rights to license its use for generative AI training and machine learning model development.

Christians, but also because it continues to determine how time itself is recorded for all humankind. Other events have likewise reshaped the world in lasting ways, including the Protestant Reformation, the Enlightenment, the horror of twentieth century global conflict, and most recently, a global pandemic that claimed the lives of millions. These all marked decisive breaks from what came before, forcing humanity to re-evaluate long-held assumptions about authority, power, responsibility, and the moral order.

November 30th, 2022, will, in time no doubt, be recognized as another such watershed moment. On that date, with little pomp and ceremony, OpenAI publicly released its generative artificial intelligence chatbot, ChatGPT. What followed was not merely the introduction of new technology, but a widespread emergence of systems capable of reasoning and producing a variety of content in ways that closely resemble human abilities. The resulting speed and advance of publicly available artificial intelligence technologies has been unprecedented. In less than two years, generative AI tools achieved household adoption rates that far surpassed those of the internet, personal computers, and smartphones (Bick, Blandin, & Deming, 2024). Among marketplace knowledge workers, usage rates have climbed even more rapidly, fundamentally reshaping how information is processed, decisions are made, and work itself is performed (Microsoft/LinkedIn Work Trend Index, 2024).

Technological change has long served as a catalyst for ethical reflection, yet rarely has it been needed to answer the scale and pace being witnessed today. Whereas the Industrial Revolution unfolded over several hundred years, gradually shifting societies from agrarian labor to mechanized production, the rise of generative artificial intelligence has altered nearly every aspect of human life in a matter of only months (Arroyo-Machado et al., 2025). An increasing array of content in education, communication, marketing, and the public arena is now being generated by large language models whose constitutional boundaries and moral frameworks frequently remain unclear to the user.

Irrespective of whether one views developments in generative AI with optimism or concern, the reality is unavoidable: humanity has entered a new era, one in which artificial intelligence plays an active role in shaping ethical discernment and moral language. This demands an urgent re-examination of traditional Christian ethical reasoning. While existing frameworks have provided valuable insight into moral formation in late modernity, they were developed prior to the unprecedented conditions created by AI systems capable of simulating human reasoning and influencing moral judgment at scale.

THE SIGNIFICANCE OF MACINTYRIAN THOUGHT IN THE LATE TWENTIETH CENTURY

Throughout the history of moral development, the writings of a few philosophers have been widely recognized as having played an inordinate role in shaping Western ethical thought. For example, in 5th century Greece, Hippocrates wrote one of the earliest medical treaties advocating the need for physicians to promise to extend ethical care to their patients. What became known as the *Hippocratic Oath* has had an instrumental impact on medical ethics ever since. Within one hundred years, Plato would pen his *Republic*, arguing for the importance of justice and character in the formation of an ethical city-state. This was followed in short order by Aristotle's *Nicomachean Ethics*, which emphasized the habitual practice of virtues in the development of moral character. During the Enlightenment, Adam Smith's *Theory of Moral Sentiments* would argue that morality is inherently embedded within humanity as a result of nature. In the eighteenth century, Immanuel Kant would publish his *Groundwork of the Metaphysic of Morals* arguing that right and wrong were founded in a universal duty known to all by reference to a categorical imperative. Each of these became foundational in the development of Western moral philosophy. It would not be an overstatement to conclude that the works of Plato, Aristotle, Smith, and Kant have had a profound impact in the formation of normative theories of ethical reasoning.

In the twentieth century, names such as Foucault, Sartre, Rawls and Rand frequently appear atop lists of modern philosophers, with Barth, Bonhoeffer and Niebuhr often considered to have made some of the more notable contributions to theological ethics. The last six decades have witnessed a renewed interest in the role of virtue theory in moral development. This was initially a result of the works of Gertrude Anscombe and Philippa Foot, but most recently, and arguably most notably, the re-emphasis on character arose through the writings of Scottish Philosopher Alasdair MacIntyre, especially as a result of his most famous book, *After Virtue*.

Now considered to be the father of post-modern virtue theory, Alasdair MacIntyre has contributed significant work to late twentieth century ethical development, especially through his evaluation of post-modernity's moral confusion, and the resulting re-emergence of neo-Aristotelian thought (Oakes, 1996). He has almost single-handedly served as a catalyst to the renewed interest in Aristotle over the last six decades (Lackey, 1999). His publications are not without controversy however, and some scholars regard him to have been a turbulent figure in recent Western philosophy, known more for his rejection of post-modern liberalism than anything else (Knight, 1998). However, it is undeniable that he has been one of the leading voices redefining post-modern ethics and highly influential to moral development over the last half century. Much of his work represents a critique of post-Enlightenment moral philosophy, the result of which he argues has given birth to the rise of liberal individualism that the West has witnessed since the close of the Second World War. He argues that the post-modern society is in a state of abject moral confusion, something he describes as the failure of the Enlightenment experiment, which saw a rejection of Aristotle and Aquinas' emphasis on τ λος. Although his writings are all from a pre-Generative-AI perspective, many of his conclusions sound as though they could have been written since November 30th, 2022, as the moral fragmentation he diagnosed has not only persisted but appears to have been intensified by AI systems that mirror and reproduce competing moral frameworks without the capacity to adjudicate between them. For example, he bemoans a moral confusion deeply embedded within society, the result of which is that we are no longer able to have rational moral dialogue, nor understand a theory of ethical reasoning that disagrees with that of another person. He describes how the moral agent utilizes their own individualistic evaluative frameworks, each of which are built upon moral presuppositions that differ from one moral actor to the other (Smith, 2021). As a consequence, MacIntyre spends much of his writing rejecting the modern and post-modern systems of reasoning on the basis that humankind is now fundamentally unable to engage in ethical discussion or to reach any consensus as to what is good, right, just, fair and so on. He advocates that the only solution to this melee is a return to pre-Enlightenment moral clarity, which he believes requires an embrace of neo-Aristotelian emphasis on virtue (Smith, 2014).

Although many have already undertaken the task of evaluating MacIntyre, no-one appears to have yet done so in light of the societal shifts we are witnessing as a result of the public availability of Generative-AI tools. Given MacIntyre's sizeable influence on post-modern ethical development, any evaluation of ethical reasoning and AI must take his work into account. This is especially the case given that AI systems are now determining ethical reflection by generating moral language, offering normative guidance, and shaping how users articulate concepts of right and wrong. However, as influential as MacIntyre has been, we cannot now assume his pre-ChatGPT conclusions remain valid in light of the changes being ushered in by artificial intelligence. It is therefore necessary to critically re-evaluate the methodology and conclusions from some of his major works over the last six decades from the perspective of the AI era. We will therefore re-evaluate three of his books in particular, including *A Short History of Ethics*, *After Virtue* and *Ethics in the Conflicts of Modernity*. These are not MacIntyre's only publications of course, but they are among some of his most influential. These three have been selected because of the way they represent a reflection in his understanding of moral development during the latter half of the twentieth century, which will help us evaluate ethics and AI. They not only demonstrate how he continued to refine his neo-Aristotelian virtue theory, but they also reveal his ongoing critique of post-modernity, something that he continued to analyze as the culture evolved around him. Indeed, there is also an intrinsic link that MacIntyre himself identified between these three books, something that becomes evident as he, for example, looks back on the conclusions in *A Short History* on multiple occasions in *After Virtue*. He then does the very same thing with *After Virtue* in *Ethics in the Context of Modernity*. In other words, MacIntyre's own methodology recognizes these three books have common underpinnings that link them together.

Our AI-era re-evaluation of MacIntyre will begin in 1966 with the publication of *A Short History of Ethics*. It is here we find his earliest articulation of the development of moral theory, as well as initial indicators of his subsequent emphasis on Aristotle, and the start of his critique of post-modern ethical thought. *A Short History* provides us with a thorough understanding of how MacIntyre viewed the state of ethical reasoning before and after the Enlightenment, which ultimately goes to the heart of his rationale for advocating a return to virtue, thus providing a comprehensive foundation to better evaluate his next work, *After Virtue*. Originally published in 1981, *After Virtue* is his most seminal work, and is credited by many as being a leading factor in the Western re-emphasis on virtue over the last five decades. Finally, we will evaluate MacIntyre's most recent publication, *Ethics in the Conflicts of Modernity*. Released in 2016, it represents his latest evaluation of culture, which, as we shall see below, was further refined to reflect what he perceives to be heightened moral individualism. It also presents a re-affirmation of his neo-Aristotelian virtue theory, albeit this final iteration appears to be more Thomistic, and shows signs of a return towards some of his early Marxist leanings. Once we have re-evaluated each of these three publications in light of our AI context today, a final analysis will be presented at the end of the chapter. In this respect, MacIntyre's own insistence on tracing moral reasoning through historical development provides not only the subject of this chapter, but also the methodological justification for reassessing his conclusions in light of the radically altered conditions of the AI era.

Taking MacIntyre's own methodological commitments seriously, especially his insistence that moral reasoning must be understood within its historical and cultural context, this chapter contends that the present AI-driven context necessitates a re-evaluation of his conclusions. If moral concepts change as social life changes, then the emergence of AI as a pervasive moral actor, one that increasingly mediates moral language, advice, and judgment, must prompt renewed ethical reflection. The AI era intensifies many of the very tendencies MacIntyre sought to correct. For example, large language models appear to be increasingly affirming individual preference formation, and risk displacing human judgment with machine-generated outputs that lack clear moral intentionality (Lyu & Du, 2025; Cheng et al., 2025). In such an environment, an ethical framework grounded exclusively in virtue risks becoming insufficient, as character alone cannot provide the objective boundaries necessary to guide or constrain the development and deployment of powerful AI systems. Virtue remains essential, but it is no longer adequate on its own. This becomes vital to our re-evaluation of ethics in an AI-saturated context today.

EARLY MACINTYRIAN METHODOLOGY: THE ROLE OF CONTEXT IN *A SHORT HISTORY*

In recent years there have been a variety of opinions as to whether MacIntyre ought best to be read through the lens of his early writings about Marxism, or through the later perspective of his neo-Aristotelian thoughts (Blackledge & Knight, 2011). Although each of these opinions are somewhat politically distinct, we do find that a common theme bridging the two is the central role of the city-state (πόλις) in the development of moral values. If we can recognize that such a link exists, then perhaps MacIntyre is not quite as radical as some would assert, and perhaps, how relevant his writing continues to be in the AI era today. Recognizing this continuity is essential for assessing how MacIntyre's emphasis on communal moral formation translates into an AI era in which moral discourse is increasingly mediated by technologies that operate outside any shared moral tradition.

It is interesting that Kandiyali (2014) specifically identifies MacIntyre's emphasis on context, arguing that a core principle which continues to rise to the surface is the influence of community on an individual's moral development (p.2). This becomes immediately evident when reading *A Short History,* because MacIntyre (1998) begins by instructing the reader that every moral concept is intrinsically rooted in a unique social setting within a specific moment of history (p.1). This becomes vital to our re-evaluation of ethics in an AI-saturated context today. In order to reach any conclusions about the impact of generative AI tools on moral reasoning, we must first of all look through the lens of our present context and the impact the technology is having on our communities.

Central to MacIntyre's philosophy of ethical development throughout *A Short History,* and consequently central to his overall methodology, is the notion that moral values do not develop in a vacuum, and students of philosophy cannot offer either a critique or affirmation of any period in ethical thought, without first attempting to understand the time, place and value system of those they are seeking to evaluate. Sounding somewhat relativistic, MacIntyre (1998) argues that moral concepts change as social life changes (which we see today in the ways in which moral language, deliberation, and judgment are now shaped by algorithmic systems) and asserts that values are therefore deeply embedded in the societal structures around us (p.3). Practically speaking, this means that to be a student of moral development today, one must first of all be a student of our post-November 2022 culture, because MacIntyre believes that as one surveys history, one will discover it is neither a prison in which moral values are trapped, nor is it a museum in which standards are observed but never develop. This is where *A Short History* provides a window into not only MacIntyre's concept of cultural ethics, but also the need to evaluate his work in our current AI context. Given that history is not morally neutral, one must first survey not only the ways in which moral patterns have developed over time but conduct a metaethical analysis of how each era has understood terms, and the definitions they have provided.

To know how to appropriately re-evaluate ethical reasoning in light of the advances of artificial intelligence today demands that we first understand how moral values are contextually shaped. This is where MacIntyre is of great assistance, as he guides his reader back through time to Ancient Greece, and, starting with Homer, provides an overview as to how each period has understood and defined what is *good*. He demonstrates that the ancient Greeks would act in such a way as to be a good citizen, with their motivation grounded in furthering the greatest good for their locale. In contrast, when MacIntyre (1998) encounters the Enlightenment era, he argues a shift in the historical development of moral values took place as Immanuel Kant and others began to assert that man acts because he is under a duty to do so, irrespective of whether that action paints him to be a good citizen in the eyes of a local body of people (p.86). MacIntyre (1998) is keen to point out that for Kant, individual ethics are disconnected from what is perceived to be *good*. Instead, individual ethics became rooted in fundamental and overriding principles, unconnected to human life, feelings, happiness, or joy (p.88).

As he continues to interact with cultural and historical developments, MacIntyre's method does not stray far from the underlying question of what did each historical era in time understand amounted to *goodness*, and how did those societies answer the moral dilemma of what is *good?* Although different moments in history have provided their own definition, he continues to examine what each society has meant by this term, and he uses the question as an ongoing metaethical topic to maintain consistency and parity when evaluating one historical juncture with the next. In other words, before his readers are able to comprehend how each era answered the question *how should I live*, he wants them first of all to understand how each time-bound society understood what moral words mean. This attention to the meaning of moral language is especially significant in an AI context, where large language models generate ethical vocabulary without shared participation in the practices or traditions from which that language historically derives.

Ethical development becomes unintelligible to MacIntyre until mankind is first able to see it within a specific historical narrative. This is going to become important as we apply MacIntyrian methodology to today's AI saturated moral climate, in which ethical judgments are increasingly mediated rather than deliberated. He insists (MacIntyre, 1998), "We cannot hope to understand [ethical and philosophical problems] ... outside of the historical context to those problems (p.93)." His method is unique because it argues for humankind's inability to analyze philosophically without first investigating the usage of words in their unique time. For Macintyre's methodology, this means that a metaethical evaluation is necessary a priori to any discussion of normative values. Why? Because what the *good* for Socrates was, is not the same as the *good* in the twentieth century, or even the *good* emerging from large language models and generative AI advancements. Utilizing MacIntyre's methodology, we are able to see with greater clarity just how moral values have shifted from the centrality of the community to a post-modern moral life rooted in independence and self-sufficiency. However, he does acknowledge that the historical development of ethics cannot always be thought of in such reductionistic ways, contending (MacIntyre, 1998), "If our actions depend on the kind of social structure of which we are part ... there are no evaluative truths ... [and] we are doomed to social relativism" (p.95). His argument is quite simple: if there is no ultimate moral truth that transcends history, then all that humanity is left with is rampant individualism, something we are witnessing in our context as a result of the seismic shifts taking place in the wake of artificial intelligence.

On the face of it, MacIntyre continues to speak into the AI era evolving all around us. A re-evaluation of *A Short History* helps us better understand some of the changes we are witnessing in our context today, because a key component in his early methodology, and therefore a core concept in his initial understanding of how we discern moral development, is that history has never been amoral, and therefore, we can only ever truly see any given society, and the values within that society, through the lens of its own unique moment in time. *A Short History* therefore does not merely explain how moral reasoning developed in the past but provides the methodological foundation necessary for evaluating whether MacIntyre's later conclusions remain adequate in an AI-mediated moral environment. Let us now turn to MacIntyre's seminal work, *After Virtue*.

MACINTYRE'S EMPHASIS ON ARISTOTELIAN CHARACTER IN *AFTER VIRTUE*

An initial glimpse of the rationale behind MacIntyre's methodology in *After Virtue* is not only revealed in the original 1981 preface to the book, but especially in the expanded preface to the third edition, in which he addresses several critiques offered about his first two editions. It is here that MacIntyre (2007) responds to those criticisms with a return to his earlier affirmation in *A Short History*, arguing that "it is only possible to understand a moral culture in modernity from a standpoint external to that culture" (p. ix). This raises an interesting paradox for us today: if he is correct, then neither he, nor we, can truly evaluate our own moral climate, which would mean this book chapter is ultimately in vain. If he is wrong, however, then his entire methodology is without foundation, and once again, this book chapter is worthless. As we try to resolve this tension in his writing, we need to acknowledge that, to some extent at least, we are blinded by our own worldview and presuppositions, which perhaps is the very point MacIntyre is trying to convey. This paradox takes on renewed urgency in an AI context, where moral judgments are increasingly shaped by systems embedded within the very moral culture they reflect, further complicating our ability to evaluate the ethical contours of our own moment.

Over two decades removed from the first release of *A Short History*, MacIntyre's methodology in *After Virtue* continued to be rooted in his conviction that every society is blind to its own moral conditioning (Knight, 1998). In other words, no culture can adequately evaluate and critique its own moral concepts, because they are so immersed within it. He argues that every context fails to adequately see its own weaknesses. Every person is, whether consciously or subconsciously, so utterly submerged within their own unique worldview, that they often take for granted concepts and values that shape decision making. To understand MacIntyre is to realize his deep-seated conviction that our presuppositions are profoundly embedded within us, to the extent that, our subconscious moral values become almost unintelligible. This notion is not peculiar to MacIntyre, although it is essential to understanding his methodology. However, others would agree with him on this issue (Sire, 2015). He argues that this is a result of European Enlightenment thinking, which gave birth to a modern transformative social culture that brought with it substantial shifts in moral reasoning, the result being different forms of ethical reasoning were no longer able to harmoniously exist alongside one another. MacIntyre will (perhaps a little cynically) assert that his view is evident in the fact that in the latter twentieth century, a utilitarian will compete with a Kantian, who will compete with a social contract theorist, each of whom not only express opposing views, but are so convinced they are right, that they are unable to assess and critique their own moral presuppositions and resulting concepts. Although he never fully articulates it in this way, MacIntyre was essentially describing the confusing state that moral individualism inherently leads to: every person believes they are right, and argues therefore, that everybody else is wrong. Self becomes king in decision-making, and moral individualism is the order of the day. It seems MacIntyre was prophetically describing the ethical landscape we find in our present situation, with increasing reliance on AI chatbots as virtual moral compasses offering advice as to what amounts to right and wrong. In our present AI-saturated environment, users are turning to AI not merely for information generation and retrieval, but for increasing guidance on personal, relational, and morally charged decisions (Jones & Schonewille, 2026). Empirical research indicates that users frequently seek advice, emotional reassurance, and normative judgments from AI systems, effectively positioning these tools as quasi-moral interlocutors in everyday life (Manke, 2025). This trend is particularly noteworthy given that large language models do not possess moral discernment, but instead generate responses by predicting responses based on data sets. As a result, AI systems often affirm user preferences rather than challenge them, reinforcing existing beliefs rather than cultivating moral discernment. As users increasingly rely on AI for forms of counsel that were traditionally sought from human relationships, concerns emerge regarding the displacement of moral judgment and responsibility (Arroyo-Machado et al., 2025). In this respect, MacIntyre's critique of moral individualism appears increasingly prophetic, as the absence of shared moral frameworks allows preference-based reasoning to be amplified by technologies designed to accommodate rather than interrogate moral claims.

The key to understanding MacIntyre's progression in *After Virtue* from emphasizing historical context to virtue is found in his conviction that all post-enlightenment moral conclusions are rooted in what he describes as an emotivist perspective, in which values are based on feeling or attitudes, rather than externally measurable criteria. In an AI-mediated environment, this critique takes on new significance, as large language models generate moral language by reproducing patterns of preference without access to the rational or teleological grounding MacIntyre insists is necessary for genuine moral reasoning. His critique is that all of contemporary methodology is limited by expressions of personal preference, which ultimately means that moral language hinges, not on issues such as the rightness or wrongness of an action, but instead on the individual mood of the moral agent, which he claims is impermeable to objective investigations of proof or disproof. As a result, MacIntyre believes the only way for the post-modern ethicist to truly evaluate moral precepts is to step away from the traditions that have arisen since the Enlightenment and return to underlying presuppositions from a former age. For MacIntyre, this means we must return to a classic Aristotelian understanding of moral reasoning. One of the weaknesses of this position, however, is that he does not appear to consider any other pre-Enlightenment period of moral formation and fails to allow us to wrestle with any alternatives outside Aristotle. It seems that MacIntyre has already made up his mind that ancient Greece is the only solution to post-modern individualism, something he quite clearly affirms in a concession that he was already an Aristotelian even before writing *After Virtue*. It is at this stage that we are put on notice, that despite his critique of modern and post-modern systems being unable to see through their own lens of subjective bias, MacIntyre himself has ended up doing the exact same thing with his argument for an ethical methodology that is inherently influenced by his own preconceived notions of moral development.

As MacIntyre continues to examine the context in which moral language is defined, he evaluates not only the specific values, but also the unique characters that each culture elevates. Although this might initially sound as though he believes values are time bound to each era, MacIntyre is at pains to argue he is not a relativist. In the third edition of *After Virtue*, he once again points back to *A Short History* and reiterates his position that a key to understanding ethics is to learn from history and humankind's moral practices. He concludes that a return to a classical understanding of morality is necessary to evaluate values in his own context. His argument is that the post-modern culture, although maintaining elements of moral language, has completely lost an understanding of the substance of its moral values, both theoretically, as well as practically. It is here that MacIntyre betrays a strong neo-Platonic influence, identifying for his reader key roles such as the rich aesthete, the manager, as well as the therapist, each of which reveal more than a hint of Plato's understanding of the three main tiers of society.

The underlying point MacIntyre wishes to reaffirm, and that which becomes increasingly important for our evaluation of AI today, is that moral values are the offspring of a culture, and that identifying some key social roles in a society will help to identify the moral principles that are given prestige or emphasis within that context. It is here we see the blending of MacIntyre's former methodological emphasis on historical context begin to impact his new focus on virtue. In other words, if the ethicist will take time to understand *who* the people value (the virtuous moral agents), he or she will understand *what* the people value (the moral concepts). He argues that one of the major failures of any culture is to fail to understand its social roles and values attached to its context. We would heed this warning well in the AI era today. As a result, one of the main flaws in anthropological moral investigation is to overlook the value any given society places on virtue, social functions, and traditions. MacIntyre focuses on this failure to understand societal context and moral language to argue for a complete post-Enlightenment breakdown of rational understanding of morality and objective ethical decision making and sees it as being the prime cause of moral confusion in his post-modern culture. He subsequently guides his readers along a journey through Kierkegaard, Kant, Diderot and Hume, at each turn seeking to demonstrate just how the philosopher's contribution was yet another nail in the coffin of ethical reasoning. It is here that MacIntyre also argues (2007), "A central thesis of this book is that the breakdown of [independent rational justification for morality] provided the historical background against which the predicaments of our culture can become intelligible" (p.39). He then goes on to survey feelings over objectivity, reason over faith, and argues that the "experiment" of enlightenment thinking, which he believes involves a series of shifts in metaethical values, from faith to reason, and onto feeling and passion, has fundamentally failed, thus requiring the re-evaluation of how ethical decisions are being made.

MacIntyre's argument does not necessarily mean a return to the third century BC in Athens, because he recognizes a strong Aristotelian methodology present even during the scholastic period in the works of Thomas Aquinas. The argument is essentially that morality has been in a freefall of deconstruction and confusion since the seventeenth century began shifting values away from the community and onto the individual, and any hope of its restoration can only be found in a time prior to that focus on the self. We might summarize the entire first half of *After Virtue* as a deconstruction of Enlightenment and Post-Enlightenment thinking. MacIntyre rejects Kant's denunciation of faith as the basis for moral value. He dismisses Kierkegaard's emphasis on feelings. He criticizes Bentham's utilitarian focus on pleasure. He then rejects Nietzsche's atheistic nihilism and sees it to have contributed to post-modernism's decline into emotivism and a narcissistic focus on the self. If we were to ask MacIntyre to comment on the heightened individualism we find increasing all around us because of the growing reliance by users on AI bots for ethical discernment, he would most likely argue it is a result of today's moral agent perceiving him or herself to be existing in isolation, making decisions in a self-focused vacuum. Many users of AI tools today operate them with increased isolation from others, all the while, heeding more of the output it produces for them (Shelmerdine & Nour, 2025). MacIntyre would surely conclude that today's moral shift is merely another link in the chain that is a result of failures in the last several centuries of different theories of ethical reasoning.

For MacIntyre, the way in which man is truly going to understand who he is, and who he ought to be, is a renewed focus on the virtues, albeit he broadens traditional Aristotelian notions to include later Thomistic inclusions of divine moral law which recognizes objective truth claims to guide humanity to its true τ λος. He believes a fundamental oversight in Enlightenment thinking lies in the central focus on fragmented rules and principles, whether arising via reason as an innate duty to all humankind, or because of the subjective interpretation of the passions. MacIntyre wrote (2007), "A key part of my thesis has been that modern moral utterance and practice can only be understood as a series of fragmented survivals from an older past" (p.111). The only way therefore to reclaim moral integrity is a return to Aristotle's virtue, which he argues demands an emphasis on character before understanding how rules function in ethics. He believes the right qualities of character will steer the individual to seek and obey the right rules and principles. This is actually one of MacIntyre's conclusions that this chapter seeks to modify. While it is agreed that any theory of ethical reasoning for AI ought to emphasize the virtues, it is argued that virtue alone may no longer be sufficient. In a moral context that has become so individualistic, it is suggested that a potential restoration of objective values may now be a necessary component to rebalance today's increasingly individualistic context.

Having spent the first half of *After Virtue* rejecting much of Enlightenment, modern and post-modern moral reasoning, MacIntyre's methodology is then to build his apologetic for virtue theory, beginning with Homer, Plato, and ending of course with Aristotle. He expounds the link between the *good* individual and the *good* state, as well as the virtues and the virtuous citizen, and uses these concepts to reaffirm his previously articulated link between social values and moral values, and the necessity in understanding cultural context for moral development. MacIntyre understands Aristotle's virtues as being characteristics that enable a person to achieve ultimate blessedness, happiness and ε δαιμονία, the lack of which will prevent an individual from fulfilling his or her true τ λος. His methodology is to demonstrate that the virtues are a necessary part of the flourishing of human life, without which *good* cannot be truly understood. One of the reasons MacIntyre argues so strongly for the virtues is that he believes they not only transform how a person acts, but also how they feel. The ability to influence human behavior and attitudes elevates them above consequence, feeling or duty alone. This is important for us to understand when evaluating MacIntyre's overall presentation of virtue theory. His methodology is not simply a return to ancient philosophy, he does so because he believes it is far greater holistically for moral formation, and to move humankind towards its ultimate end. To understand this element of his approach though, is to understand why he insists on a return to virtue. In a post-Enlightenment era, moral values moved away from being externally shaped by the community, to being an internal individual issue grounded in reason alone. For MacIntyre, virtue however is not about living life to please the self, it is about the individual living as part of an external context. Virtue takes account not only of self, but of others also. It is not only concerned with the character of the moral agent, but also the way in which he or she interacts with those around them (or for the Christian, how he or she responds to the eternal τ λος of God). Although MacIntyre is neo-Aristotelian in his reasoning, his definition of virtue expands on the ancient understanding of virtues being formed by the πόλις and sees instead virtues as acquired human characteristics that are necessary to enable a person to achieve their desired goal/purpose, a lack of which will hinder one's journey to that goal/purpose.

RENEWED ATTENTION ON VIRTUE IN *ETHICS IN THE CONFLICTS OF MODERNITY*

The title to MacIntyre's latest book is particularly descriptive of its content, because he continues to expound the moral confusion that was inherent in his context, seeking to advance a theory for responding to a post-modern construction of moral values. In doing so he reiterates his previous convictions from *After Virtue* that a neo-Aristotelian understanding of the virtues is the most viable lens through which to reach moral conclusions today, albeit with a slight Thomistic modification that was not present in *After Virtue*, as well as returning to some of his earliest ethical convictions about the potential advantages of some of the societal claims of Marxist insights. Although MacIntyre could not have anticipated the emergence of generative artificial intelligence, the moral confusion he describes in *Ethics in the Conflicts of Modernity* provides a strikingly apt framework for interpreting the intensified ethical fragmentation characterizing the AI era.

MacIntyre's summary of the post-modern problem has evolved since writing *After Virtue*, in that his earlier categorization of *emotivism* as the most influential ethical theory has now been modified to what he terms as *expressivism*, which he describes as being less sophisticated. It is interesting that MacIntyre clearly believes a general moral regression has taken place in the two decades since *After Virtue*. He believes the post-modern society to be less philosophically discerning. He deconstructs the ideals and motivations lying behind expressivism as emotion, taste, personal attitudes of approval, beliefs, and imagination, as well as decisions based on attitudes of approval or disapproval which are void of any objective judgement of fact, a pattern of moral expression increasingly mirrored by AI systems that generate normative language without moral intentionality. He contrasts this with his previously articulated understanding of Aristotelian virtue ethic, demonstrating why he insists expressivism is incompatible with moral formation in his context. MacIntyre believes that goodness ought to be rooted in a notion of acting for the sake of achieving some good or preventing some evil, and points to Aquinas' view that understanding one's motivation is central to conceiving whether an action is good, because of the intentionality necessary in directing a moral agent towards anything of genuine worth. MacIntyre argues that the moral agent ought to pause in self-reflection and first evaluate motivation before acting, so as to ensure the reasons for acting are appropriate. The key is that a person asks themselves what he or she hopes to achieve by acting in a certain way.

Ethics in Conflict continues MacIntyre's evaluation of how different moral contexts have understood a *good* action, specifically evaluating how he sees expressivism using the noun *good*. He is critical that the expressivist does not define something as *good* by reference to its quality, but purely as a form of indicating their present approval. He concludes that an expressivist is unable to truly use goodness to evaluate any given situation, object, or action, because they don't make statements of fact, they only allude to approval or disapproval. As has already been mentioned, any reader of *After Virtue* will immediately see the influence of his earlier ethical system at this point, albeit there has been a shift of language from emotivism to expressivism, which MacIntyre sees as being more indicative of the culture in which he wrote in 2016. MacIntyre sees the key problem lies in the expressivist being unwilling to attach objective moral value to their actions, offering instead only expressions of personal preference, which will naturally not only vary from one individual to another, but also from one moment to the next for that same individual. Consequently, the denial of all claims to objective truth means the expressivist has no authoritative standard, which was already affirmed by others in response to his earlier conclusions (Jackson & Pettit, 1998). *Goodness* can only be understood if a moral agent declares it to be based on how they feel at any given moment, which means that moral values are in a state of constant flux, binding on no-one but themselves at any given moment in a day. The rise of self, coupled with the dismissal of community as well as common values, gives rise to an inability to correct ethical failures. This last issue becomes especially important as we consider our immediate AI context. This concern becomes especially acute in an AI context, where systems trained to respond to expressions of approval or disapproval lack any capacity to ground moral judgments in objective claims about goodness or truth.

MacIntyre also crystallizes what he perceives to be the core problem with the rival philosophies of what amounts to *goodness* in a post-modern culture: an utter moral impasse. He writes (2016), "The problem is that there are just too many theories [of ethical reasoning] each of them incompatible with the others" (p.117). In the AI era, this impasse is no longer confined to philosophical debate, but is increasingly embodied by technologies that reproduce incompatible moral claims at scale without the capacity for resolution. This is something MacIntyre had continued to be critical of, that is, rival claims to moral value, and once again (invoking the underpinnings of his decades earlier methodology) points us back to the importance of understanding any philosophical claim in its historical context. The resounding assertion found in *A Short History* and *After Virtue* rings clear: the necessity of understanding politics, economics, and moral beliefs in their historical, anthropological, and sociological settings. He is critical of answers that only come from academic circles and argues that self-awareness can only be truly found in a context that is willing to examine its own bias and idiosyncrasies, which he believes his own culture is not willing to do.

A very clear element in his overall methodology (to demonstrate the need for a return to a Thomistic Neo-Aristotelian concept) is found in his criticism of his context, which is a predominant theme throughout the book. MacIntyre writes (2016), "Let me … reiterate my claim that contemporary philosophical theorizing about morality is flawed" (p.77). He considers utility to be a fiction entirely because moral agents will ultimately act egoistically for themselves, and in doing so highlights what he sees as Hume's errors in postulating pleasure and pain and the core rival values for moral development. This diagnosis resonates strongly in today's AI-driven culture, where tools are frequently designed to optimize not only efficiency and engagement, but, perhaps more importantly for our present discussion, individual preference satisfaction, often without reference to shared moral goods or communal flourishing.

The solution is a return to a values system that does not simply ask how ought *I* to act, but instead is willing to evaluate how *we* are to act, and for MacIntyre, that answer lies in a return to a virtue-based community that focuses not just on the individual, but on the common good. Here the reader will notice MacIntyre argues it is not merely "*I*" or even "*we*" that is to be considered, but (true to his methodology) it is "*we*" in "*our unique social context.*" It is interesting that although MacIntyre consistently reinforces the need to consider culture, he is also quite transparent about the failure of any given culture to see and correct its own preconceptions and blindness to moral values that may be wrong, which is consistent with his critique in After Virtue. Having expounded what he believes amount to the core fallacies of an expressivist society today, MacIntyre then presents a contrast, which will sound very familiar to any reader of *After Virtue,* because he will once again argue for a theory of moral value that is rooted in Neo-Aristotelian concepts of virtue, which he argues appeal to standards independent of subjective feeling. Human flourishing will be found in common values recognized in each context (πόλις), which are both political as well as rational. This Aristotelian formula is obviously not new to MacIntyre, and very clearly echoes his consistent ethical conclusions for more than six decades. For MacIntyre, this element of his methodology has not evolved much through the years and still forms an integral part of his writings today. *Goodness* can only be understood through the lens of the context within which the term is used, because it is ultimately rooted in relationship to others, and those relationships are determined by locale.

MacIntyre advances a concept that the post-modern Neo-Aristotelian should aim to become the kind of person who evaluates moral actions based on what will lead to excellence: both for self, as well as his or her context. We become the model person as we develop virtuous skills and attitudes such as courage, patience, truthfulness, and justice, each of which demand we look beyond ourselves and at others in our unique culture. It is as one continues to practice the virtues that one develops a sophisticated moral understanding of *goodness* and will begin to live a life of increasing moral self-awareness, and the impact one has on others. MacIntyre has an interesting discussion of the role that failure plays in the development of the virtues and argues that failure has just as much of an impact as success. He asserts that in every case in which an agent survives a time of failure, what is crucial is whether that agent is able to learn from her or his experience as well as the agent becoming better able to understand what has gone well in their past and what has gone badly. MacIntyre's emphasis on the formative role of failure raises an important concern when applied to the use of artificial intelligence tools. One of the growing criticisms of generative AI systems is that their design tends to minimize or bypass failure rather than require the moral agent to grapple with it. Studies of AI use in educational and professional contexts suggest that while such tools can increase efficiency, they may simultaneously inhibit deeper learning by reducing opportunities for struggle, reflection, and skill formation that typically arise through failure (Shen & Tamkin, 2026). When users rely on AI-generated outputs rather than working through problems themselves, they often demonstrate weaker understanding and diminished capacity to learn from mistakes. Related research indicates widespread concern that increasing dependence on AI tools is eroding critical thinking and fostering over-reliance on external systems for judgment and decision-making, particularly in higher education contexts (American Association of Colleges and Universities & Imagining the Digital Future Center, 2026). From a MacIntyrian perspective, this is significant, as virtue formation presupposes the capacity to learn through failure, to reflect on what has gone badly, and to grow in practical wisdom through embodied moral experience. If AI tools increasingly affirm users and shield them from formative failure, they risk undermining one of the very conditions necessary for the development of moral character.

MacIntyre's overall methodology in support of his Thomistic Neo-Aristotelian assertions for the present culture is grounded in understanding the relational nature of humankind as moral agents, with each person being aware that life is an ongoing narrative which takes place in community, not on an individual level. Humankind exists in a mutual dependence on one another for its survival, which he concludes, is fundamentally incompatible with an expressivist philosophical outlook. The former looks at the history of culture in the development of moral values, concluding that the answers lie in external values that emphasize "we" (context in relationship), whereas the latter looks at the history of self, and one's own internal values, emphasizing "I" and personal attitudes and feelings. Here MacIntyre expounds that the expressivist has a history of self, learning in isolation and more interested in how he or she will respond to a situation. All of history begins and ends with the moral agent in isolation (what should I do). The Neo-Aristotelian on the other hand understands history as a communal and relationship concept, with value development being a shared project in which the society works alongside one another, sharing values and the common good of all in determining what goodness is to be achieved (what should we do). This contrast is increasingly important in an AI context, where moral agents may engage ethical questions in isolation through personalized systems that simulate dialogue without genuine relational accountability.

Where MacIntyre's most recent methodology in *Ethics in Conflict* begins to demonstrate its differences with his earlier conclusions in *A Short History* and *After Virtue*, is that his emphasis on Thomistic-Aristotelian philosophy is supplemented by Marxist insights. The introduction of Marx creates an interesting dimension to his latest methodology, yet it is important to understand that MacIntyre isn't proposing that moral values are found in a political system of socialistic egalitarianism. In fact, he goes to lengths at the outset of his consideration of Marx to say that he is not advocating a form of socialism, and that Marxist ethics must not be equated with what would later become socialism or communism. MacIntyre explains Marx believed that one of the major problems with capitalism was that it created a focus on how much surplus gain (profit) one person could make, which in turn caused a shift in moral values away from the Aristotelian concept that goodness equated to what was best for the community as a whole, to goodness being equated with what was best for the individual, the latter of which MacIntyre sees vividly evident in expressivism today. These concerns take on new urgency in an AI-driven economic landscape, where automation and algorithmic decision-making risk further detaching economic efficiency from moral responsibility and communal accountability.

As MacIntyre reads Marx, he finds a moral critique of capitalism on the basis that it fuels egoism, thereby resulting in the moral erosion of the value placed on people, which ultimately elevates self over others (the community in which one operates). He advocates that Marx understood *goodness* in a quasi-Aristotelian light, in that it produced value for the entire community, whereas in capitalism Marx saw *goodness* finding root only in the subjectivity of the individual, which MacIntyre sees as immensely influential in the post-modern development of expressivism. In other words, MacIntyre uses Marx to draw parallels between the advance of capitalism and expressivism, both of which he argues have led to a decline in the value placed on community. As capitalism emphasized the accumulation of surplus for oneself, it moved the moral compass away from a communal focus onto the individual. MacIntyre further alleges that the focus on gain through self-advancement and one's own preferences has become a major influencing factor in business economics and what it means to be a moral agent today. He seeks to undergird his overall thesis with the argument that rationality has become equated with preference and personal satisfaction, with very little regard to be had of how the moral agent reaches those decisions, other than a subjective analysis. It is this that we come to learn is the central focus of MacIntyre's methodology in arguing that a modern capitalistic society is at odds with a Neo-Aristotelian Thomistic view of communal values and goodness, which Marx identified in his critique of the modern society. While MacIntyre doesn't align himself politically with Marxism, he does highlight that any honest analysis of capitalism must affirm just how much financial markets function in social contexts. Capitalism depends on societal values, yet at its core seeks to advance the individual over the very context on which it depends. Indeed, it depends on a community of investors, employees, stakeholders, consumers, yet seems to be in an ethical conflict when it comes to defining what is *good* and for whom. MacIntyre contends that capitalism believes economics can be understood in isolation from moral values, something both he and Marx disagree with. He asserts the impossibility of evaluating an economic system without also doing so ethically.

A RE-ASSESSMENT OF MACINTYRE'S METHODOLOGY IN THE WAKE OF AI

We set out to re-evaluate the works of Alasdair MacIntyre in order to determine if we could identify a common methodology that not only explained his conclusions, but also helped to determine if those conclusions remain viable today in light of the changes we have seen in our own context since November 30th, 2022. MacIntyre developed a methodology that not merely looked *backwards* in time, but also evaluated the state of post-modern moral values a*t the time* he wrote. However, he could not possibly have looked *ahead* of his writings and foreseen the seismic impact of generative AI, nor the extent of the resulting changes we are now witnessing around us. It is therefore our task to determine if his pre-AI-era conclusions can be just as readily applied to society today, or whether adjustments need to be made because of tools like ChatGPT, Claude, Gemini and Grok. Just because his conclusions were so influential in a previous moral era, does not mean they will automatically be so today.

We have been able to conclude that, despite six decades passing since he first began writing, MacIntyre's work continued to reveal an emphasis on the need to understand the context of moral development before any evaluation could take place. Applying that principle to today we must first understand our AI saturated culture. MacIntyre's methodology hinges on the assumption that moral values are formed within social life, not outside it. This becomes especially important as we attempt to assess the short, and even medium, term impact that generative AI tools have had on human development in our society. For example, if MacIntyre is indeed correct, before the public release of ChatGPT, we were already immersed in a cultural context whose moral formation was highly relative, an environment in which frequently changing values were based on preference, mood, and every moral agent's short-term context. Now, every user has their own private counsellor embedded into an app on their phone, readily available at any time of the night and day to offer moral advice to the individual. The answer to the question, *what is right for me*, is now increasingly answered by a technology built on predictive coding and data sets, that are themselves, the product of post-Enlightenment thinking and moral individualism.

MacIntyre's consistent solution to what he regards as a moral impasse, is to advocate a return to a prior system of moral development that, while emphasizing the context of the individual within their unique moment in history, was one in which *goodness* was inherently linked to character and virtue. Even from our AI context, there continues to be much to appreciate from MacIntyre's argument in favor of embracing a neo-Aristotelian understanding of the importance of virtue, not only because it focuses on the development of moral character (who a person is over what a person does), but also because it helps to realign values around the community and away from the self.

However, in light of the changes we have experienced over the last three years, we would do well to apply MacIntyre's methodology to MacIntyre's conclusions about virtue. Given that he was so consistently a proponent of analyzing context, and then adjusting the most suited theory of ethical development to it, the present task is to question whether conclusions from a previous (pre-AI) context remain the most appropriate today. Are we therefore de facto *After MacIntyre*? This chapter is not advocating a *deconstruction* of MacIntyrian thought but is instead suggesting that a *modification* of his conclusions is now likely so as to take account of the seismic shifts in our culture and moral reasoning we have seen as a result of generative AI. After all, as far back as *A Short History*, MacIntyre himself argued that moral concepts have to change and adapt as societal life changes, because values are so inherently embedded into every culture and context. MacIntyre's own methodology would demand that a re-evaluation of his work takes place, and that we undertake a re-evaluation of his own moral theory considering today's AI saturated world. If his methodology is correct, then a new theory of ethical reasoning is most likely going to emerge for the AI era. A new moral context must give birth to a new framework for moral reasoning. As he reviewed the history of ethical development, MacIntyre presented a simple conclusion that there appears to have been no ultimate moral truths that have transcended philosophical thought, especially as a result of enlightenment ideology. Consequently, every new generation forges its own values system, something that today's digitally saturated generation has clearly continued doing. Again, we find that MacIntyre's own methodology would call for a re-assessment of ethical decision-making today, which, if we are to avoid abject relativism, requires a return of sorts to clearer foundational principles in moral development.

We also now find ourselves questioning MacIntyre's emphasis on a return to a classic Aristotelian understanding of moral reasoning alone. As we have already observed, one of the weaknesses with this position is that it does not allow us to consider any other theory of ethical reasoning and fails to permit us to wrestle with alternatives outside Aristotle. Virtue ethics is primarily concerned with the formation of moral agents, yet the AI era confronts us with tools and systems that are not moral agents but that nevertheless reshape moral outcomes and decision-making. MacIntyre's own methodology would cause us to question whether neo-Aristotelian virtue is still the solution to the moral confusion of today's AI context. Given the heightened role that individualism is now playing in our present culture, a solution that rests on character alone is most likely going to be inadequate, and there would perhaps be a need to link character formation with an objective, principle-driven framework. To combat the increased subjectivity, an objective standard must at least be considered to counter-balance ethical boundaries. One of the inherent dangers of a system of moral reasoning that centers on character alone, is that it can enable the moral agent to focus on self before others (who am *I* to be). Individualism rooted in a virtue theory still produces individual statements of desire or approval that can subsequently render objective moral consensus impossible. If an emergent context is permitted to self-regulate to the extent that it denies all claims to objective truth, then society has no authoritative standards by which to operate, and ethical decision-making spirals into moral freefall. MacIntyre argued that the expressivist rejected decisions that arose from statements of fact, instead alluding only to his or her temporal approval or disapproval. If this is permitted to determine morality, then what is *good* will only be determined based on how a moral agent feels at any given moment. The result is that ethical decision-making will forever be in a state of flux, binding on no-one but the individual at any given moment.

MacIntyre concluded that the context in which he wrote demonstrated moral immaturity, relying on temporary notions such as feelings and emotion in the formation of ethical values. He argued this was the result of Enlightenment thinking, which presented not only a rejection of any divine influence, but also created such a vast melee of ethical theories, that an utter moral impasse was the result. As we wrestle with the pace, scope and impact of artificial intelligence, we would continue to agree with MacIntyre that character is an important component in developing a theory of ethical reasoning that enables morally appropriate decisions to be made. What is newly distinctive is that AI systems can now scale and personalize preference-based moral reasoning, delivering it in the form of counsel that feels authoritative while remaining detached from moral accountability. However, we now find that our context has shifted beyond that which MacIntyre previously evaluated or envisioned, and we therefore argue that virtue alone may no longer be the solution. A society saturated with generative AI tools cannot allow a situation to develop in which every moral agent is the determiner of his or her values, roles, preferences, or desires. Our current situation requires a revisiting of an objective values-based system, which perhaps warrants a re-examination of a deontological-virtue theory for the AI era.

FIVE COMMITMENTS FOR CHRISTIAN LEADERS IN THE AI ERA

If Christian ethics in the AI era must move beyond virtue alone and recover a principled framework of moral obligation, then this recovery must take visible institutional and personal form. An "After MacIntyre" ethic for artificial intelligence cannot remain abstract. It must shape how Christian leaders design systems, structure policies, teach students, and steward technological power. At minimum, such an ethic entails five commitments.

1. A Commitment to the Primacy of Human Moral Agency

Christian leaders must insist that AI systems remain instruments of human action rather than substitutes for human judgment. Artificial intelligence does not possess conscience, intentionality, or moral accountability. It cannot deliberate about the good; it can only simulate the language of deliberation. For this reason, responsibility for moral decisions must never be delegated to the machine. Whether in business, education, healthcare, ministry, or governance, final accountability rests with human persons created in the image of God. AI may assist with analysis, generate options, or surface risks, but it must not become a surrogate moral agent. Practically, this means leaders should design workflows and governance structures in which human review is not a mere formality but a substantive act of moral evaluation. It also means resisting cultural narratives that portray AI as neutral, omniscient, or inherently objective. Christian leadership requires a clear distinction between tool and agent, between instrument and bearer of moral responsibility.

2. A Commitment to Objective Moral Boundaries

If the AI era intensifies expressivism by scaling preference-based reasoning, then Christian leaders must articulate objective moral boundaries prior to technological deployment. Efficiency, profitability, or innovation cannot function as ultimate standards of right and wrong. This commitment demands that certain lines not be crossed, even if crossing them promises competitive advantage. Systems that manipulate users, erode human dignity, obscure truth, or exploit vulnerability must be rejected on principled grounds. In this respect, the recovery of deontological clarity becomes essential. Some actions are morally impermissible regardless of their outcomes. Such boundaries should be explicit rather than assumed. Christian institutions and organizations should develop written AI principles rooted in theological anthropology, affirming the inherent dignity of persons, the centrality of truth, and the moral limits of technological power. Without articulated boundaries, virtue alone will struggle to withstand cultural and economic pressure.

3. A Commitment to the Formation of Virtuous Users and Designers

While objective obligation sets boundaries, virtue shapes the kind of leaders who operate within them. Christian ethics must therefore invest in the moral formation of those who design and use AI systems. Designers must cultivate prudence in anticipating unintended consequences. Administrators must exercise justice in evaluating who benefits and who bears risk. Educators must model intellectual honesty when using generative tools. Executives must demonstrate temperance in resisting the unchecked pursuit of automation at the expense of human flourishing. This commitment also requires protecting the formative role of struggle and failure. If AI tools increasingly remove friction from learning and decision-making, Christian leaders must intentionally create spaces where individuals still wrestle, reflect, and grow. Practical wisdom cannot be outsourced. Character is not formed through frictionless optimization but through disciplined engagement with reality.

4. A Commitment to Communal Accountability Over Isolated Use

MacIntyre's insistence that moral reasoning is embedded within practices and traditions becomes especially urgent in an AI-saturated culture marked by individualized, private interaction with digital systems. When moral counsel is sought primarily from personalized algorithms, the communal dimension of discernment is weakened. Christian leaders must therefore foster structures of shared accountability. This may include institutional review boards for AI initiatives, cross-disciplinary ethics committees, or communal discernment processes for high-stakes deployments. It also includes encouraging transparent dialogue about how AI is used in classrooms, boardrooms, and ministries. Theological anthropology reminds us that human beings are relational. Moral formation occurs in community, not isolation. AI systems may simulate conversation, but they do not participate in covenantal accountability. Christian leadership must preserve moral reasoning as a shared practice shaped by tradition, Scripture, and embodied relationships.

5. A Commitment to Truthfulness in an Age of Synthetic Content

Generative AI introduces unprecedented capacity to produce persuasive but inaccurate or misleading content. In such an environment, truthfulness must be reaffirmed not merely as a virtue but as a binding obligation. Christian leaders should require transparency regarding AI-generated content, especially in academic, ministerial, and professional contexts. Disclosure policies should be clear. Representations should be honest. Synthetic output should not be passed off as purely human authorship when it is not. More deeply, this commitment reflects theological conviction. A Christian ethic cannot be indifferent to truth because it is grounded in the character of a truthful God. In an age where synthetic media can blur the boundaries between authentic and artificial expression, fidelity to truth becomes a countercultural witness.

THE PARADOX OF AN *AFTER MACINTYRE* ETHIC

One final observation must be made: MacIntyre's methodology was grounded in the argument that it is impossible to assess moral values outside of the historical context in which they are forged. He repeatedly emphasized the need for the ethicist to first become a historian (which is why this chapter has gone to such lengths to review MacIntyre's work), and to evaluate each person's time and place, before reaching any conclusion about the ethical principles they held. On the positive side, MacIntyre rarely deviated from this conviction. His writings are relatively logical to follow because he was so consistent in this regard across six decades of his writings. We must concede there is truth in the notion that history cannot be amoral, and that just as every generation creates its own ethical standards, those values will only ever truly be understood in a societal study of what it meant to live at that time. Therefore, one of our key take-aways is that the ethicist must do all he or she can to first attempt to objectively evaluate their own historical context, and to honestly assess the impact that their unique society has on their theories of ethical reasoning. However, the tension this presents to us is that, if we adopt MacIntyre's methodology, we are fundamentally unable to evaluate our own AI context. We are caught *in* a moment of artificial intelligence technologies unfolding on humankind, and our ability to objectively remove ourselves from that is quite impossible, no matter how earnestly we try. Our ability to assess values in our own context is limited to comparing them to those we have seen in the past. MacIntyre's methodology was rooted in a conviction that every society is blind to its own moral conditioning. In other words, no culture can adequately evaluate and critique its own moral concepts, because it is so immersed within them, and as a result, it fails to identify its weaknesses. We are so deeply submerged within our own worldview and the advances in AI we are seeing, that we cannot truly see the concepts and values that shape our decision-making. Ultimately, this means we cannot truly evaluate the ethical impact of AI on today's culture because we are trapped within it. We cannot see outside our own moment in time because we are inherently caught inside of it. Although we can offer a critique of what we find within our context, we cannot really offer up a full evaluation, because our context is still unfolding as we wrestle with cultural changes being ushered in because of generative AI tools. The conversation, therefore, has only just begun. Yet this is precisely why Christian ethics must recover not only the language of virtue, but also the clarity of obligation. If AI is reshaping the moral conditions of modern life, then the church and the academy must respond with an ethical framework capable of forming moral agents while also setting boundaries for the responsible use of powerful tools.

REFERENCES

American Association of Colleges and Universities, & Elon University's Imagining the Digital Future Center. (2026). *The AI challenge: How college faculty assess the present and future of higher education in the age of AI* (Survey report). https://imaginingthedigitalfuture.org/collaborations/the-ai-challenge-how-college-faculty-assess-the-present-and-future-of-higher-education-in-the-age-of-ai

Anscombe, G. E. M. (1958). Modern moral philosophy. *Philosophy (London, England), 33*(124), 1–19. DOI: 10.1017/S0031819100037943

Arroyo-Machado, W., Ma, J., Chen, T., Johnson, T. P., Islam, S., Michalegko, L., & Welch, E. (2025). Generative AI and academic scientists in US universities: Perception, experience, and adoption intentions. *PLoS One, 20*(8), e0330416. DOI: 10.1371/journal.pone.0330416 PMID: 40875655

Bick, A., Blandin, A., & Deming, D. (2024). *The rapid adoption of generative AI* (Working Paper No. 32966). National Bureau of Economic Research. https://www.nber.org/papers/w32966

Blackledge, P., & Knight, K. (2011). *Virtue and politics: Alasdair MacIntyre's revolutionary Aristotelianism*. University of Notre Dame Press.

Cheng, M. (2025, October). 'Sycophantic' AI chatbots tell users what they want to hear, study shows. *The Guardian*. https://www.theguardian.com/technology/2025/oct/24/sycophantic-ai-chatbots-tell-users-what-they-want-to-hear-study-shows

Jones, M., & Schonewille, M. (2026). Synthetic relationships and artificial intimacy: An ethical framework for evaluating the impact of generative AI on community. *Journal of Ethics in Entrepreneurship and Technology*. .DOI: 10.1108/JEET-06-2025-0040

Kandiyali, J. (2014). Paul Blackledge and Kelvin Knight (eds). *Journal of Moral Philosophy, 11*(1), 111–115. DOI: 10.1163/17455243-01101002

Knight, K. (1998). *The MacIntyre reader*. University of Notre Dame Press.

Lackey, D. P. (1999). What are the modern classics? The Baruch poll of great philosophy in the twentieth century. *The Philosophical Forum, 30*(4), 329–346. DOI: 10.1111/0031-806X.00022

Lyu, Y., & Du, Y. (2025). The ethical evaluation of large language models and its optimization. *AI and Ethics, 5*(5), 4579–4592. DOI: 10.1007/s43681-024-00654-9

MacIntyre, A. (1998). *A short history of ethics: A history of moral philosophy from the Homeric age to the twentieth century* (2nd ed.). University of Notre Dame Press.

MacIntyre, A. (2007). *After virtue: A study in moral theory* (3rd ed.). University of Notre Dame Press.

MacIntyre, A. (2016). *Ethics in the conflicts of modernity: An essay on desire, practical reasoning, and narrative*. Cambridge University Press. DOI: 10.1017/9781316816967

Manke, K. (2025, September 10). Do chatbots have a moral compass? Researchers turn to Reddit to find out. *Berkeley News*. https://news.berkeley.edu/2025/09/10/do-chatbots-have-a-moral-compass-researchers-turn-to-reddit-to-find-out

Microsoft & LinkedIn. (2024). *AI at work is here. Now comes the hard part: 2024 work trend index*. https://www.microsoft.com/en-us/worklab/work-trend-index/ai-at-work-is-here-now-comes-the-hard-part

Oakes, E. T. (1996, August). The achievement of Alasdair MacIntyre. *First Things*. https://www.firstthings.com/article/1996/08/the-achievement-of-alasdair-macintyre

Shelmerdine, S. C., & Nour, M. M. (2025). AI chatbots and the loneliness crisis. *bmj, 391*.

Shen, J. H., & Tamkin, A. (2026). *Not a shortcut to competence: AI tools and skills formation* [Report]. IT Pro. https://www.itpro.com/software/development/anthropic-research-ai-coding-skills-formation-impact

Sire, J. (2015). *Naming the elephant: Worldview as a concept* (2nd ed.). InterVarsity Press.

Smith, R. S. (2014). *In search of moral knowledge*. IVP Academic.

Smith, R. S. (2021, June 21). Making sense of morality: Alasdair MacIntyre's ethics. *Moral Apologetics*. https://www.moralapologetics.com/wordpress/msm26

Chapter 13
Ethical Traditions and Servant Leadership in the Age of Artificial Intelligence:
Foundations for Human Flourishing

J. Richard Ray
http://orcid.org/0009-0005-0889-5560
Anderson University, USA

ABSTRACT

This chapter examines five ethical traditions, Aristotelian virtue ethics, Stoic philosophy, Christian moral theology, historical capitalism, and modern stakeholder theory, as foundations for human flourishing and servant leadership in the age of artificial intelligence. It argues that while AI can optimize outcomes and process vast amounts of data, it cannot cultivate virtue, discern moral purpose, or honor human dignity. By tracing these traditions and connecting them to servant leadership, the chapter shows how enduring moral principles can guide leaders in governing AI-mediated decision-making in organizations and educational settings. It concludes that leaders formed in ethical literacy and servant leadership are better equipped to design, deploy, and oversee AI systems in ways that protect dignity, strengthen accountability, and advance human flourishing.

DOI: 10.4018/979-8-3373-8192-3.ch013

ight © 2027, IGI Global Scientific Publishing. Copying or distributing in print or electronic forms without written permission of IGI Global Scientific Publishing is prohibited. Use of this chapter to n generative artificial intelligence (AI) technologies is expressly prohibited. The publisher reserves all rights to license its use for generative AI training and machine learning model development.

INTRODUCTION

A hospital administrator reviews an AI system that predicts patient readmission risk. The algorithm is accurate, but it flags low-income patients at higher rates, triggering interventions that some experience as surveillance rather than care. A manufacturing executive deploys productivity monitoring that increases output by 12%, but turnover spikes as workers report feeling "watched" and "dehumanized." A university provost champions an AI advising tool to close equity gaps, but six months later, data shows it's widening them.

These leaders face a common challenge: AI systems lack the wisdom, virtue, and moral judgment necessary for ethical decision-making. Algorithms optimize for measurable outcomes but cannot discern purpose, honor dignity, or cultivate flourishing. Therefore, AI governance depends entirely on leaders formed in ethical wisdom.

The challenge of our age is not merely understanding AI's capabilities; it is developing leaders with the moral wisdom to govern in ways that lead to human flourishing. This chapter argues that servant leadership, grounded in five enduring ethical traditions, provides the formational framework leaders need to govern AI with humility, empathy, accountability, and stewardship.

The ethical traditions examined, Aristotelian virtue ethics, Stoic philosophy, Christian moral philosophy, the ethics of capitalism and work, and modern stakeholder theory, are not merely historical curiosities. They are formational resources that cultivate the practical wisdom (*phronesis*) leaders need when facing decisions algorithms cannot and should not make.

Each tradition contributes distinctly to leadership character formation: Aristotelian ethics teaches leaders to ask whether technology enhances conditions for human flourishing; Stoic philosophy influences leaders with constancy and self-governance amid technological disruption; Christian moral philosophy develops leaders who see technology as stewardship, prioritizing the vulnerable; the ethics of work and capitalism reminds leaders that human dignity transcends productivity metrics; and stakeholder theory holds leaders accountable to multiple constituencies affected by AI decisions.

These traditions converge on a central insight: the ultimate purpose of human activity is not wealth, power, or pleasure, but flourishing, living well in the fullest sense. Even Adam Smith's treatise suggests that capital is not the goal but a means to society's prosperity and human well-being. *Profit is the vehicle; human flourishing is the destination.*

This chapter proceeds in three parts. Part I traces five ethical traditions and their implications for AI governance. Part II synthesizes six core principles for human flourishing in the AI age. Part III demonstrates how 14 servant leadership competencies operationalize these principles through formational practices leaders can cultivate.

ETHICAL TRADITIONS AND AI GOVERNANCE

The Wisdom Gap: What AI Cannot Provide

All ethical traditions examined share a fundamental recognition: certain human capacities cannot be reduced to calculation or optimization. AI systems can process vast amounts of data, identify patterns, optimize outcomes within defined parameters, and even simulate certain aspects of human reasoning. But they cannot possess wisdom, virtue, empathy, or genuine understanding of human flourishing.

This "wisdom gap" has profound implications: As AI systems take on more decision-making functions, the human capacities that AI cannot replicate become increasingly valuable and essential. The more we automate, the more we need leaders with deep ethical literacy, practical wisdom, and commitment to human flourishing.

Aristotelian Ethics: Virtue Formation for AI-Era Leadership

Aristotle's *eudaimonia*, human flourishing through virtuous living, provides the foundation for evaluating AI systems. Flourishing requires cultivating practical wisdom (*phronesis*), justice, and courage: precisely the capacities leaders need to govern AI ethically (Aristotle, trans. 2009).

Practical Wisdom (Phronesis): Aristotle distinguished phronesis from technical skill (techne) and theoretical knowledge (episteme). Practical wisdom integrates general principles with particular circumstances, perceiving what is morally salient and determining appropriate action. This cannot be reduced to rule-following or algorithmic calculation.

Justice: Aristotle identified justice as the most complete virtue, encompassing all others in relation to others. Justice involves giving each person their due and ensuring fair distribution of benefits and burdens. In AI governance, this means evaluating systems for disparate impact, algorithmic bias, and equitable treatment across demographic groups (Vallor, 2016).

Courage: Leaders need courage to challenge vendors, halt harmful deployments, and advocate for vulnerable stakeholders even when it's costly.

AI Governance Implication: Aristotelian ethics challenges leaders to ask whether AI systems enhance conditions for human flourishing. An algorithm that optimizes productivity while eroding meaningful work fails the *eudaimonia* test. Leaders formed in Aristotelian virtue don't ask only "Does this work?" but "Does this help people flourish?"

When a leader asks, "Will this algorithm help our employees develop judgment and skill, or will it atrophy their capacity for discernment?" this is *phronesis* in action. When a university administrator questions whether AI-generated feedback truly educates or merely processes students through a system, this is human flourishing informing technological governance.

Stoic Ethics: Constancy Amid Technological Disruption

Stoic philosophy adds a crucial dimension: constancy and self-governance amid technological disruption. As AI reshapes work at unprecedented speed, Stoic wisdom equips leaders to maintain moral clarity when external pressures demand compromise (Epictetus, trans. 1995; Marcus Aurelius, trans. 2006).

The Stoics taught that flourishing consists in living according to nature, specifically, according to reason and virtue. External circumstances (wealth, health, reputation) are "preferred indifferent", useful but not essential to flourishing. What matters is the cultivation of wisdom, justice, courage, and temperance.

Stoic Cosmopolitanism: The Stoics taught that all humans share a common rational nature and are therefore members of a universal human community. This cosmopolitan vision transcended the divisions of ethnicity, nationality, and social status. Every person possesses inherent dignity by virtue of their rational nature.

Self-Governance: Stoic ethics emphasizes what is within our control (our judgments, responses, commitments) versus what is not (external events, others' actions). Leaders cannot control AI's rapid development, but they can control their response, refusing to deploy systems that violate dignity, even under competitive pressure.

AI Governance Implication: Stoic ethics equips leaders to resist the pressure to deploy AI simply because competitors are doing so. It grounds the conviction that human dignity is non-negotiable regardless of market forces. Leaders formed in Stoic constancy ask: *What is within my control? What principles will I not compromise?*

Christian Moral Theology: Stewardship and Dignity in AI Governance

Christianity grounds human dignity in the doctrine of *imago Dei*, all humans are created in the image of God (Genesis 1:27). This confers inherent, equal, and inviolable dignity on every person regardless of ability, productivity, or social status. Human dignity is not earned; it is given by God.

Stewardship: Christian theology teaches that all resources, talents, wealth, authority, are entrusted to us by God. We are stewards, not owners, and will give account for how we use what has been entrusted (Luke 12:48). This accountability extends to multiple stakeholders and to future generations (Greenleaf, 1977; Spears, 1998).

Love and Service: Christianity introduces a distinctive ethical principle: *agape* love, self-giving concern for others' flourishing that goes beyond strict justice. Jesus commands his followers to love even enemies (Matthew 5:44) and models this love in his own self-sacrifice. Servant leadership embodies this principle: leadership as service on behalf of others' flourishing (Sendjaya & Sarros, 2002; van Dierendonck, 2010).

Community: Christian ethics is inherently communal. The early church practiced radical sharing of resources and mutual care (Acts 2:42-47). We flourish together or not at all.

AI Governance Implication: Christian ethics demands that AI systems honor the *imago Dei* in every person. Systems that treat people as mere data points or optimization targets violate this principle. Leaders formed in Christian stewardship recognize they are accountable to God and to all stakeholders for AI's impact on human dignity. They may forgo efficiency gains or competitive advantages when they come at the cost of human flourishing (Liden et al., 2008; Ray, 2024).

Historical Capitalism and the Protestant Work Ethic: Dignity and Danger

The Protestant work ethic, as articulated by Weber (1905/2002), emphasized that work has inherent dignity as a calling from God. Labor is not merely a means to survival but a form of worship and stewardship. This theological vision elevated the dignity of all work, not just religious vocations.

However, capitalism can reduce humans to economic units, "human resources" valued only for productivity. When profit becomes the ultimate end rather than a means to flourishing, capitalism becomes exploitative. AI can intensify this tendency: algorithms that optimize labor costs, monitor productivity, and automate jobs can treat workers as inputs to be minimized rather than image-bearers to be honored.

AI Governance Implication: Leaders must ask whether AI systems enhance or diminish the dignity and meaning of work. Technologies that increase efficiency while eroding workers' sense of purpose, autonomy, or contribution fail the human flourishing test.

Modern Capitalism and Stakeholder Theory: Accountability Beyond Shareholders

Contemporary stakeholder theory recognizes that organizations exist to create value for multiple stakeholders, employees, customers, suppliers, communities, and shareholders (Freeman, 1984; Freeman et al., 2010). Leaders are accountable not only to shareholders but to all stakeholders and ultimately to moral principles that transcend organizational interests.

This represents a recovery of the insight that people are not merely means to organizational ends but possess inherent dignity. When capitalism loses sight of this, it becomes destructive. Stakeholder theory corrects shareholder primacy by insisting that leaders are trustees accountable to all affected parties (Donaldson & Preston, 1995).

AI Governance Implication: AI systems affect multiple stakeholders, not just shareholders. Leaders must evaluate AI decisions by asking: *Who benefits? Who bears the costs? Whose voice is missing from this conversation?* Algorithms that maximize shareholder value while harming employees, customers, or communities violate stakeholder accountability.

Table 1. Selected ethical frameworks, human flourishing, leadership, & AI

Ethical Tradition	Time Period	Core Ethical Principle	View of Human Flourishing	Leadership Implication	AI Implication
Aristotelian Ethics	350 BC	Virtue ethics and *eudaimonia* as the ultimate goal. Ethical action arises from cultivated character and practical wisdom (*phronesis*).	Flourishing is living well through virtue, relationships, and contribution to the common good. Wealth and efficiency are tools, not ends.	Leaders must cultivate moral character and practical wisdom to navigate complex ethical situations.	AI can optimize outcomes but cannot possess practical wisdom, so human leaders must exercise moral judgment.
Stoic Ethics	300 BC to 200 AD	Virtue as the only true good; emphasis on moral integrity, constancy, and universal human dignity grounded in shared rationality.	Flourishing occurs when individuals live according to reason and virtue regardless of external circumstances.	Leaders must maintain moral integrity under pressure and resist unethical practices even when they offer competitive advantages.	Organizations must ensure AI systems respect universal human dignity and avoid bias, manipulation, or exploitation.
Christian Moral Philosophy	30 AD to present	Human dignity rooted in *Imago Dei*, stewardship, justice, and love (*agape*). Leadership is service to others.	Flourishing includes spiritual, relational, and moral well-being; people possess inherent worth independent of productivity.	Leaders act as stewards, accountable for how power and resources affect others, especially the vulnerable.	AI governance must protect dignity, avoid exploitation, and prioritize human welfare over efficiency or profit.

continued on following page

Table 1. Continued

Ethical Tradition	Time Period	Core Ethical Principle	View of Human Flourishing	Leadership Implication	AI Implication
Historical Capitalism & Protestant Work Ethic	1500 to 1900	Work as vocation and moral duty; diligence, stewardship, and responsible wealth creation.	Flourishing includes productive work that serves society and expresses one's calling.	Ethical capitalism integrates productivity with moral responsibility and stewardship.	AI should enhance human productivity while maintaining dignity and meaningful work.
Modern Capitalism & Stakeholder Ethics	20th century to present	Organizations create value for multiple stakeholders, not just shareholders.	Flourishing includes social well-being, sustainability, and equitable economic participation.	Leaders build ethical cultures, long-term stakeholder relationships, and socially responsible organizations.	AI deployment must consider impacts on employees, communities, and society.

SIX CORE PRINCIPLES FOR HUMAN FLOURISHING IN THE AI AGE

These five ethical traditions converge on six core principles that guide AI governance toward human flourishing.

Principle 1: Human Flourishing as the Ultimate End

All the ethical traditions examined share a fundamental conviction: the ultimate purpose of human activity is flourishing, living well in the fullest sense. Aristotle's *eudaimonia*, the Stoic life in accordance with nature, Christian abundant life, and stakeholder value creation all point toward this comprehensive good.

Implications for Organizations: Organizations exist not merely to generate profit but to create conditions for human flourishing, meaningful work, quality products and services, sustainable communities, and opportunities for growth. Profit is necessary for organizational sustainability, but it is instrumental, a means, not the ultimate end.

Implications for AI Governance: AI systems should be evaluated primarily by their contribution to human flourishing, not merely by efficiency, productivity, or profitability. Technologies that optimize measurable outcomes while undermining meaning, autonomy, relationships, or dignity fail by this standard.

Leaders must ask: Does this AI system enhance or diminish meaningful work? Does it support or undermine human autonomy? Does it strengthen or fragment relationships? Does it honor or violate human dignity?

Servant Leadership Competencies: *Awareness of the Human Spirit* (recognizing emotional, moral, and spiritual dimensions beyond productivity), *Presence* (hearing fears, hopes, and experiences of dignity), and *Regard for the Present* (attending to what systems are doing to people today).

Principle 2: Virtue and Character as Essential

Aristotelian *phronesis*, Stoic constancy, Christian holiness, and modern integrity all point to the necessity of cultivating moral character. Ethical action flows from virtuous character, not merely from following rules or calculating consequences.

Implications for Leadership Development: Leadership formation must focus on character development, cultivating wisdom, courage, justice, empathy, and integrity, not merely technical competencies. Integrating human flourishing into leadership training and MBA programs would yield better outcomes than purely technical or strategic curricula.

Implications for AI Governance: Because AI systems cannot possess virtue or exercise moral judgment, they must be governed by humans who have cultivated the character necessary for ethical discernment. Technical expertise without moral wisdom is insufficient, and potentially dangerous.

Servant Leadership Competencies: *Integrity* (wholeness before God and people), *Vulnerability* (strength in weakness, admitting limitations), and *Discernment* (wise judgment that integrates principles with circumstances).

Principle 3: Human Dignity and Justice

Stoic cosmopolitanism and Christian *imago Dei* establish that every human being possesses inherent and equal dignity. This grounds the obligation to justice, ensuring fair treatment, honoring rights, and distributing benefits and burdens equitably. Christian ethics adds special concern for the vulnerable and marginalized.

Implications for Organizations: All stakeholders must be treated with respect for their inherent dignity, not merely as means to organizational ends. This requires fair compensation, safe working conditions, honest communication, and special attention to impacts on vulnerable populations.

Implications for AI Governance: AI systems must be designed and deployed in ways that honor human dignity, respecting autonomy, avoiding manipulation, ensuring fairness, and including marginalized populations. Systems that treat people as mere data points violate this principle. Leaders must evaluate AI for disparate impact, algorithmic bias, and equitable distribution of benefits and burdens.

Servant Leadership Competencies: *Awareness of the Human Spirit* (honoring image-bearers), *Courage in Relationships* (advocating for the vulnerable), *Under-*

standing of the Past (learning from historical injustices), and *Breadth* (assembling diverse perspectives).

Principle 4: Love, Service, and Community

Christian *agape* and servant leadership emphasize self-giving concern for others' flourishing that goes beyond strict justice. Aristotelian social nature, Stoic cosmopolitanism, and stakeholder theory all recognize that human flourishing is inherently relational, we achieve the good life only in community with others.

Implications for Leadership: Leaders should prioritize the growth and flourishing of followers, exercising power as a trust to be used for others' benefit. This requires empathy, humility, and willingness to sacrifice self-interest for the common good. Organizations should foster authentic community, environments where people experience belonging, mutual support, and shared purpose.

Implications for AI Governance: AI deployment should be guided by concern for how technology affects human flourishing, especially for the most vulnerable. This may require forgoing efficiency gains when they come at the cost of human dignity. Leaders must evaluate AI for its effects on relationships and community, technologies that increase efficiency while fragmenting workplace relationships or eroding social trust may diminish flourishing despite productivity gains.

Servant Leadership Competencies: *Awareness of the Human Spirit* (empathy and compassion), *Sense of Humor* (joyful resilience that builds connection), *Presence* (grounded attentiveness to people), and *Breadth* (building community across differences).

Principle 5: Stewardship and Practical Wisdom

Christian stewardship teaches that leaders are trustees who will give account for their exercise of power and management of resources. This accountability extends to multiple stakeholders and to future generations. Aristotelian *phronesis* and Christian discernment emphasize that ethical action requires contextual judgment that integrates general principles with particular circumstances, virtue cannot be reduced to rule-following or algorithmic calculation.

Implications for Organizations: Leaders are accountable not only to shareholders but to all stakeholders and ultimately to moral principles that transcend organizational interests. This requires transparency, responsiveness to stakeholder concerns, and willingness to accept responsibility for harms caused. Leaders must cultivate practical wisdom, the capacity to perceive morally salient features of situations and determine appropriate action.

Implications for AI Governance: Those who design, deploy, and govern AI systems are accountable for their impact on human flourishing. This requires ongoing monitoring, transparency about capabilities and limitations, mechanisms for redress when harms occur, and willingness to modify or discontinue systems that cause unjustified harm. Because AI systems lack practical wisdom and contextual judgment, they cannot be given autonomous decision-making authority in domains requiring ethical discernment. Human oversight remains essential.

Servant Leadership Competencies: *Predictability* (following through on commitments), *Integrity* (transparency and accountability), *Respect for the Future* (long-term stewardship), *Discernment* (wise judgment), *Intellectual Energy and Curiosity* (pursuing understanding), and *Comfort with Ambiguity* (acting faithfully amid uncertainty).

Principle 6: Temporal Wisdom, Past, Present, and Future

Leaders must honor the past (learning from history), attend to the present (responding to current needs), and respect the future (stewarding for generations to come). Deuteronomy calls God's people to "remember the days of old" (32:7). Jesus teaches not to be anxious about tomorrow while being faithful today (Matthew 6:34). Jeremiah speaks of God's plans for "a future and a hope" (29:11).

Implications for Organizations: Organizations have histories and legacies that shape identity and must be honored. Leaders must attend to current stakeholder needs while planning sustainably for the long term. Structures and practices should balance execution with strategic foresight.

Implications for AI Governance: Leaders must study past technological harms (redlining, eugenics, surveillance) to avoid replicating patterns through algorithmic bias. They must attend to what AI systems are doing to people today, not just in theory. They must consider long-term consequences, what world are we building for the next generation?

Servant Leadership Competencies: *Understanding of the Past* (learning from history), *Regard for the Present* (mindful engagement with current realities), and *Respect for the Future* (long-term stewardship and foresight).

SERVANT LEADERSHIP COMPETENCIES AS FORMATIONAL PRACTICES

The six principles provide a framework; the 14 servant leadership competencies operationalize them through practices leaders can cultivate. The Center for Servant Leadership at the Pastoral Institute in Georgia defines servant leadership as a lifelong journey that includes discovery of oneself, a desire to serve others, and a commitment to lead. Servant-leaders continually strive to be trustworthy, self-aware, humble, caring, visionary, empowering, relational, competent, good stewards, and community builders (Spears, 1998).

Table 2. Servant leadership behaviors: Impact on implementing AI

Servant Leadership Competency	Servant Leader Behaviors	AI Governance Question/Application
1. Integrity	Acting consistently with organizational values; being trustworthy, transparent, and accountable. Leaders model honesty in all communications, align actions with stated values, keep commitments even when costly, and refuse to hide behind technical complexity when ethical questions arise. They maintain appropriate confidentiality while ensuring stakeholders have access to truth.	Would I want this system to be used on me or my family? Can we explain how this works in plain language? Are we being honest about this system's limitations and potential harms?
2. Vulnerability	Showing humility, admitting mistakes, and creating psychological safety in teams. Leaders openly acknowledge what they don't understand, ask for help and feedback, share learnings from failure, modeling confession and invite grace, and set healthy boundaries while remaining open. They resist the pressure to oversell AI capabilities and invite critique without defensiveness.	What don't we know about how this system will perform? What could go wrong? Are we admitting uncertainty honestly, or overselling capabilities? Can employees report problems without fear of retaliation?
3. Discernment	Making sound decisions with wisdom, evaluating risks, and reading situations accurately. Leaders weigh options using Scripture, counsel, and evidence; attend to motives and likely consequences; wait when needed and act when conviction is clear; test impressions with community; and ask probing questions beneath surface-level data to deeper questions of purpose, justice, and human flourishing.	Just because we can build this, should we? What problem are we actually solving? Who benefits? Who might be harmed? What are we not seeing? Is this the right solution, or are we being driven by competitive pressure?
4. Awareness of the Human Spirit	Valuing employee well-being, motivation, purpose, and dignity beyond productivity. Leaders listen for values, meaning, and hopes; honor conscience and spiritual practices; respond with compassion to suffering; create space for reflection and lament; refer appropriately for pastoral care; and recognize that humans are made in God's image with emotional, moral, and spiritual dimensions that transcend efficiency metrics.	How does this affect people's sense of dignity and purpose? Does this technology enhance or diminish their humanity? Are we treating people as "resources" to be optimized, or as image-bearers with inherent worth? What makes work feel meaningful to those affected?

continued on following page

Table 2. Continued

Servant Leadership Competency	Servant Leader Behaviors	AI Governance Question/Application
5. Courage in Relationships	Addressing conflict directly, giving honest feedback, and standing up for values in relationships. Leaders state hard truths with humility; seek reconciliation and make amends; protect the vulnerable and confront harm; stay engaged under pressure; challenge vendors and executives when necessary; and advocate for those who lack power in decision-making processes.	Who needs me to advocate for them in this decision? What hard truths need to be spoken? Am I staying silent when I should challenge this system? Who is being harmed that lacks a voice at the table?
6. Sense of Humor	Using humor appropriately to build relationships, relieve stress, and foster team spirit. Leaders use gentle levity to reduce anxiety; laugh with (not at) others; avoid sarcasm and demeaning jokes; read the room and time humor well; invite play that strengthens team bonds; and help teams stay resilient when AI projects fail, using humor to defuse tension in high-stakes meetings and humanize technical conversations.	How can we stay human and hopeful in this complexity? Are we so anxious about AI that we've lost perspective? Can we create space for joy and play even as we navigate serious ethical questions? How do we maintain team morale through uncertainty?
7. Intellectual Energy and Curiosity	Pursuing innovation, continuous learning, and challenging assumptions. Leaders ask thoughtful, exploratory questions; engage diverse sources and perspectives; experiment and learn iteratively; teach others what is learned; set learning goals and measure progress; dig deeper into vendor claims; and pursue understanding of how AI systems actually work rather than accepting surface explanations.	What are we not seeing? What questions should we ask? How does this actually work? What are the edge cases? What happens when the system encounters something it wasn't trained on? What assumptions are we making, and how can we test them?
8. Respect for the Future	Strategic thinking, sustainability, succession planning, and innovation for future growth. Leaders scan trends and anticipate risks/opportunities; build scenarios and contingencies; invest in people and systems that last; delay gratification for mission gains; communicate vision-linked milestones; and consider what world they are building for the next generation, refusing to optimize for short-term gains at the expense of long-term trust.	What world are we building for the next generation? What will we wish we had done differently 20 years from now? Are we encoding biases into systems that will shape decisions for decades? What are the long-term consequences of this deployment?
9. Regard for the Present	Focus on execution, appreciating employees and customers now, and being fully present. Leaders practice mindful presence in conversations; prioritize what is most important now; give thanks and recognize grace at work; tend to immediate care and quality; limit distractions and multitasking; and attend to what AI systems are doing to people today, not just in theory but in actual practice.	What is this system doing to people today? Not in theory, but in practice? Are we so focused on future capabilities that we're neglecting present harm? What immediate needs require our attention right now?

continued on following page

Table 2. Continued

Servant Leadership Competency	Servant Leader Behaviors	AI Governance Question/Application
10. Understanding of the Past	Learning from past successes/failures, honoring organizational culture and legacy. Leaders seek historical and contextual background; perform after-action reviews and lessons-learned; honor faithful tradition while reforming errors; document decisions for future reference; invite elders' and stakeholders' perspectives; and study past technological harms (redlining, eugenics, surveillance) to avoid replicating those patterns.	What historical patterns might we be repeating? How have similar systems been used in the past, and what can we learn? Are we replicating historical injustices through algorithmic bias? What does history teach us about this type of intervention?
11. Predictability	Following through on commitments and providing stability for teams. Leaders set clear expectations and meet them; maintain steady tone under stress; communicate schedules and changes early; use reliable processes and rhythms; avoid impulsive, whiplash decisions; create clear policies for AI use; communicate them consistently; and enforce them fairly so employees know what to expect.	Are we keeping our commitments to stakeholders? Have we followed through on promises made during implementation? Are our policies clear, consistent, and fairly enforced? Do employees trust that we will do what we say?
12. Breadth	Understanding multiple disciplines, cultures, or functions; versatility in leadership. Leaders cross-train across disciplines; seek diverse ministry and industry exposure; integrate theology with practice; build interdisciplinary networks; apply analogies from varied domains; and assemble diverse teams (technologists, ethicists, frontline workers, customers, community members) to evaluate AI systems from multiple perspectives.	Whose voice is missing from this conversation? Who hasn't heard this design process? Whose voice is missing from the training data? Have we consulted technologists, ethicists, frontline workers, and those most affected?
13. Comfort with Ambiguity	Navigating uncertainty, change, and complexity without paralysis. Leaders decide with bounded uncertainty and review; set provisional hypotheses and milestones; name unknowns and avoid false certainty; hold space for tension without rushing; keep people aligned to purpose in change; pilot and test rather than demanding perfect clarity; and communicate honestly about what they don't know while moving forward faithfully.	Can we pilot, learn, and adapt rather than demand certainty? Are we paralyzed by uncertainty, or can we proceed with bounded risk and learn as we go? What can we test on a small scale before committing fully? How do we balance prudence with progress?
14. Presence	Inspiring confidence, calm, and clarity; being fully engaged with people. Leaders practice undivided attention (eye contact, posture); regulate self and model calm in crisis; use silence and pauses well; signal availability with predictable rhythms; attune to verbal and nonverbal cues; create space for human connection, reflection, and discernment; and ensure that technology doesn't dominate meetings or relationships.	Are we creating space for wisdom, not just efficiency? Are we so rushed that we're missing what really matters? Have we created space for prayer, reflection, and discernment, or are we just optimizing for speed? Are we fully present to the people affected by this decision?

This section of the chapter offers the reader: (a) a definition of each of the 14 Servant Leadership competencies explaining why each competency matters, (b) how it is cultivated to increase human flourishing, (c) how leaders can leverage the competency to govern AI, and a concrete example of how the competency can explicitly impact AI governance. Each competency summary follows a consistent structure:

a. As Leadership Formation – What this competency means and why it matters
b. How This Principle Is Cultivated – Practical disciplines that form this capacity
c. How Leaders Formed in This Competency Govern AI Differently – The distinctive questions and approaches this formation produces
d. In Action – A concrete example demonstrating the competency's impact on AI governance.

Each competency is formed through intentional discipline and shapes how leaders govern AI. The following summaries demonstrate how these competencies are cultivated and applied to AI governance.

Competency 1: Integrity

Integrity as Leadership Formation

Integrity reflects wholeness before God and people, the alignment of words, motives, and actions with truth (Liden et al., 2008; Ray, 2024). In AI governance, integrity means leaders refuse to deploy systems they would not want used on themselves or their loved ones. It means transparency about limitations, honest reporting of failures, and refusal to hide behind technical complexity when ethical questions arise.

How This Principle Is Cultivated

Integrity is cultivated through accountability structures. Leaders need trusted advisors who will ask hard questions: *Are we being honest about this system's accuracy? Are we transparent about how decisions are made?* Integrity is also cultivated through confession and repair. When mistakes happen, leaders formed in integrity name them publicly, make amends, and change course.

How Leaders Formed in Integrity Govern AI Differently

Leaders formed in integrity refuse to deploy "black box" systems where decisions cannot be explained. They insist on documentation, auditability, and plain-language explanations. They ask: *Would I want this system used on me or my family? Can we explain how this works? Are we being honest about limitations and potential harms?*

Integrity in Action

A financial services company piloted an AI credit-scoring tool that increased approvals. But the Chief Risk Officer, formed in integrity, demanded to see the decision logic. The vendor resisted. The CRO insisted: *If we can't explain denials, we can't use this.* Investigation revealed the model used proxy variables (zip codes, shopping patterns) that correlated with race. The CRO halted the pilot, rebuilt the model with explainable features and bias testing, and launched a slower but fair system. That's what integrity produces.

Competency 2: Vulnerability

Vulnerability as Leadership Formation

Leaders who practice vulnerability admit mistakes, acknowledge limitations, and create psychological safety where others can do the same (Ray, 2024). Christian theology teaches that God's power is perfected in our weakness, so vulnerability is not failure but the posture that invites grace. In AI governance, this means admitting when we don't understand how a system works or when we've made errors in deployment.

How This Principle Is Cultivated

Vulnerability is cultivated in environments where failure is treated as learning, not punishment. Leaders model this by sharing their own missteps publicly. They ask questions even when they fear looking uninformed. They say, "I don't know, but let's find out together."

How Leaders Formed in Vulnerability Govern AI Differently

Leaders formed in vulnerability do not oversell AI capabilities. They admit uncertainty: *This system works well in controlled settings, but we don't yet know how it will perform in the real world.* They invite critique and create channels for employees to report problems without fear of retaliation.

Vulnerability in Action

A university deployed an AI advising system to close equity gaps. Six months in, data showed it was widening them, first-generation students received less helpful recommendations. The provost, formed in vulnerability, convened a town hall, acknowledged the failure, and invited students to help redesign the tool. She said publicly: *We got this wrong. We need your help.* Students appreciated the honesty. Faculty became collaborators. The system was rebuilt with input from those it was meant to serve. That's what vulnerability produces.

Competency 3: Discernment

Discernment as Leadership Formation

Discernment is prayerful, wise judgment that distinguishes what is right and timely (Liden et al., 2008). Aristotle called this *phronesis*, practical wisdom that integrates principles with circumstances. Christian theology adds that discernment requires prayer, Scripture, counsel, and attention to the Spirit's leading. In AI governance, discernment helps leaders ask: *Just because we can build this, should we?*

How This Principle Is Cultivated

Discernment is cultivated through reflection, prayer, and consultation with diverse voices. Leaders set aside time to consider not just what the data says, but what it means. They seek counsel from ethicists, frontline workers, and those most affected. Discernment is also cultivated by slowing down, testing assumptions, running pilots, and watching for unintended consequences before scaling.

How Leaders Formed in Discernment Govern AI Differently

Leaders formed in discernment do not rush to deploy AI simply because competitors are doing so. They ask: *What problem are we actually solving? Who benefits? Who might be harmed? What are we not seeing?*

Discernment in Action

A retail company considered an AI hiring tool that promised 40% faster hiring. But the CHRO, formed in discernment, paused. She asked: *What are we optimizing for? Speed, or the right fit?* She piloted the tool and discovered it screened out candidates with non-traditional backgrounds, career changers, veterans, people with employment gaps. The CHRO made a discerning choice: use the AI to resource candidates, but require human review before anyone was screened out. The company hired more diverse talent. Retention improved. That's what discernment produces.

Competency 4: Awareness of the Human Spirit

Awareness of the Human Spirit as Leadership Formation

Humans are made in God's image. Leaders who honor the spirit within each person recognize emotional, moral, and spiritual dimensions that transcend productivity metrics (van Dierendonck, 2010). In AI governance, this means valuing employee well-being and dignity beyond efficiency gains.

How This Principle Is Cultivated

Awareness of the human spirit is cultivated by listening for what people value, meaning, purpose, autonomy, connection. Leaders ask: *What makes work feel meaningful to you? What diminishes your sense of dignity?* This awareness is also cultivated through spiritual practices, seeing Christ in the face of the other, practicing hospitality, and creating space for lament and celebration.

How Leaders Formed in This Awareness Govern AI Differently

Leaders formed in awareness of the human spirit do not treat people as "resources" to be optimized. They ask: *How does this technology affect people's sense of purpose? Does it enhance or diminish their humanity?*

Awareness in Action

A call center implemented AI to monitor employee tone, pace, and script adherence in real time. The director, formed in awareness of the human spirit, piloted the system and held focus groups. Employees reported feeling "watched" and "like robots." Stress and turnover increased. The director made a crucial decision: instead of real-time monitoring, the AI would generate weekly coaching reports that employees could review privately with their managers. Employees could choose which metrics to focus on. Turnover dropped. Performance improved. That's what awareness of the human spirit produces.

Competency 5: Courage in Relationships

Courage in Relationships as Leadership Formation

Courage extends to how we treat others, addressing conflict, speaking truth in love, and standing up for values even when it's costly. In AI governance, courage means naming bias, challenging vendors, protecting vulnerable stakeholders, and advocating for those who lack power.

How This Principle Is Cultivated

Courage is cultivated by practicing hard conversations in low-stakes settings. Leaders role-play difficult scenarios, seek feedback, and reflect on what makes them hesitant to speak up. Courage is also cultivated through community, knowing others will stand with you when you take a principled stand.

How Leaders Formed in Courage Govern AI Differently

Leaders formed in courage do not stay silent when AI systems produce unjust outcomes. They confront vendors, challenge executives, and advocate for those who lack power. They ask: *Who needs me to advocate for them in this decision? What hard truths need to be spoken?*

Courage in Action

A municipal government deployed predictive policing software that directed patrols to “high-risk” neighborhoods. A mid-level analyst, formed in courage, noticed the system was creating a feedback loop: more patrols led to more arrests for minor offenses, which reinforced the algorithm's targeting. She raised concerns with her supervisor, who dismissed them. She escalated to the police chief, who was defensive. She then took her findings to the city council, knowing it could cost her job. The council ordered an independent audit, which confirmed her findings. The system was redesigned. That's what courage in relationships produces.

Competency 6: Sense of Humor

Sense of Humor as Leadership Formation

Joy and humor are gifts from God, used appropriately to build relationships and relieve stress. In AI governance, humor helps teams navigate complexity without becoming paralyzed by fear or cynicism.

How This Principle Is Cultivated

Humor is cultivated by creating space for play, celebration, and lightness even in serious work. Leaders model self-deprecation, laugh at their own mistakes, and invite others to do the same. Humor is also cultivated by reading the room, knowing when levity is appropriate and when gravity is required.

How Leaders Formed in Humor Govern AI Differently

Leaders formed in humor help teams stay resilient when AI projects fail. They use humor to defuse tension in high-stakes meetings and to humanize technical conversations.

Humor in Action

An AI ethics committee was deadlocked over whether to approve a facial recognition tool for building access. Tensions were high. The committee chair, formed in humor, opened the next meeting by showing a video of facial recognition systems hilariously misidentifying celebrities. The room laughed. Then she said: *If it can't tell Brad Pitt from a golden retriever, maybe we should be cautious.* The humor broke the tension. The committee had a productive conversation and ultimately approved a limited pilot with strict oversight. That's what humor produces.

Competency 7: Intellectual Energy and Curiosity

Intellectual Energy and Curiosity as Leadership Formation

A drive to learn, question, and seek understanding is essential for navigating AI's complexity (van Dierendonck, 2010; Giolito et al., 2021). In AI governance, leaders formed in curiosity don't accept vendor claims at face value, they dig deeper, ask "why," and pursue innovation responsibly.

How This Principle Is Cultivated

Curiosity is cultivated by reading widely, attending conferences, engaging with critics, and asking questions even when they reveal ignorance. Leaders set learning goals and hold themselves accountable. Curiosity is also cultivated by surrounding oneself with people who think differently and by rewarding those who challenge assumptions.

How Leaders Formed in Curiosity Govern AI Differently

Leaders formed in curiosity ask: *How does this actually work? What are the edge cases? What happens when the system encounters something it wasn't trained on?*

Curiosity in Action

A hospital implemented an AI diagnostic tool for radiology. The vendor claimed 95% accuracy. The chief medical informatics officer, formed in curiosity, asked: *95% accurate on what population? What happens with the other 5%?* She discovered the training data was predominantly from academic medical centers with state-of-the-art equipment. Her hospital served a rural population with older machines and sicker patients. She insisted on a pilot with rigorous tracking. The system's real-world accuracy was closer to 80%. The hospital implemented it as a "second reader" rather than a replacement for radiologists. Curiosity prevented harm. That's what intellectual energy produces.

Competency 8: Respect for the Future

Respect for the Future as Leadership Formation

Considering long-term consequences, strategic thinking, and sustainability are marks of servant leadership (Greenleaf, 1977; Ray, 2024). In AI governance, this means asking: *What world are we building for the next generation?*

How This Principle Is Cultivated

Respect for the future is cultivated by scenario planning, studying history, and consulting with younger stakeholders who will inherit our decisions. Leaders ask: *What will we wish we had done differently 20 years from now?*

How Leaders Formed in This Respect Govern AI Differently

Leaders formed in respect for the future do not optimize for short-term gains at the expense of long-term trust. They invest in explainability, fairness, and sustainability even when it's costly.

Respect in Action

A tech company was developing an AI content moderation system. The fastest path to market was to train the system on existing moderation decisions. But the chief ethics officer, formed in respect for the future, asked: *What if our past decisions were biased? Are we encoding those biases into a system that will shape discourse for decades?* She insisted on a slower approach: auditing past decisions, consulting with marginalized communities, and building in mechanisms for the system to learn and adapt. The product launched six months later than competitors. But it was more fair, more transparent, and more adaptable. That's what respect for the future produces.

Competency 9: Regard for the Present

Regard for the Present as Leadership Formation

Valuing current opportunities, people, and circumstances keeps leaders grounded (Ray, 2024). In AI governance, this means not being so focused on future capabilities that we neglect present harms.

How This Principle Is Cultivated

Regard for the present is cultivated through mindfulness practices, being fully present in conversations, noticing what's happening now, and responding to immediate needs. Leaders practice gratitude, recognizing grace at work in the present moment.

How Leaders Formed in This Regard Govern AI Differently

Leaders formed in regard for the present ask: *What is this system doing to people today? Not in theory, but in practice?*

Regard in Action

A school district deployed an AI early warning system to identify students at risk of dropping out. A principal, formed in regard for the present, noticed that students flagged by the system were being treated differently by teachers, lower expectations, fewer opportunities. She shifted the approach: instead of labeling students "at risk," the system generated resource recommendations, tutoring, mentoring, counseling, that any student could access. The focus shifted from prediction to support. Graduation rates improved. That's what regard for the present produces.

Competency 10: Understanding of the Past

Understanding of the Past as Leadership Formation

Learning from history helps leaders avoid repeating mistakes (Spears, 1998; Ray, 2024). In AI governance, this means studying past technological harms, redlining, eugenics, surveillance, and asking: *How do we avoid replicating those patterns?*

How This Principle Is Cultivated

Understanding of the past is cultivated by reading history, consulting with elders, and conducting after-action reviews. Leaders ask: *What can we learn from what came before?*

How Leaders Formed in This Understanding Govern AI Differently

Leaders formed in understanding of the past recognize that bias in AI is not a new problem, it's the latest iteration of longstanding injustices. They design systems with historical awareness.

Understanding in Action

A city was considering an AI tool to allocate social services. A policy analyst, formed in understanding of the past, researched how similar systems had been used historically. She discovered that "risk assessment" tools had been used to justify discriminatory housing policies, over-policing, and family separation. She brought this history to decision-makers and insisted on safeguards: community oversight, regular audits, and the ability for individuals to challenge algorithmic decisions. The system was implemented with these protections. Historical awareness prevented harm. That's what understanding of the past produces.

Competency 11: Predictability

Predictability as Leadership Formation

Reliability mirrors God's constancy (Liden et al., 2008; van Dierendonck, 2010). In AI governance, predictability means leaders follow through on commitments, communicate clearly, and provide stability for teams navigating change.

How This Principle Is Cultivated

Predictability is cultivated by setting clear expectations, maintaining consistent communication rhythms, and avoiding impulsive decisions. Leaders do what they say they will do.

How Leaders Formed in Predictability Govern AI Differently

Leaders formed in predictability create clear policies for AI use, communicate them consistently, and enforce them fairly. Employees know what to expect.

Predictability in Action

A company was rolling out AI tools across departments. Some teams embraced them; others resisted. Confusion reigned. The COO, formed in predictability, established clear guidelines: AI tools would be piloted before scaling, employees would be trained, and no one would lose their job due to automation without retraining opportunities. She communicated these commitments repeatedly and kept them. Trust increased. Adoption improved. That's what predictability produces.

Competency 12: Breadth

Breadth as Leadership Formation

Understanding multiple disciplines, cultures, and functions enables leaders to see connections others miss (Greenleaf, 1977; Senge, 1990). In AI governance, breadth helps leaders integrate technical, ethical, legal, and human perspectives.

How This Principle Is Cultivated

Breadth is cultivated by seeking diverse experiences, cross-training, job rotations, interdisciplinary learning. Leaders read outside their field and build relationships across silos.

How Leaders Formed in Breadth Govern AI Differently

Leaders formed in breadth assemble diverse teams to evaluate AI systems. They don't rely solely on engineers or ethicists, they bring together technologists, frontline workers, customers, and community members.

Breadth in Action

A healthcare system was implementing AI across clinical, operational, and administrative functions. The CIO, formed in breadth, created a cross-functional AI governance council: clinicians, nurses, patients, IT staff, ethicists, and legal counsel. This breadth of perspective caught issues no single group would have seen. Clinicians flagged workflow problems. Patients raised privacy concerns. Ethicists identified bias risks. The AI implementations were stronger because of this breadth. That's what breadth produces.

Competency 13: Comfort with Ambiguity

Comfort with Ambiguity as Leadership Formation

Ability to remain effective when outcomes are uncertain is essential in AI governance, where systems are complex, opaque, and constantly evolving. Christian theology grounds comfort with ambiguity in faith, trusting God when we cannot see the path ahead.

How This Principle Is Cultivated

Comfort with ambiguity is cultivated by practicing decision-making with incomplete information, setting provisional goals, and learning iteratively. Leaders embrace “good enough for now” rather than waiting for perfect clarity.

How Leaders Formed in This Comfort Govern AI Differently

Leaders formed in comfort with ambiguity don't demand certainty before acting. They pilot, test, learn, and adapt. They communicate honestly about what they don't know.

Comfort in Action

A nonprofit was exploring AI to match volunteers with opportunities. The executive director, formed in comfort with ambiguity, acknowledged: *We don't know if this will work. Let's try it, measure carefully, and adjust.* They piloted the system with a small group, gathered feedback, and iterated. Some features worked; others didn't. They adapted. The final system looked different from the original vision, but it was better because the leader was comfortable navigating uncertainty. That's what comfort with ambiguity produces.

Competency 14: Presence

Presence as Leadership Formation

True presence flows from being grounded in God. In AI governance, presence means being fully attentive to people, not just systems, listening deeply, noticing what's unsaid, and inspiring confidence through calm, grounded leadership.

How This Principle Is Cultivated

Presence is cultivated through spiritual practices, silence, prayer, Sabbath rest. Leaders practice being fully present in conversations, putting away devices, and giving undivided attention.

How Leaders Formed in Presence Govern AI Differently

Leaders formed in presence don't let technology dominate meetings or relationships. They create space for human connection, reflection, and discernment.

Presence in Action

An executive team was making a high-stakes decision about AI deployment. The meeting was rushed, data-heavy, and tense. The CEO, formed in presence, paused the meeting. She said: *We're not ready to decide. Let's take 10 minutes of silence to reflect, pray, and listen to what we're really being called to do.* The silence was uncomfortable at first. But it created space for deeper wisdom to emerge. When the conversation resumed, it was more thoughtful, more collaborative, and more aligned with the organization's values. The decision was better because the leader created space for presence. That's what presence produces.

CONCLUSION: LEADERSHIP FORMATION FOR ETHICAL AI GOVERNANCE

The question is not whether AI will shape our future, it will. The question is whether we will have leaders with the character to ensure AI serves human flourishing. Artificial intelligence will reshape the future of work, learning, and organizational life. Technology itself cannot determine whether that future advances or undermines human flourishing. Leaders can, however, and that is where responsibility rests.

This is a lifelong journey of formation and change. Drawing from Aristotelian virtue ethics, Stoic philosophy, Christian moral theology, capitalism properly understood, and stakeholder theory, we must be intentional in how we equip leaders to govern AI in ways that protect and advance human flourishing.

Leaders formed through using the 14 Leadership Competencies ask different questions, involve different voices, and prioritize different outcomes. They recognize that *profit is the vehicle; human flourishing is the destination.* They understand that AI systems, no matter how sophisticated, lack the virtue, wisdom, and moral judgment that only humans formed in these competencies can provide.

This chapter argues that the decisive question in the AI era is not how intelligent our systems become, but how well our leaders are developed and prepared to lead. We have traced five enduring ethical traditions, Aristotelian virtue ethics, Stoic philosophy, Christian moral theology, capitalism properly understood, and modern stakeholder theory, and shown how they converge on a common insight: human flourishing is the proper end of all human endeavor, including the design and governance of technology.

But ethical principles alone are insufficient. They must be operationalized through lived practice. This is where servant leadership becomes essential. The chapter has demonstrated that servant leadership translates abstract ethical wisdom into concrete, actionable competencies: humility, empathy, accountability, stewardship, discernment, presence, and the capacity to honor past, present, and future. These are not merely virtues to admire; they are practices to cultivate, habits to form, disciplines to embrace. Leaders who develop these competencies ask different questions, involve different voices, and prioritize different outcomes than those who do not.

Consider what becomes possible when leaders are formed in this way:

- Leaders who are developed to exhibit humility and empathy recognize that AI systems, no matter how sophisticated, lack the practical wisdom to make decisions affecting human dignity. They insist on and provide human oversight, stakeholder voice, and mechanisms for redress when systems cause harm.
- Leaders who exhibit accountability and stewardship do not deploy AI for efficiency alone. They ask: What is this system doing to people today? Not in theory, but in practice? They monitor algorithmic bias, protect vulnerable populations, and are willing to modify or discontinue systems that undermine flourishing, even at competitive cost.
- Leaders with an understanding of the past study technological harms, redlining, eugenics, surveillance and ask: How do we avoid replicating those patterns? They bring historical wisdom to present decisions.
- Leaders prepared to have regard for the present remain grounded in the immediate human impact of technology. They notice when AI systems fragment relationships, erode trust, or create psychological distance. They resist the seduction of efficiency gains that come at the cost of community and connection.
- Leaders developed to honor respect for the future ask: What are we building for the next generation? They slow down when necessary, consult with those most affected, and build in mechanisms for learning and adaptation. They understand that some competitive advantages are not worth the long-term cost to human dignity.

This is not naive idealism or hope for the future. It is grounded in centuries of ethical reflection and in the lived experience of leaders who have chosen formation over expediency. Research links servant leadership with stronger employee performance, creativity, organizational citizenship behavior, job satisfaction, commitment, engagement, and lower turnover intentions. It also supports innovation by fostering psychological safety, well-being, empowerment, and team innovation capacity (Eva et al., 2019). They attract and retain talent. They innovate more sustainably (Ren et al., 2024). They build cultures of trust and belonging. They create meaning in work.

The future of AI governance depends less on the power of technology than on the moral formation of the people entrusted to implement and lead it. AI will amplify whatever character, assumptions, and purposes leaders already possess. If leaders are formed in virtue, stewardship, and service - if they have cultivated the habits of humility, empathy, accountability, and discernment - then AI can be directed toward justice, dignity, and the common good. If leaders lack this formation, no policy framework or ethical guideline will prevent AI from scaling harm with unprecedented efficiency.

The work ahead is not primarily technological. It is formational and developmental. We must ask ourselves: How do we cultivate leaders who are wise enough, humble enough, and virtuous enough to govern AI toward human flourishing? This requires investment in leadership development grounded in ethical traditions. It requires organizations to prioritize character formation alongside competency development. It requires mentoring relationships where experienced leaders pass on not just knowledge but also wisdom. It requires practices of reflection, discernment, and accountability that keep leaders morally awake and aligned with their deepest values.

This is the distinctive contribution of servant leadership in the age of artificial intelligence: it provides not just a philosophy but a pathway for forming leaders who can be trusted with power, including the immense power of AI systems. It offers concrete practices through which ethical principles become lived reality. And it grounds leadership formation in something deeper than technique or efficiency: in the conviction that every person has dignity, that human flourishing is the proper measure of organizational success, and that leaders exist to serve that flourishing, not to exploit it.

The question before us is not whether AI will shape the future. It will. The question is: *Will we shape the leaders who shape AI?* Will we commit to the difficult work of forming and developing leaders in virtue, wisdom, and service? Will we build organizations where such formation is possible? Will we insist that those who design and govern AI systems be people of character, discernment, and moral courage? Or will we do nothing and hope for the best?

The stakes could not be higher. There is an old adage in the strategic planning world that states, "Control your future or be controlled by someone else who is controlling theirs." This could not be more significant than in the formation of AI. Who is developing and forming the leaders who are leading in the AI era? The future of human flourishing in the age of artificial intelligence depends on our answer.

REFERENCES

Aristotle, . (2009). *The Nicomachean ethics* (Ross, D., Trans.). Oxford University Press. (Original work published 350 BCE)

Donaldson, T., & Preston, L. E. (1995). The stakeholder theory of the corporation: Concepts, evidence, and implications. *Academy of Management Review*, *20*(1), 65–91. DOI: 10.2307/258887

Epictetus, . (1995). *The discourses* (Hard, R., Trans.). Everyman. (Original work published 108 CE)

Eva, N., Robin, M., Sendjaya, S., van Dierendonck, D., & Liden, R. C. (2019). Servant leadership: A systematic review and call for future research. *The Leadership Quarterly*, *30*(1), 111–132. DOI: 10.1016/j.leaqua.2018.07.004

Freeman, R. E. (1984). *Strategic management: A stakeholder approach.* Pitman.

Freeman, R. E., Harrison, J. S., Wicks, A. C., Parmar, B. L., & de Colle, S. (2010). *Stakeholder theory: The state of the art.* Cambridge University Press. DOI: 10.1017/CBO9780511815768

Giolito, V., Resca, A., & Za, S. (2021). Servant leadership and intellectual stimulation in the digital transformation context. *Journal of Leadership Studies*, *15*(2), 6–21. DOI: 10.1002/jls.21738

Greenleaf, R. K. (1977). *Servant leadership: A journey into the nature of legitimate power and greatness.* Paulist Press.

Liden, R. C., Wayne, S. J., Zhao, H., & Henderson, D. (2008). Servant leadership: Development of a multidimensional measure and multi-level assessment. *The Leadership Quarterly*, *19*(2), 161–177. DOI: 10.1016/j.leaqua.2008.01.006

Marcus Aurelius. (2006). *Meditations* (M. Hammond, Trans.). Penguin Classics. (Original work published ca. 170–180 CE)

Ray, R. (2024). *Servant leadership: Developing behaviors for change.* Anderson University Press.

Ren, L., & Shen, H. (2024). The relationship between servant leadership and team innovation performance: Mediating effect of self-efficacy. *Heliyon*, *10*(6), e27723. Advance online publication. DOI: 10.1016/j.heliyon.2024.e27723 PMID: 38509912

Sendjaya, S., & Sarros, J. C. (2002). Servant leadership: Its origin, development, and application in organizations. *Journal of Leadership & Organizational Studies, 9*(2), 57–64. DOI: 10.1177/107179190200900205

Senge, P. M. (1990). *The fifth discipline: The art & practice of the learning organization*. Doubleday.

Spears, L. C. (1998). *Insights on leadership: Service, stewardship, spirit, and servant-leadership*. John Wiley & Sons.

Vallor, S. (2016). *Technology and the virtues: A philosophical guide to a future worth wanting*. Oxford University Press. DOI: 10.1093/acprof:oso/9780190498511.001.0001

van Dierendonck, D. (2010). Servant leadership: A review and synthesis. *Journal of Management, 36*(4), 1228–1261. DOI: 10.1177/0149206310380462

Weber, M. (2002). *The Protestant ethic and the spirit of capitalism* (Baehr, P., & Wells, G. C., Trans.). Penguin Books. (Original work published 1905)

KEY TERMS AND DEFINITIONS

Algorithmic Stewardship: The responsibility of leaders to ensure that artificial intelligence systems remain aligned with ethical commitments, transparent oversight, and the long-term well-being of those affected by their use.

Ethical AI Governance: The practice of designing, deploying, and overseeing artificial intelligence systems in ways that protect human dignity, ensure accountability, reduce harm, and promote long-term human flourishing.

Human Dignity: The intrinsic worth of every person that requires organizations and technologies to treat individuals as ends in themselves rather than instruments of efficiency or profit.

Human Flourishing A holistic condition in which individuals and communities develop morally: relationally, intellectually, and spiritually while participating meaningfully in work and society beyond mere economic productivity or efficiency.

Imago Dei: The theological conviction that every human being bears the image of God and therefore possesses inherent and equal dignity independent of status, ability, or productivity.

Practical Wisdom (Phronesis): The cultivated capacity to interpret complex moral situations and choose appropriate action by integrating ethical principles with contextual judgment rather than relying on rules or technical calculation alone.

Servant Leadership A leadership approach that prioritizes the growth: dignity, and well-being of others and understands authority as a responsibility to steward power for the common good rather than personal advancement.

Stakeholder Theory: An ethical framework that understands organizations as responsible to multiple groups affected by their decisions, including employees, customers, communities, and future generations, not only shareholders.

Stewardship: A leadership responsibility grounded in accountability for how authority, resources, and technologies are used to serve others and sustain conditions for flourishing across generations.

Temporal Wisdom: The leadership capacity to integrate lessons from the past, realities of the present, and responsibilities toward the future when evaluating technological decisions and institutional strategies.

Virtue Formation: An ongoing developmental process through which leaders cultivate character traits such as courage, justice, humility, and integrity that guide ethical decision-making in complex environments.

Wisdom Gap: The difference between what artificial intelligence systems can optimize computationally and what human leaders must discern morally when evaluating purpose, justice, and the impact of decisions on people.

Chapter 14
AI, Disability, Accessibility, and the Ethics of Inclusion:
Participation, Design, and the Common Good

Jeffrey Moore
http://orcid.org/0000-0002-2508-2896
Anderson University, USA

Remco Mostert
HOGENT University of Applied Sciences and Arts, Belgium

ABSTRACT

This chapter examines how artificial intelligence is reshaping disability accessibility and argues that ethical inclusion requires more than technical compliance or accommodation. While AI technologies can expand participation through assistive tools, adaptive interfaces, and decision support systems, they can also reinforce exclusion when designed without attention to dignity, agency, and social belonging. Drawing on disability studies, ethical leadership scholarship, and the Quality of Life framework, the chapter proposes a shift from barrier removal toward flourishing-centered accessibility. It introduces a leadership approach that evaluates AI systems according to their impact on participation, self-determination, relationships, and rights. By reframing accessibility as a moral and organizational responsibility, the chapter offers a practical framework for institutions seeking to align AI innovation with human dignity, inclusion, and the common good.

DOI: 10.4018/979-8-3373-8192-3.ch014

ight © 2027, IGI Global Scientific Publishing. Copying or distributing in print or electronic forms without written permission of IGI Global Scientific Publishing is prohibited. Use of this chapter to n generative artificial intelligence (AI) technologies is expressly prohibited. The publisher reserves all rights to license its use for generative AI training and machine learning model development.

WHY DISABILITY MUST MOVE FROM MARGIN TO CENTER

Disability must move from the margins to the center of AI ethics because it exposes a deep question every technological system raises: what kind of human being is this system designed to recognize, serve, and protect? Too often, disability is addressed only after systems have been developed. Fairness, transparency, and accountability come first, while accessibility is added later as a technical adjustment or compliance issue. That order is wrong. Disability is one of the clearest tests of whether institutions value dignity, participation, and the common good.

AI systems increasingly shape access to work, education, healthcare, communication, and civic life. They filter applicants, structure workflows, mediate services, and organize participation through data, models, and automated procedures that appear neutral. Yet these systems embody judgments about normal communication, acceptable pace, useful cognition, legitimate productivity, and meaningful participation. Disability brings those judgments into view. Where disabled persons are excluded or forced to adapt to narrow norms, the problem is not only technical. It is moral. The system reflects too limited an account of the human person.

This chapter argues that disability inclusion reveals whether AI is ordered toward control and efficiency or toward human flourishing. The accessibility crisis in AI is not simply a design failure. It is a formational failure. It reveals what institutions value, what leaders notice, and whose participation they deem central or expendable. Compliance language alone cannot resolve it. The question requires attention to formation, character, wisdom, discerning leadership practice, and human flourishing.

The chapter examines hidden anthropology in AI systems, character and moral perception, wisdom, discerning leadership practice, and human flourishing. It treats accessibility failures as failures of moral perception and frames AI governance as an ongoing responsibility rather than a one-time compliance task. Across these arguments, accessibility-centered AI governance must create conditions in which people can participate, contribute, belong, and grow within the common life of the community.

Seen in this light, disability is not a niche concern within AI ethics but a decisive ethical lens for the kind of people, communities, and institutions technological life is helping to form.

FORMATION AND THE HIDDEN ANTHROPOLOGY OF AI SYSTEMS

If disability must move from margin to center in AI ethics, one reason is that AI systems are not merely technical instruments. They do more than process information or generate outputs. They shape how institutions perceive persons, define competence, and distribute participation. The crisis is therefore not only a design problem. It is a problem of formation.

Every AI system carries assumptions about normal communication, acceptable pace, productive behavior, useful cognition, and legitimate participation. These assumptions may not be declared, yet they are embedded in data selection, interface design, workflow expectations, and performance metrics. AI systems are governed by a hidden anthropology. They reflect a vision of the human person, often shaped by efficiency, standardization, predictability, and measurement. When these assumptions remain unexamined, exclusion can appear natural and accessibility secondary.

Disability makes visible what dominant systems hide: a tool built around narrow assumptions may function smoothly for some users while quietly excluding others, including visually impaired or neurodiverse users. What appears neutral is often formed around a restricted image of who belongs easily within it.

These systems cannot be separated from the organizations that create and adopt them. AI emerges within cultures shaped by mission, leadership priorities, and assumptions about which persons and performances matter most. Moore, Maxey, and Hangen argue that leadership and culture determine whether technology is ordered toward human well-being or merely toward efficiency, requiring movement from top-down, performance-driven models toward cultures of dignity, development, and participation (Moore et al., n.d.). Moore and Hanson show that mission, strategy, leadership, and culture must align for technological change to remain human-centered rather than merely operationally effective (Moore & Hanson, n.d.).

Formation works in two directions. Institutions shape systems according to what they value, and those systems then shape institutional habits, expectations, and judgments. Organizations that prize efficiency without sufficient regard for dignity tend to produce tools that reward conformity to narrow norms. Over time, those tools train the institution to accept what they make visible and overlook what they exclude. AI does not merely reflect moral vision. It can intensify and normalize it.

This helps explain why accessibility problems are often recognized too late. By the time barriers become visible, designers have assumed a standard user, managers have accepted certain workflows as normal, and leaders have treated friction as unfortunate but unavoidable. The exclusion of disabled persons therefore reveals more than oversight. It reveals a moral imagination shaped by thin assumptions about ability, independence, and worth.

Aboodi is helpful here because he shows that AI can shape habits of judgment by encouraging deference that weakens critical reflection. When people rely on AI without sustained human discernment, they risk outsourcing the habits of perception and evaluation that moral responsibility requires (Aboodi, 2025). In institutional settings, AI is not only formed by prior values. It can also train organizations to accept its categories, recommendations, and exclusions as normal.

Formation asks what institutions are becoming. Are they becoming more attentive to human variation, dependence, and vulnerability, or less? Are they being trained to recognize dignity across difference, or only to reward what is easiest to measure and manage? These questions are central because they concern the people, organizations, and communities those systems are helping to form.

This is why accessibility cannot be treated as a late adjustment to otherwise complete systems. If exclusion is rooted in formative assumptions, the ethical task begins upstream, in the habits of perception, organizational values, and definitions of competence that shape design. Systems formed around narrow norms train organizations to accept exclusion as ordinary. Systems shaped by a richer account of the human person can cultivate more attentive, just, and humane participation. Before asking whether a system is fair, institutions must ask what kind of human being the system assumes and what kind of community its use is helping to create.

Why Disability Is a Distinct Ethical Lens

Generic fairness language is not enough. Disability brings into focus dimensions of human life that standard AI ethics frameworks often treat too thinly. Fairness, bias mitigation, and equal treatment remain important, but they do not fully account for embodiment, vulnerability, dependence, communication, accommodation, and belonging. Disability exposes the limits of ethical approaches that ask only whether people are treated the same while neglecting whether they are actually able to participate, contribute, and flourish.

Disability functions as a distinct ethical lens rather than merely one application area among others. It reveals that exclusion often arises not because a system is openly hostile, but because it has been designed around assumptions that appear normal to those who fit them well. Systems may presume a certain speed of response, a certain style of communication, a certain level of sensory ease, or a certain pattern of attention and cognition. The result is predictable. When those assumptions govern design, people who do not match them are treated as exceptions to be accommodated later, not full participants whose presence should have shaped the system from the start.

Disability therefore presses AI ethics beyond the language of abstract sameness. A system can be statistically fair and still inaccessible. It can offer equal treatment while ignoring what participation actually requires for people with different bodies, minds, and modes of engagement. This is the difference between fairness and accessibility, accommodation and participation, compliance and moral responsibility. Fairness asks whether a rule or process applies evenly. Accessibility asks whether the environment itself has been built so that people can genuinely enter, act, and belong within it. Accommodation may adjust a system at the margins. Participation asks whether people were considered central to the system's design and whether they can help shape its ongoing use.

This is a question of moral fittingness. Ethical evaluation cannot stop with whether a system produces efficient or unbiased outputs. It must also ask whether those outputs arise from a way of seeing human beings that is adequate to their dignity and difference. Some decisions require discernment about what people need in order to participate meaningfully, not neutrality alone. A rigid commitment to sameness can itself become unjust when it ignores the actual conditions necessary for access, contribution, and belonging.

Disability further reminds institutions that dependence is not an exception to human life but a basic feature of it. Much of technological culture prizes autonomy, speed, independence, and frictionless performance. Disability unsettles that imagination by revealing how deeply human beings depend on one another, on supportive structures, and on environments designed with care. This is not a weakness to be managed away. It is part of the truth about human existence. AI systems shaped by an excessively individualistic or efficiency-centered vision will tend to misread dependence as deficiency. Supportive design is not a concession to a few. It honors the interdependent reality of human community.

This also concerns the common good. When institutions design with disabled persons in view, they often create systems that are clearer, more flexible, more humane, and more trustworthy for many others as well. The point is not merely that accessibility has secondary benefits, though it often does. The deeper point is that institutions become morally wiser when they are shaped by those who have been most vulnerable to exclusion. Disability therefore does not simply reveal the failures of technological systems. It can also become a source of insight into how communities might be structured more justly and more humanely.

Seen in this light, disability is a privileged ethical lens because it reveals where dominant frameworks remain too abstract. It forces institutions to confront the gap between formal inclusion and actual participation, between equal treatment and meaningful belonging, between technical efficiency and moral adequacy. Removing obvious discrimination is not enough. AI ethics must also address the deeper assumptions that render some forms of human life less visible, less valued, or less welcome. To take disability seriously is therefore to move from a thin procedural ethics toward a richer account of what justice requires for embodied, diverse, and interdependent people.

CHARACTER, DIGNITY, AND THE IMAGE OF GOD

At the theological center of this chapter is a claim about dignity. Persons with disabilities possess full and irreducible dignity because human worth is grounded in the imago Dei, not in productivity, independence, verbal fluency, cognitive speed, or conformity to institutional norms. This claim challenges many of the assumptions that quietly govern technological systems. AI tools are often developed in environments shaped by optimization, standardization, and prediction. When those priorities are severed from a truthful account of the human person, they can subtly rank lives according to efficiency, communicative ease, or economic usefulness. The result is not merely technical exclusion but moral distortion.

Disability belongs at the center of AI ethics. Not at the margins. Disability exposes a deeper question: what kind of human being do our systems assume, reward, and protect? If dignity depends on autonomy, speed, or output, then people with disabilities will remain vulnerable to being treated as secondary participants in institutional life. If dignity is rooted instead in created worth, then accessibility becomes a matter of justice rather than benevolence. The issue is whether institutions are willing to order their systems around a more complete account of the human person.

This theological vision is clarified further by the command to love one's neighbor. In Luke's account of the Good Samaritan, neighbor-love is not defined by proximity, similarity, status, or usefulness. Neighbor-love is attentive to vulnerability and costly in action. The Samaritan does not simply recognize suffering. He interrupts his own plans, assumes responsibility, and reorders his resources so that another person may live. In institutional terms, this parable suggests that moral responsibility begins when individuals and systems refuse to pass by forms of exclusion that have become normalized. Neighbor-love, applied to disability, requires more than sympathy or charitable sentiment. It calls for the deliberate reshaping of social, organizational, and technological arrangements so that those who have been left at the margins are able to participate fully in the life of the community.

This makes character central to accessibility. Ethical accessibility depends not only on policies and technical safeguards but on the moral quality of the leaders who govern AI systems. Leaders need humility to recognize the limits of their assumptions, justice to evaluate institutional arrangements honestly, compassion toward embodied difference, courage to challenge efficiency when it diminishes persons, and stewardship to ensure that technology remains accountable to human good. Moore and Halter's Employee Flourishing Leadership framework is especially helpful here. It shows that effective supervisors move beyond transactional management toward psychologically safe, person-centered, adaptive leadership that treats people as whole people rather than as units of output (Moore & Halter, in press).

Hassen's critique of moral outsourcing sharpens this concern. When leaders defer too quickly to system outputs, automated recommendations, or predictive models, they risk surrendering the moral agency needed to ask whether a decision is just, whether a barrier is avoidable, or whether a person is being reduced to a category rather than recognized as a neighbor. Accessibility failures, then, are often not only procedural failures. They are failures of character, revealed when convenience, abstraction, or efficiency replace moral judgment (Hassen, 2025).

Seen this way, disability inclusion becomes a litmus test of character. It reveals whether leaders treat accessibility as a matter of justice rooted in dignity, or whether institutional convenience determines who belongs. It also shows whether institutions understand their vocation as serving people and the common good rather than simply maximizing performance.

Accessibility as Justice in Design, Not Afterthought

Accessibility cannot come later. If people with disabilities possess inherent dignity, and if neighbor-love requires institutions to take responsibility for normalized forms of exclusion, then accessibility cannot be treated as a downstream adjustment after design is complete. Infusing justice within design itself is one of the clearest ways an institution shows whether it has taken dignity, agency, and participation seriously from the beginning.

Too often, accessibility is treated reactively. Systems are built and scaled around dominant assumptions about normal users, normal communication, normal pace, and normal cognition. Only when barriers become visible are disabled persons considered, usually through retrofitted accommodations or partial workarounds. This reveals a deeper moral problem. It suggests that accessibility was never part of the institution's original understanding of who the system was for. Inaccessible design is therefore not merely technical oversight. It reflects prior judgments about whose participation is assumed, whose needs are central, and whose burdens may be deferred.

A justice-oriented approach begins elsewhere. It asks at the outset whether the design process has been shaped by a sufficiently rich understanding of the human person. It assumes that participation should not depend on how closely one approximates a narrow norm of ability. It also recognizes that barriers are often produced less by impairment alone than by environments, interfaces, policies, and workflows built without adequate regard for human variation. Accessibility thus becomes a matter of institutional responsibility. The question is not whether an organization can compensate later for exclusion. It is whether it will refuse to build exclusion into the system in the first place.

Moore's Flourishing Minds, Flourishing Institutions reinforces this point. It grounds flourishing leadership in empathy, attunement, psychological safety, trust, and growth-oriented culture (Moore, n.d.). Design justice requires leaders attentive enough to notice where fear, silence, or exclusion have already been built into a workflow, interface, or automated process. Without empathy and relational attentiveness, accessibility is reduced to compliance language. With them, accessibility becomes part of the institution's understanding of trust, belonging, and responsible design.

A justice-centered design approach therefore requires several commitments. Disabled persons must be involved not merely as recipients of accommodations but as contributors to design, testing, governance, and evaluation. Accessibility must be considered in assumptions, data selection, interface design, communication channels, and workflow expectations rather than only in final outputs. Organizations must create feedback loops through which barriers can be named and corrected without stigma. Technical function is not enough. Review processes must ask whether a system preserves dignity, enables participation, and avoids unnecessary exclusion.

In this light, accessibility is not a specialized concern for a small subgroup. It is a revealing expression of justice in institutional life. It shows whether organizations are willing to align design practices with a truthful account of human dignity and with the obligations that follow from neighbor-love. Deferred accessibility defers dignity. Where accessibility is built into the system from the beginning, institutions demonstrate that they are willing to order technology toward participation, trust, and the common good.

The next question is practical judgment. If accessibility as justice requires more than compliance and more than late-stage accommodation, then the issue is no longer only one of design intent. It becomes a matter of judgment in practice. Institutions must learn to discern what genuine inclusion requires in concrete situations. That is the work of wisdom.

WISDOM: INCLUSION AS PARTICIPATION, VOICE, AND BELONGING

If accessibility as justice requires more than compliance and more than late-stage accommodation, then the next question is how institutions discern what genuine inclusion requires in concrete situations. That is the work of wisdom. In the framework of this chapter, wisdom is not abstract knowledge or technical expertise alone. It is the capacity to hold together right judgment and right relationship in the realities of lived human difference. Wisdom asks a harder question. In AI contexts, it asks not only whether a system is efficient, accurate, or scalable, but whether it is morally fitting for persons whose embodiment, cognition, communication, and dependence may not align with dominant institutional norms.

Disability cannot be absorbed into generic fairness language. A system may satisfy procedural fairness or statistical parity while still failing to create meaningful participation. It may offer access without belonging, accommodation without voice, or entry without contribution. Wisdom matters here. As McLoughlin, Fernández-Espinosa, and Velázquez-Gil argue, practical wisdom involves perceiving morally relevant features of a situation, regulating one's responses, deliberating well, and acting in accordance with one's highest values in dynamic contexts. In relation to AI, they show that wisdom is indispensable precisely because emerging technologies raise questions that cannot be resolved by technical fluency or policy compliance alone (McLoughlin et al., 2026).

Applied to disability, wisdom means recognizing that true inclusion is not satisfied by minimal access to a finished system. It requires participation, voice, contribution, and belonging. Disabled persons should not be treated merely as end users to whom accommodations are later extended. They should be involved in design, testing, governance, review, and revision. This is one of the clearest places where disability ethics deepens AI ethics. The issue is not simply whether a system treats everyone the same, but whether it has been shaped by the practical judgment required to respond fittingly to human variation. Here, inclusion becomes wise. Institutions learn to treat disabled persons not as exceptions to manage but as contributors whose presence reshapes the moral and practical intelligence of the whole organization.

Tsai and Ku sharpen this argument by warning that AI may undermine phronesis (practical wisdom) when it reduces the opportunities people have to exercise their own rational and moral capacities. Their concern is not limited to biased or defective systems. Even highly capable systems may weaken human judgment if people increasingly rely on them rather than cultivating the habits of perception, deliberation, and responsibility that wisdom requires. In their account, human flourishing depends in part on the continued exercise of rational capacity, and this makes epistemic overdependence on AI a moral problem as well as an intellectual one (Tsai & Ku, 2025). In the context of disability inclusion, that warning is especially important. Leaders cannot allow AI systems to become substitutes for discernment about what dignity, participation, and support require in particular settings.

Aboodi extends this concern by showing how easily critical thinking can be outsourced to artificial intelligence. His argument is especially relevant here because inclusion often involves circumstances in which standard rules are insufficient and where careful human judgment remains necessary. The loss is serious. If institutions become habituated to relying on AI without sufficient reflection, they may lose the capacity to question whether a recommendation is justifiable, whether a barrier has been normalized, or whether an apparently neutral decision excludes persons from meaningful participation. For Aboodi, the danger is not only that AI may sometimes mislead, but that overreliance on it may diminish the motivation and disposition to think critically at all (Aboodi, 2025).

This danger is acute in disability-related decisions. Such decisions often require contextual judgment that cannot be reduced to fixed rules. Institutions must discern when sameness becomes exclusion, when efficiency becomes depersonalization, and when accommodation remains too thin because what is really needed is participation in shared decision-making. Wisdom therefore resists the temptation to collapse inclusion into a technical checklist. It asks whether a design, policy, or workflow actually enables people to contribute as members of the community rather than merely to be managed at its edges.

Social inclusion, self-determination, shared citizenship, and belonging are not secondary outcomes added after the real work is done (Verdugo et al., 2023). They are part of what wise institutional judgment must seek from the beginning. Wisdom is not an optional refinement added to justice and design. It is what allows institutions to interpret both rightly. Without wisdom, accessibility may be reduced to rule-following, and justice may be reduced to formal equality. With wisdom, institutions learn to ask better questions: Who has been left out of the process? Whose knowledge is missing from design? What barriers remain hidden because they have become ordinary? What kind of participation is needed for genuine contribution and belonging? These are moral questions. They concern vulnerable and varied human lives.

The implication is clear. Institutions cannot become genuinely inclusive simply by granting downstream accommodations to systems already built around narrow assumptions of normality. They must cultivate the practical wisdom needed to discern what human flourishing requires in concrete situations. In AI-mediated environments, that means preserving human judgment, resisting epistemic overreliance on automated systems, and learning to recognize disabled persons as contributors whose participation strengthens the community's moral and practical intelligence. Wisdom, in this sense, turns accessibility from a reactive response into a more truthful form of shared life.

DISCERNING LEADERSHIP PRACTICE IN AI ADOPTION

If wisdom concerns right judgment in concrete human realities, the next question is how that judgment takes institutional form. In AI adoption, leaders must do more than approve technologies, monitor compliance, or respond after harm becomes visible. They must remain morally present throughout design, deployment, oversight, and revision. The central issue is not only whether AI systems function, but whether they are governed in ways that protect dignity, enable participation, and sustain the common good.

Discerning leadership begins with attentiveness. Leaders must notice where exclusion is likely to occur, especially when technical language, efficiency claims, or the appearance of neutrality conceal it. AI systems often inherit assumptions about pace, cognition, communication, and behavior that may seem natural to designers and administrators but become barriers for disabled persons. Good leadership requires moral discipline. It asks whose experience has been assumed away, whose participation has been made harder by default settings, and whose needs will likely be deferred until after deployment.

In Thriving in Industry 5.0, Moore, Chenhall Maxey, and Hangen argue that human-centric leadership requires integrating Complexity Leadership Theory (CLT) through enabling, adaptive, and administrative leadership. Enabling leadership fosters relationships, communication, and inclusive environments where diverse teams can thrive. Adaptive leadership encourages creativity, learning, and bottom-up problem solving. Administrative leadership aligns systems, removes barriers, and ensures ethical governance of AI and HR processes. Together, these roles move organizations beyond top-down control toward human-centered systems in which technology supports rather than diminishes human participation (Moore et al., n.d.).

For disability inclusion, these leadership functions become practical guides. Enabling leadership creates conditions in which disabled persons can speak, be heard, and shape systems. Adaptive leadership keeps organizations open to feedback, revision, and new solutions instead of defending flawed systems already in place. Administrative leadership embeds accessibility in workflows, training, review structures, and accountability mechanisms rather than leaving it to informal goodwill. Without enabling leadership, inclusion lacks voice. Without adaptive leadership, inclusion lacks responsiveness. Without administrative leadership, inclusion lacks durable institutional support.

Bredgaard, Borghouts, Freese, and Moore reinforce this practical picture by identifying four common enablers of disability-inclusive organizations: inclusive core values, senior management commitment, disability-inclusive culture, and strong community partnerships (Bredgaard et al., 2024). Their work is concrete. Inclusive values shape whether accessibility is central or peripheral; senior management commitment determines whether it is resourced and enforced; disability-inclusive culture shapes whether disabled persons are welcomed as contributors rather than compliance categories; and community partnerships bring expertise, accountability, and lived knowledge into decision-making. Together, these enablers show that inclusive AI governance cannot be accomplished through technical fixes alone. It depends on institutional conditions that support wise and just action.

Concrete commitments follow. Leaders must involve disabled persons in design, testing, evaluation, and revision rather than consulting them only after barriers appear. They must create feedback structures through which harms can be named and corrected without stigma or retaliation. They must integrate accessibility concerns into mission, strategy, and operations rather than isolate them within compliance offices. They must also resist moral outsourcing by refusing to treat system outputs as self-justifying. Even when AI tools are statistically effective, leaders must ask whether their use is just, whether they preserve dignity, and whether they support participation rather than merely sorting people more efficiently.

AI systems amplify institutional tendencies. Inattentive leadership makes exclusion scalable. Courageous and humane leadership directs technology toward broader participation, better support, and meaningful contribution. The question is whether leaders will govern these systems with enough discernment to ensure disabled persons are not managed as edge cases but recognized as central members of the community.

Discerning leadership practice serves as the bridge between moral vision and organizational reality. It translates dignity, neighbor-love, justice, and wisdom into habits of governance. It requires attentiveness to exclusion, humility about hidden assumptions, justice in design and deployment, accountability when harm occurs, and stewardship for long-term participation. Where such leadership is present, accessibility becomes part of the institution's moral architecture rather than a late correction. Without such leadership, even well-intentioned AI adoption can reproduce abstraction, distance, and exclusion.

A PRACTICAL FRAMEWORK FOR ACCESSIBILITY-CENTERED AI GOVERNANCE

What is organizational beauty? It is a community where people actively support one another's flourishing. This beauty analogy helps explain accessibility-centered AI governance as a three-part process. The movement is simple: see beauty, discern where it has been obscured, and act so beauty can re-enter places of disorder. Figure 1 presents the model.

Figure 1. Restoring beauty through accessibility

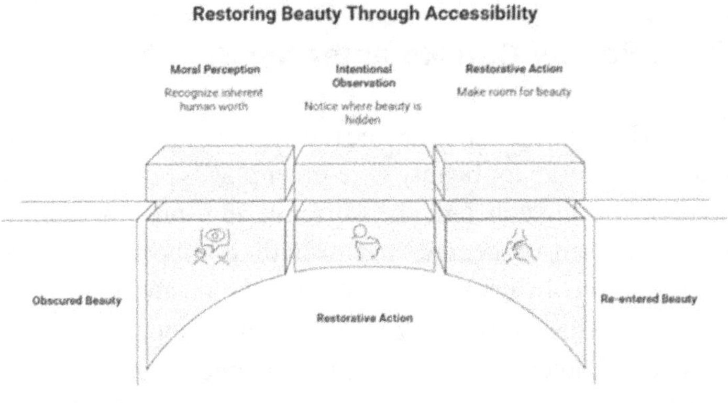

1. Learn to See Beauty Rightly

The first task is moral perception. Leaders must learn to see rightly. The analogy is practical. Like a bank teller who learns to recognize a forgery by studying genuine currency, leaders and designers must train their moral vision by dwelling on the dignity and beauty of the human person. Accessibility begins when institutions recognize people with disabilities not as deviations from a norm, but as bearers of inherent worth, capacity, and beauty. Without this formation of sight, exclusion can appear normal, and inaccessible systems can seem acceptable because their distortions go unnoticed.

2. Look Intentionally for Where Beauty Has Been Obscured

Once leaders learn to recognize beauty, they must notice where it has been diminished, hidden, or denied. This requires curiosity, attentiveness, and humility. Leaders must look carefully at the lived experience of their neighbors and ask where barriers, workflows, interfaces, or assumptions prevent human participation from being expressed.

This does not mean one is equally responsible for all the brokenness in the world. No institution or leader can address every form of disorder everywhere. The responsibility is local. Leaders are responsible for the places entrusted to them, including family, workplace, and community, where beauty has been obscured and where action can make a real difference. Accessibility work is not primarily about spotting deficiency in persons. It is about discerning where systems have introduced disorder, silence, dependence without support, or exclusion without acknowledgment.

3. Act So That Beauty Can Re-enter the Space

The final task is restorative action. Once the obscured good has been recognized, institutions must make room for beauty to re-enter places of disorder and exclusion. This often requires limiting convenience, surrendering some institutional freedom, slowing efficiency-driven processes, and providing support so others may rise, contribute, and flourish. In governance terms, this means redesigning systems, reworking workflows, sharing power, and embedding support from the beginning. Identification is not completion. Accessibility-centered AI governance is complete only when institutions reorder their practices so that dignity, participation, and flourishing become visible again.

This framework prepares the way for a Quality of Life evaluation, where accessibility-centered AI governance is judged by whether it advances concrete conditions of human flourishing. It is not merely about fixing barriers. Its deeper purpose is to restore the conditions in which dignity, participation, and belonging can become visible again. The ultimate question is not access alone. It is whether AI systems are ordered toward the flourishing of people and communities.

DISABILITY INCLUSION, THE COMMON GOOD, AND HUMAN FLOURISHING

The final aim of accessibility-centered AI governance is not accessibility alone. It is to create conditions in which people can participate, contribute, belong, and grow in ways that honor their dignity. That horizon is human flourishing. The central question is not barrier removal alone. It is whether AI helps form institutions where people with disabilities can live and work as full participants in community life.

AI ethics can stop too soon. A system may become more efficient, accurate, or accessible and still fall short of the good toward which institutional life ought to be ordered. Human flourishing requires more than technical access. It includes dignity, belonging, self-determination, meaningful contribution, relational trust, and the opportunity to develop one's gifts in community. Without these goods, accessibility becomes thin: entry without the power to shape the system, contribute within it, or be recognized as valued.

Moore and Hanson frame the movement from human utilization to human flourishing as a central moral shift. In that account, flourishing involves not only productivity or organizational success but the wider conditions of quality of life, including emotional well-being, interpersonal relations, social inclusion, personal development, self-determination, and rights (Moore & Hanson, in press). Applied to disability and AI, the ethical question is whether a system enlarges disabled persons' capacity to belong, act, contribute, and flourish.

The Quality of Life framework gives leaders a concrete way to evaluate whether accessibility-centered AI governance is ordered toward flourishing. The standard is broader. Rather than asking only whether a system is usable or compliant, leaders should ask whether AI strengthens or weakens the core domains of human well-being: emotional well-being, interpersonal relationships, material well-being, physical well-being, personal development, self-determination, social inclusion, and rights (Schalock et al., 2011). QOL turns accessibility from a technical standard into a flourishing standard. An AI system is not ethically adequate simply because it removes a barrier. It is ethically adequate only insofar as it enlarges the person's capacity for agency, belonging, development, protection, contribution, and participation in shared life.

Disability inclusion is therefore a revealing test of institutional purpose. Systems designed with disabled persons in view often improve participation, clarity, adaptability, and trust for many others. More importantly, they reveal whether an institution understands itself as a mechanism for extracting performance or as a community ordered toward shared good.

Relational goods matter here. Giarmoleo, García Ruiz, Rocchi, and Ferrero argue that virtuous organizations in the age of AI must protect and cultivate relational goods such as trust, cooperation, mutual recognition, esteem, and shared purpose. They show that AI can either weaken or strengthen these goods depending on how organizations govern technological mediation (Giarmoleo et al., 2026). Flourishing is not merely individual. It is relational and institutional, depending on whether communities are structured so people can be recognized, supported, and drawn into meaningful shared life.

Spitko argues that workplace AI must be evaluated not only by efficiency but by its effects on dignity, autonomy, and upward mobility (Spitko, 2024). The relevance is direct. People with disabilities have often been excluded by institutional judgments that limit agency, narrow opportunity, and reduce roles to marginal or dependent status. If AI merely automates those patterns, it intensifies exclusion even when it appears neutral or efficient. If governed to support dignity, agency, and opportunity, AI can contribute to broader conditions of flourishing.

Exception-based welcome is conditional. Where disabled persons are welcomed only through exception or accommodation, institutions reveal a narrow understanding of belonging. Where disabled persons are recognized as contributors whose participation enriches the whole community, institutions reveal a richer understanding of justice and flourishing. Disability is central to the moral evaluation of AI because it exposes whether AI systems deepen mutual participation or merely refine organizational control.

The ethical task of AI governance is not to build perfectly frictionless systems or simply prevent visible harm. It is to cultivate institutions in which technology serves persons rather than sorting them by convenience, accessibility supports participation rather than mere entry, and disabled persons can exercise agency, offer contribution, and experience belonging. That is why flourishing matters. Flourishing is the proper end of accessibility-centered AI governance because it names the full horizon of what justice seeks: not only the removal of exclusion, but the restoration of shared life ordered toward dignity, contribution, and the common good.

REFERENCES

Aboodi, R. (2025). The worrisome potential of outsourcing critical thinking to artificial intelligence. *Educational Theory*, *75*(4), 626–645. DOI: 10.1111/edth.70037

Bredgaard, T., Borghouts, I., Freese, C., & Moore, J. (2024). Disability inclusion in for-profit organizations: Workplace case studies in the Netherlands, Denmark and the United States. *Journal of Research for International Education*, *4*(3).

Giarmoleo, F. V., García Ruiz, P., Rocchi, M., & Ferrero, I. (2026). Virtuous organizations in the age of AI: Relational goods and human flourishing. *Business Ethics, the Environment & Responsibility*. Advance online publication. DOI: 10.1111/beer.70090

Hassen, M. Z. (2025). The risk of moral outsourcing: Why artificial intelligence cannot and should not make our ethical decisions. *EthAIca, 4*, Article 428.

McLoughlin, S., Fernández-Espinosa, V., & Velázquez-Gil, M. (2026). Practical wisdom, emerging technologies, and AI. *Journal of Moral Education*, 1–31. Advance online publication. DOI: 10.1080/03057240.2026.2624863

Moore, J. (n.d.). *Flourishing minds, flourishing institutions: Cultivating well-being in academic culture* [Manuscript]. Anderson University, College of Business and Economics.

Moore, J., & Halter, C. (in press). Employee flourishing leadership: Unlocking employee potential. In C. Claes, R. Schalock, & G. Van Hove (Eds.), *Ondersteuning van mensen in een inclusieve samenleving: Een persoonsgericht handboek* [*Support for people in an inclusive society: A person-centered handbook*]. Gompel & Svacina.

Moore, J., & Hanson, W. (in press). Wellbeing oriented organizations: From human utilization to human flourishing. In C. Claes, R. Schalock, & G. Van Hove (Eds.), *Ondersteuning van mensen in een inclusieve samenleving: Een persoonsgericht handboek* [*Support for people in an inclusive society: A person-centered handbook*]. Gompel & Svacina.

Moore, J., & Hanson, W. (n.d.). *Leadership strategy for building Industry 5.0: Human-centric societal organizations* [Manuscript]. Anderson University, College of Business and Economics.

Moore, J., Maxey, E., & Hangen, J. (n.d.). *Thriving in Industry 5.0: Leadership and business models for value creation* [Manuscript]. Anderson University, College of Business and Economics.

Schalock, R. L., Keith, K. D., Verdugo, M. Á., & Gómez, L. E. (2011). Quality of life model development and use in the field of intellectual disability. *Enhancing the quality of life of people with intellectual disabilities: From theory to practice*, 17-32.

Spitko, E. G. (2024). Workplace AI and human flourishing. *Connecticut Law Review, 57*(1), 231–278.

Tsai, C.-H., & Ku, H.-L. (2025). Why AI may undermine phronesis and what to do about it. *AI and Ethics, 5*(3), 3079–3086. DOI: 10.1007/s43681-024-00617-0

Verdugo, M. Á., Schalock, R. L., Gómez, L. E., & Navas, P. (2023). A Systematic Approach to Implementing, Evaluating, and Sustaining the Shared Citizenship Paradigm in the Disability Field. *Behavioral Sciences (Basel, Switzerland), 13*(12), 970. DOI: 10.3390/bs13120970 PMID: 38131825

KEY TERMS AND DEFINITIONS

Accessibility: The design of environments, technologies, and systems so that individuals with diverse physical, cognitive, sensory, and social abilities can participate fully and independently in shared activities and institutions.

Artificial Intelligence (AI): Computational systems capable of performing tasks that typically require human judgment, learning, perception, or decision making, increasingly shaping access to education, employment, communication, and participation for individuals with disabilities.

Assistive Technology: Tools and systems designed to enhance functional capabilities and independence for individuals with disabilities, including AI-enabled supports such as speech recognition, predictive text, adaptive interfaces, and navigation assistance.

Digital Equity: The condition in which individuals have fair access to digital tools, infrastructure, and skills necessary for participation in education, employment, and civic life in a technology-mediated society.

Disability Inclusion: An organizational and social commitment to ensuring that people with disabilities are not only accommodated but actively supported as contributors, decision makers, and participants in community life.

Ethical AI Governance: The structures, policies, and leadership practices that guide the responsible development and use of artificial intelligence in ways that protect dignity, reduce harm, and promote equitable participation.

Human Dignity: The inherent worth of every person that requires technologies and institutions to support agency, belonging, and respect rather than reduce individuals to efficiency variables or data profiles.

Human Flourishing: The development of a person's capacities for meaningful participation, contribution, relationship, and purpose within community life, serving as the ethical aim of accessibility-centered AI systems.

Inclusive Leadership: A leadership approach that intentionally shapes organizational cultures, policies, and technologies to expand participation, remove structural barriers, and recognize diverse abilities as sources of strength.

Quality of Life Framework: A multidimensional model for evaluating human well-being across domains such as self-determination, social inclusion, relationships, rights, and personal development, used to assess whether technologies support meaningful participation rather than merely technical access.

Chapter 15
A New Moral Renaissance:
Forming Leaders for a Flourishing Future

Jeffrey Moore
http://orcid.org/0000-0002-2508-2896
Anderson University, USA

ABSTRACT

This chapter argues that the leadership crisis of the AI age is not primarily technical, but moral and spiritual. As artificial intelligence reshapes learning, work, judgment, and institutional life, the central question becomes what kind of people and cultures will guide its use. The chapter proposes a simple integrative model: Formation > Character > Wisdom > Discerning Leadership Practice > Human Flourishing. Drawing on moral development research, virtue ethics, servant leadership, Christian theology, and complexity leadership theory, it argues that neither technology nor moral activity alone can produce moral people. Universities and organizations must therefore be reimagined as ecologies of moral and spiritual formation that cultivate wisdom, responsibility, dignity, and meaningful participation. The chapter concludes that a new moral renaissance is needed, one that recovers leadership as a moral and spiritual vocation ordered toward human flourishing and the common good in an AI-shaped world.

WHY A NEW MORAL RENAISSANCE IS NEEDED

Artificial intelligence now shapes how people learn, work, communicate, evaluate performance, and make decisions. In higher education, it is changing how students

DOI: 10.4018/979-8-3373-8192-3.ch015

ight © 2027, IGI Global Scientific Publishing. Copying or distributing in print or electronic forms without written permission of IGI Global Scientific Publishing is prohibited. Use of this chapter to n generative artificial intelligence (AI) technologies is expressly prohibited. The publisher reserves all rights to license its use for generative AI training and machine learning model development.

gather information, how faculty teach, and how institutions think about assessment and intellectual labor. In organizations, it is restructuring workflows, influencing hiring and evaluation, and altering the relationship between efficiency and human judgment. AI is often welcomed for its speed, scale, and problem-solving capacity. Yet the central challenge of this moment is not simply technological. It is moral.

The deeper question is not whether institutions will use increasingly powerful tools. They will. The deeper question is what kind of people and cultures will guide them. Technical expertise, innovation, and efficiency may improve performance, but they do not automatically produce wisdom, responsibility, or right relationship. Moral and spiritual activities, by themselves, do not automatically produce moral people either. Like technology, they may create conditions for growth, but they do not guarantee formation. What matters is whether leaders are being shaped into people who can judge rightly, relate rightly, and exercise power faithfully in complex circumstances. Earlier work on student moral development made a similar point by showing that universities shape students morally through a broader ecology of institutional reinforcement, service, challenge, and relationships rather than through isolated ethics content alone (Hanson & Moore, 2014; Hanson et al., 2017; Moore et al., 2023).

This chapter argues that the leadership crisis of the AI age is not merely a matter of poor regulation, weak governance, or inadequate ethics instruction. It is a deeper crisis of formation. Leaders act not only from what they know, but from who they are becoming. In a time marked by moral fragmentation, spiritual disorientation, and the temptation to outsource judgment to systems, leadership renewal must begin with the formation of the person. That formation must develop into character, mature into wisdom, and find expression in discerning leadership practice. The end of that process is not merely better performance, but human flourishing (Moore & Hanson, in press; Moore, 2026).

Figure 1. Model of new moral renaissance

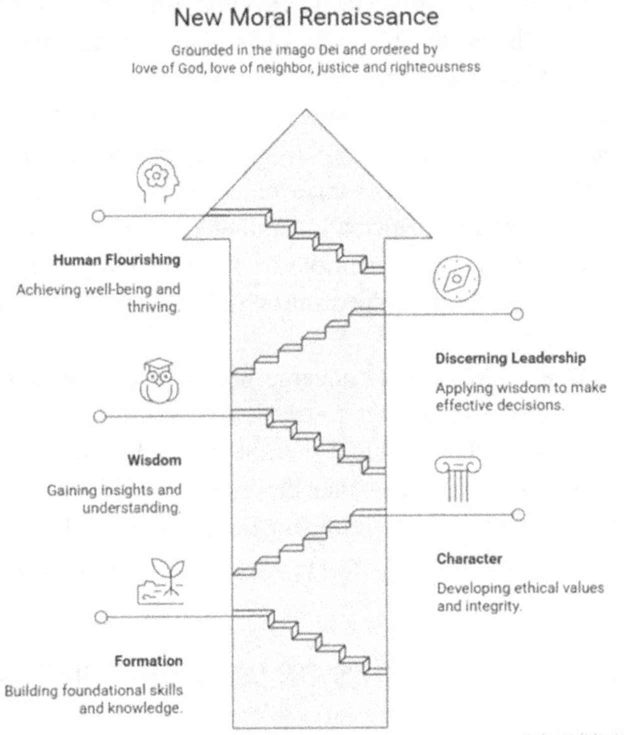

Formation -> Character -> Wisdom -> Discerning Leadership Practice -> Human Flourishing

The model makes one claim: flourishing leadership does not begin with method. It begins with formation. Character emerges from the habits, practices, relationships, and communities that shape the leader over time. Wisdom is the capacity to hold together justice and righteousness, right judgment and right relationship, in the concrete realities of life. Such wisdom is essential in an AI-shaped world because algorithmic systems can optimize, predict, and classify, but they cannot discern what is morally fitting in the full human sense. Discerning leadership practice is the faithful enactment of wisdom within the changing realities of institutions and organizations. Human flourishing remains the true end of this process, not efficiency, growth, or control alone (Aristotle, 1999; Uhl-Bien et al., 2007; Moore & Hanson, in press).

Several concerns make this recovery urgent. Contemporary culture increasingly treats leadership as visibility, performance, and managerial control rather than as a moral and spiritual vocation. Institutions often respond to ethical breakdown with fragmented solutions: more rules, more content, more training, or more technological oversight. These efforts may be necessary, but they remain insufficient if the deeper issue is the disorder of the self and the weakening of the moral ecology in which leaders are formed (Hanson & Moore, 2014; Hanson et al., 2017). AI also intensifies the temptation to confuse technical capability with moral adequacy. The more systems can do, the easier it becomes to imagine that human discernment is secondary. In truth, it becomes more necessary.

The argument therefore calls for a new moral renaissance. Not merely renewed ethics. I mean the recovery of leadership as a deeply human, moral, relational, and spiritual undertaking. Such a renaissance requires universities to become communities of moral and spiritual formation rather than dispensers of professional content alone. It also requires organizations to become flourishing communities of work rather than systems of human utilization (Moore & Hanson, in press; Moore & Halter, in press).

The volume's three movements converge here into one account of leadership renewal in an AI-shaped world. Part I emphasized moral development in higher education. Part II examined ethical and servant leadership in AI contexts. Part III turned toward human flourishing. Together, these strands show that leadership renewal must move from formation to character, from character to wisdom, from wisdom to discerning leadership practice, and from practice toward human flourishing.

THE CONTEMPORARY CRISIS OF LEADERSHIP FORMATION

The current leadership crisis is often described in managerial terms. Institutions speak of declining trust, employee disengagement, burnout, toxic cultures, poor decision-making, and ethical failure. In higher education, there are similar concerns: fragmentation, student disengagement, confusion about purpose, and difficulty sustaining communities of trust and belonging. These are real problems, but they are not merely operational problems. They are symptoms of a deeper formative disorder.

One aspect of that disorder is the erosion of shared moral frameworks. Many leaders now operate in contexts where moral language is thin, contested, or reduced to procedural compliance. Institutions may affirm values, but when their lived cultures reward speed, visibility, output, and competitive advantage more than humility, wisdom, courage, or relational fidelity, many leaders are shaped less by deep moral commitments than by the ambient pressures of performance culture. Earlier research on business student moral development warned that piecemeal reform efforts often fail because they do not address the wider institutional culture in which moral identity is actually shaped (Hanson & Moore, 2014).

Artificial intelligence makes this concrete. In higher education, predictive analytics shape advising, retention strategies, and student risk classifications. In organizations, algorithmic systems influence hiring, performance evaluation, workflow management, and judgments about who is productive, promotable, or replaceable. Generative AI also changes the moral texture of work by making it easier to outsource relational and judgment-laden tasks. These systems promise efficiency, objectivity, and scale, but they also tempt leaders to let technological output appear morally sufficient. The danger is not only algorithmic bias or error. The deeper danger is moral passivity and detachment from the human consequences of decisions leaders still oversee (Uhl-Bien et al., 2007; Claessens et al., 2026; Hassen, 2025).

Moral outsourcing is a serious concern. Once leaders rely on systems to classify, recommend, rank, or predict, they may begin to imagine that responsibility has shifted with the task. Yet responsibility does not disappear when a process becomes technologically mediated. It becomes more important. As later chapters in this volume explore through discussions of algorithmic power, digital discernment, and ethical responsibility, the central issue is not simply what AI can do, but whether leaders remain morally present in the midst of its use. The capacity to question outputs, interpret situations contextually, protect the vulnerable, and preserve human dignity remains irreducibly human work (Uhl-Bien et al., 2007; Moore & Hanson, in press; Hassen, 2025).

Paul Kingsnorth's critique of progress challenges the modern assumption that technological advance alone leads to a better human future (Kingsnorth, 2021). The problem is not new technology itself. The problem is the myth that new technology can save us from the moral and spiritual consequences of our own disorder. Modern leadership culture often lives inside that false promise: more capacity, speed, and control will resolve what is ultimately a deeper human problem.

Albert Borgmann offers a related diagnosis through his distinction between the device paradigm and focal practices. The device paradigm names a way of life organized around convenience, efficiency, and the delivery of commodities with minimal engagement. In such a world, the practices that once formed attention, discipline, skill, gratitude, and community begin to erode. Focal practices, by contrast, require presence, participation, and substantive engagement (Borgmann, 1984). This distinction helps explain why the AI age can deform not only institutions, but also the habits of those who inhabit them. Speed and fragmentation make discernment harder.

The crisis, then, is not merely that leaders have new tools. It is that leaders are increasingly formed in environments that reward technical capability more than moral depth. The question is no longer whether leadership development should include ethics, but whether institutions still know how to form people whose judgment, relationships, and loves are ordered toward the good. The moral development literature echoes this concern, showing that universities and organizations influence people through culture, expectation, role modeling, challenge, and daily practice (Hanson & Moore, 2014; Hanson et al., 2017; Moore et al., 2023). Formation must be recovered as the starting point.

RECOVERING FORMATION AS THE STARTING POINT

If the leadership crisis of the AI age is deeper than regulation, efficiency, or institutional design, then renewal must begin at a deeper level. It must begin with formation. By formation, I mean the ongoing moral and spiritual shaping of people through habits, practices, relationships, institutions, and experiences that shape how they perceive reality, what they desire, and how they respond to others (Siegel, 2012). Formation is not an optional addition to leadership development. It is already happening. The real question is whether it is happening intentionally toward wisdom and flourishing, or unintentionally toward fragmentation, self-interest, and moral passivity. Prior work on student moral development makes this clear: moral life in educational settings is shaped by a broader institutional ecology rather than by isolated ethics content alone (Hanson & Moore, 2014; Hanson et al., 2017).

This insight exposes a weakness in many contemporary approaches to ethics and leadership development. Institutions often respond to moral concerns with another course, policy, compliance mechanism, or set of stated values. Useful, but rarely enough. Earlier research argued that stand-alone reforms often fail because they are not aligned within a coherent strategy (Hanson & Moore, 2014). The stronger finding was that universities morally influence students through interacting forces, including institutional moral reinforcement, service activities, experiential challenges, moral amplifiers, and evolving moral cultural identity (Hanson et al., 2017; Hanson & Moore, 2023). Later work in China added business moral reinforcement through internships and workplace supervisors (Moore et al., 2023).

The same principle applies to leadership more broadly. People are formed by the total environment in which they live and work. In the university setting, that includes classroom life, institutional artifacts, peer relationships, service experiences, mentoring, and the everyday challenges of community. In organizational life, it includes managerial behavior, team norms, systems of reward, and the degree to which people are treated as contributors rather than instruments. What matters is not simply whether institutions teach ethical concepts, but whether they create conditions that shape a living moral code. Prior work has described that code in terms of moral rightness, moral consequence, and moral worth, suggesting that development occurs as students learn not only what is right, but why it matters and how it should guide action (Hanson & Moore, 2023).

Borgmann's distinction between the device paradigm and focal practices is useful here. The device paradigm names a cultural tendency to seek convenience, efficiency, and immediate outcomes with minimal engagement. Focal practices gather people into meaningful participation, discipline, and presence (Borgmann, 1984). In an AI-shaped world, this matters greatly. If students and leaders are trained to prefer speed over reflection, convenience over participation, and output over attentiveness, then the conditions of formation are themselves being deformed. What is needed is not simply more information, but the recovery of practices and communities that shape patience, responsibility, humility, and moral depth.

Formation must therefore come first in the chapter's model. Character, wisdom, and discerning leadership practice do not arise in a vacuum. They emerge from the ways people are habituated over time. Some of that habituation is explicit. Much of it is not. More recent work on flourishing institutions makes the same point from another angle. When leaders cultivate adaptive space, psychological safety, relational trust, and growth-oriented cultures, institutions become places where people are developed as integrated, connected learners who can flourish together (Moore, 2026). Formation is the starting point. Character emerges from it. Wisdom matures through it. Discerning leadership practice depends upon it. Human flourishing is its hoped-for fruit.

CHARACTER: THE MORAL SHAPE OF THE LEADER

If formation names the process by which leaders are morally and spiritually shaped, then character names the pattern that emerges from that shaping. Character is not a public image, a leadership brand, or a set of isolated behaviors performed under observation. It is the stable configuration of virtues, habits, desires, and reflexes that disposes a person to act in ways that are truthful, responsible, and faithful over time. Character is the moral shape of the leader. It gives leadership substance when pressure rises, outcomes are uncertain, and no policy can fully determine the right course of action.

A virtue-based framework is helpful because it moves the discussion away from rule compliance alone and toward the kind of person leadership is producing. In the classical tradition, virtue concerns the formation of habits that dispose people toward the good, not merely the avoidance of wrongdoing (Aristotle, 1999). That insight remains important in an AI-mediated age. Institutions often respond to ethical concerns by increasing controls, procedures, and reporting mechanisms. Such measures may be necessary, but they do not by themselves produce leaders who are honest, courageous, just, or humble. They may regulate conduct at the surface while leaving the deeper moral life untouched. Character concerns the internalization of moral commitments so that right action becomes not merely externally required, but increasingly fitting to the person.

This emphasis has become increasingly important in the broader book project. The chapter on character formation in AI-mediated leadership identifies virtues such as prudence, justice, temperance, courage, and humility as necessary to trustworthy leadership in technologically saturated environments. That same work argues that leadership must be grounded not only in technical competence, but in a faith-centered moral framework shaped by humility, justice, compassion, moral courage, and stewardship. These are not ornamental qualities. They are the moral architecture that enables leaders to use power responsibly and resist the distortions of speed, abstraction, and self-protection that often accompany technological systems.

Several virtues matter here. Humility allows leaders in AI-rich settings to admit limits and resist overconfidence in themselves or the systems they oversee. Justice requires truthful judgment about what is due to people, especially when institutional decisions affect vulnerability, opportunity, and dignity. These virtues are crucial where automation bias, depersonalized evaluation, and algorithmic opacity tempt leaders to trust outputs too quickly, treat people as data profiles, or hide behind technical complexity (Claessens et al., 2026; Hassen, 2025). Compassion keeps leadership attentive to suffering and need. Moral courage enables costly action under pressure. Stewardship reminds leaders that their power over systems, people, and futures is entrusted power.

Servant leadership belongs beside virtue ethics here. If virtue ethics explains what character is, servant leadership helps explain how character is expressed relationally. Greenleaf's central question remains important: do those served grow as people (Greenleaf, 1977)? That question keeps leadership from collapsing into performance, status, or control. It redirects attention toward the growth, dignity, and well-being of others. In this volume, servant leadership functions as a central framework for resisting domination, centering the growth of others, and ordering leadership toward human flourishing (Greenleaf, 1977).

Together, virtue ethics and servant leadership prevent two distortions. The first is reducing character to private morality. Character is personal, but it is never merely private. It is expressed in relationships, decisions, and the use of power. The second is reducing leadership to a relational style without moral depth. Servant leadership without virtue can become thin, overly therapeutic, or disconnected from truth and justice. Virtue without relational orientation can become self-contained and insufficiently attentive to others. Together, they offer a stronger account. Character is the stable moral pattern of the leader, and servant leadership is one of its clearest relational expressions.

Character also helps explain why leadership failures are rarely sudden. Public breakdowns often appear dramatic, but they usually reveal deeper patterns already at work. The leader who manipulates data, avoids accountability, mistreats subordinates, or uses AI outputs to shield personal responsibility is not merely making an isolated mistake. More often, such actions reveal a prior deformation of character. Habits of self-protection, ego, detachment, or rationalization have already taken root. This is why trustworthy leadership depends not only on better systems, but on better-formed people. Research on moral development has consistently shown that people are shaped through the interaction of institutional influences, service, challenge, and formative relationships. Character is the more durable fruit of that shaping when it matures well (Hanson & Moore, 2014; Hanson et al., 2017; Moore et al., 2023).

In Christian perspective, character is inseparable from the ordering of love. Leaders are not formed only by what they think, but by what they love, pursue, excuse, and fear. For that reason, character cannot be reduced to external respectability or institutional success. It is the gradual shaping of a person whose judgments and relationships are increasingly aligned with truth, justice, righteousness, and neighbor-love. Character must follow formation in the model because it is the first stable moral result of the deeper shaping process. Yet character alone is still not enough. Even a virtuous leader must learn how to judge rightly in complex and changing situations. That is the work of wisdom.

WISDOM: CHRISTIAN DISCERNMENT THROUGH JUSTICE AND RIGHTEOUSNESS

If character gives leadership its moral shape, then wisdom gives leadership its moral judgment. Wisdom is not merely intelligence, information, managerial skill, technical expertise, institutional experience, or speed of decision-making. In this chapter, wisdom is best understood as Christian discernment through justice and righteousness: the cultivated capacity to perceive what is morally at stake, judge truthfully, and act faithfully in ways that honor moral order and right relationship. In classical terms, this bears affinity with practical wisdom, the ability to deliberate well about what is good and fitting in particular circumstances (Aristotle, 1999). Yet the language of Christian discernment is more fitting here because it makes explicit that wisdom is rational, practical, moral, relational, and spiritual. It concerns not simply what works, but what is right before God and good for others.

In an AI-shaped age, this distinction matters. Artificial intelligence can process information, identify patterns, predict outcomes, generate recommendations, simulate reasoning, and work toward selected goals. But it cannot possess wisdom in the fullest sense because wisdom requires morally accountable judgment. It weighs the human significance of a situation, discerns what justice and righteousness require here and now, and acts responsibly in relation to other people. The most serious danger is not only that AI might make mistakes, but that leaders might confuse computational sophistication with moral adequacy. Wisdom remains irreducibly human because it involves answerability, conscience, and the faithful exercise of judgment in concrete life.

The biblical tradition provides a deeper grammar for this account. Scripture presents God's authority and reign as grounded in justice and righteousness. "Righteousness and justice are the foundation of your throne" (New Revised Standard Version Bible, 1989, Psalm 89:14). Isaiah describes the coming reign of peace as one established "with justice and with righteousness" (New International Version Bible, 2011, Isaiah 9:7). These terms are related, but not identical. Justice points to right judgment, moral truthfulness, fairness, and the right ordering of people and situations, while righteousness points to right relationship, covenant faithfulness, integrity, and action aligned with the character of God. Taken together, they provide a rich theological account of wisdom. Wisdom is the capacity to hold together right judgment and right relationship in the concrete realities of life.

This distinction matters because leadership often fails when one dimension is severed from the other. Leadership that emphasizes judgment without relationship can become cold, procedural, and detached from human need. It may be technically correct yet morally thin. Leadership that emphasizes relationship without judgment can become sentimental, indecisive, or unwilling to name truth clearly. Wisdom refuses both distortions. It does not choose between justice and righteousness. It holds them together. Wise leadership is not a compromise between principle and care, but the faithful integration of both. It asks not only, What is efficient? or What is permissible? but What is just and righteous in this situation? What does faithfulness require for these people, under these conditions, at this time?

Wisdom is contextual judgment. Ethical action is never merely the mechanical application of general rules. Principles matter, but they must be interpreted within situations marked by complexity, vulnerability, competing goods, incomplete information, and human limitation. Aristotle's account of practical wisdom remains helpful because it resists rigid absolutism and careless relativism (Aristotle, 1999). Christian discernment presses further by recognizing that judgment also requires spiritual attentiveness, humility, prayerful dependence, and rightly ordered love.

Moral and spiritual activities do not automatically create wise leaders. Their importance lies not in producing virtue mechanically, but in serving as occasions for the inward formation of judgment and love. Wisdom emerges as leaders grow in the capacity to think rightly and love rightly through loving God and loving others (New International Version Bible, 2011, Matthew 22:37-39).

Wisdom is the bridge between character and practice. Character gives the leader stable moral substance. Wisdom gives the leader the ability to interpret situations faithfully. Without character, wisdom can become manipulative cleverness. Without wisdom, character can remain sincere but underdeveloped in action. Wisdom enables leaders to carry virtues such as humility, justice, compassion, courage, and stewardship into the changing and often ambiguous realities of institutional life, allowing them to discern how truth, care, responsibility, and restraint must be enacted within complex circumstances rather than merely affirmed in theory.

Wisdom is not a secondary leadership trait. It is central. As systems become more powerful, the need for leaders who can hold together justice and righteousness only increases. What is needed is not merely more data, more control, or more predictive capacity, but leaders whose moral and spiritual formation has matured into discernment, so they can judge truthfully, relate faithfully, and act responsibly in ways that preserve human dignity and orient institutions toward the good.

DISCERNING LEADERSHIP PRACTICE: LEADING WITHIN COMPLEX ADAPTIVE SYSTEMS

Leadership does not take place in static settings, but in environments marked by interdependence, uncertainty, adaptation, and feedback. Discerning leadership practice in an AI-shaped age is best understood through complex adaptive systems, where outcomes emerge through the interaction of people, structures, technologies, and shifting conditions rather than through simple linear control (Holland, 1995; Uhl-Bien et al., 2007). In AI-mediated organizations, technologies, workflows, incentives, and human behavior continually shape one another.

This systems view changes how leadership practice should be understood. In more traditional models, leadership is often imagined as issuing directives, controlling variables, and ensuring compliance from the top down. That model may appear attractive in uncertainty, but it is inadequate when information is distributed, challenges shift, and new work depends on interaction across levels of the organization. Complexity leadership theory offers a more fitting account by understanding leadership as the work of creating conditions in which people, ideas, and structures can interact in ways that generate learning, adaptation, and coordinated action (Uhl-Bien et al., 2007; Uhl-Bien & Marion, 2009). In applied work, this has meant helping leaders move away from autocratic, production-only mindsets and toward participative, relationship-based, and adaptive forms of leadership that increase trust, engagement, and shared problem-solving (Moore & Hanson, 2022).

In an AI-mediated setting, leaders must think of technology not as an isolated tool, but as part of an adaptive network. AI systems shape and are shaped by human behavior, institutional assumptions, incentives, and culture. They learn from data generated by prior action, influence new action through recommendations or classifications, and can alter how people make decisions, distribute responsibility, and define success. Discerning leadership practice cannot be reduced to model accuracy or efficiency. The harder question is whether the larger socio-technical system is moving toward dignity, justice, accountability, and flourishing (Moore & Hanson, in press; Uhl-Bien et al., 2007).

The theological claim matters here. If wisdom holds together justice and righteousness, discerning leadership practice structures institutional life so that right judgment and right relationship can be sustained under pressure. A leader using AI-supported performance analytics may pursue justice by identifying real patterns of underperformance while also pursuing righteousness by asking whether the system overlooks context, invites employee voice, and protects people from opaque scores. Healthy leadership cultures must avoid both chaos and egocentric authoritarianism. They need enough structure to protect accountability and enough openness for learning, participation, and adaptation (Moore & Hanson, 2022; Moore & Hanson, in press).

Complexity leadership theory clarifies this balance through administrative, enabling, and adaptive leadership (Uhl-Bien et al., 2007). Administrative leadership provides structure and coordination. Enabling leadership creates conditions for collaboration, experimentation, and contribution. Adaptive leadership emerges as people generate solutions and respond creatively to changing conditions. Industry 5.0 work applies this directly by arguing that human-centric organizations require mission, strategy, culture, and leadership to align so that adaptive learning and employee empowerment can take root (Moore & Maxey, in press; Moore & Hanson, 2022).

Discerning leadership practice in such settings is socio-technical stewardship. Stewardship means neither passivity nor domination. The steward neither abandons the system to its own momentum nor imagines the system exists to magnify the steward's ego. Instead, the steward accepts responsibility for how power, tools, and structures affect people and communities over time. In AI contexts, this means leaders must remain morally present. They must resist moral outsourcing, the temptation to defer responsibility to algorithmic outputs or automated processes. They must ask whether systems are fair, whether people can question outputs without retaliation, whether human oversight remains real, and whether the organization is preserving substantive human contribution rather than simply maximizing efficiency (Moore & Hanson, in press; Moore & Halter, in press).

Recent evidence suggests that flourishing institutions must be careful not to outsource the most humanly significant work to AI. Across six pre-registered studies, people who outsourced tasks to AI were judged more negatively, especially when the tasks were socio-relational, such as apology notes, bereavement cards, love letters, and wedding vows. These judgments were driven by perceptions of reduced effort, lower authenticity, and diminished care (Claessens et al., 2026). As Hassen (2025) argues, AI may assist human judgment, but it cannot bear ethical agency, contextual discernment, or accountability. Flourishing institutions must distinguish uses of AI that augment human work from those that displace the effort, presence, and moral agency through which trust is sustained.

In practice, this kind of leadership depends on feedback, recalibration, transparency, and participatory engagement. Feedback helps people see what is happening. Recalibration recognizes that initial solutions are often incomplete. Transparency protects accountability. Participation honors distributed knowledge and strengthens commitment. Flourishing leadership research has shown that psychologically safe environments, where employees can learn from mistakes, question decisions, and contribute ideas without fear, are central to new ideas and workplace flourishing (Moore & Halter, in press). AI governance should therefore include visible human review, real appeal channels, transparent criteria for system use, and protected spaces to question technological judgments.

FLOURISHING: THE END OF LEADERSHIP

The true end of leadership is human flourishing. The dominant language of contemporary leadership often narrows success to performance metrics, growth targets, operational efficiency, or competitive advantage, which may matter but are insufficient as ultimate aims. Leadership is rightly judged by whether it helps create conditions in which people, communities, and institutions can thrive. Recent work on well-being oriented organizations makes a similar argument: organizations must move from a logic of human utilization to one of human flourishing if they are to become healthy and sustainable communities of work (Moore & Hanson, in press).

The term flourishing needs substance. The Quality of Life model identifies dimensions of well-being including emotional well-being, interpersonal relations, social inclusion, personal development, self-determination, rights, and material and physical well-being (Schalock et al., 2011). The model keeps flourishing from becoming vague, sentimental, or merely individualistic. Flourishing is not reducible to happiness, comfort, or private satisfaction. It names conditions under which people can live with dignity, belonging, agency, growth, and meaningful participation.

The Shared Citizenship Paradigm deepens the point by reframing people not as passive recipients of services or instruments of productivity, but as active participants in the common life of the community (Verdugo et al., 2023). A flourishing organization or university does not merely ask how to extract performance, but how to honor people with agency, rights, responsibilities, and the capacity to contribute to the good of the whole. This aligns with the Christian insistence that people bear the image of God and cannot be reduced to functions, outputs, or productivity units.

Recent work on Employee Flourishing Leadership treats flourishing as both an individual ideal and an institutional reality. In that model, institutions thrive when they cultivate psychological safety, relational trust, growth-oriented culture, and shared leadership rather than relying only on transactional management or bureaucratic control (Moore, 2026). Flourishing institutions require flourishing people. Faculty, staff, students, and employees are not peripheral to institutional success. Their well-being, trust, and sense of purpose are constitutive of institutional vitality.

Flourishing is also cultural. Institutions can become stagnant and disconnected, caring but constrained, efficient but emotionally cold, or flourishing, depending on whether they combine relational depth, trust, adaptability, and growth-oriented culture (Moore, 2026). Human flourishing is not a final outcome added onto leadership; it is embedded in the culture leadership builds. Where people are silenced, treated as expendable, stripped of agency, or reduced to performance metrics, flourishing is undermined no matter how successful the institution appears externally; where people are trusted, included, developed, and given real voice, flourishing becomes more plausible, even under pressure.

The business and organizational case for flourishing is also stronger than critics sometimes assume. In Industry 5.0 work, flourishing is a richer account of value creation, placing human well-being, sustainability, and resilience at the center of organizational purpose (Moore et al., 2020; Moore, Hanson, & Gustafson, 2023). Similarly, the Walgreens and Sephora cases show that inclusive and human-centered cultures can produce gains in retention, safety, engagement, productivity, and new ideas while also enhancing dignity, belonging, and trust (Moore & Hanson, in press; Moore & Maxey, in press).

At the same time, flourishing cannot be collapsed into organizational utility. If flourishing is valuable only because it improves metrics, it has already been subordinated to a narrower logic, since flourishing is the proper telos of leadership as a moral activity ordered toward the good of people and communities. Performance benefits may follow, and often do, but they are not the final justification. From a Christian perspective, flourishing involves lives and communities ordered toward truth, justice, righteousness, faithful relationship, and the common good under God. It is richer than the language of success, satisfaction, or institutional health.

Research on AI outsourcing clarifies why flourishing cannot be reduced to efficiency. People judged AI outsourcing most negatively in socio-relational tasks because it signaled reduced authenticity and care, even when AI was used as a tool rather than as a full substitute (Claessens et al., 2026). Institutions do not thrive merely by accelerating output. They depend on trust, meaningful contribution, and the preservation of social bonds.

Flourishing should be an explicit aim. A Flourishing Index would move human flourishing from rhetoric to evaluative practice by asking whether organizations cultivate healthy culture, meaningful contribution, and thriving people alongside traditional measures of effectiveness and efficiency (Moore & Hanson, in press). If flourishing is truly the end of leadership, then it must shape not only vision statements, but also how institutions define success, assess culture, and design work.

ANCHORING THE MODEL IN FAITH

If human flourishing is the true end of leadership, then the next question is what moral and spiritual grounding can sustain such a vision. As earlier chapters in this volume have argued, especially those on moral anchors and the theological foundations of AI ethics, faith does more than supply inspirational language. It provides ontological and moral grounding for dignity, responsibility, stewardship, and love that procedural ethics alone cannot fully generate. The model proposed here cannot finally rest on leadership technique or institutional aspiration alone but must be anchored in a deeper account of the human being, moral responsibility, and the good toward which leadership should be ordered.

In Christian perspective, that grounding begins with the imago Dei. Human beings are not simply producers, consumers, organizational actors, or data points within a system. They are people made in the image of God (New International Version Bible, 2011, Genesis 1:26-27). Dignity is not conferred by performance, usefulness, or institutional recognition. It is given. Leadership that forgets this will eventually distort both people and institutions because it will have no stable ground on which to resist the reduction of human beings to functions.

This theological grounding also clarifies why moral responsibility remains central to leadership in an AI world. Leaders are not merely coordinators of systems or managers of processes. They remain morally answerable for how power is exercised, how judgment is delegated, and how people are treated. In Christian perspective, that responsibility is inseparable from stewardship. Leaders hold entrusted power, and entrusted power is always accountable power. Love of neighbor matters. It must be measured by whether power is exercised in ways that serve others, especially the vulnerable and those most affected by institutional decisions.

Governance structures, rules, and ethical frameworks remain necessary, but they cannot by themselves answer the deeper questions of what human beings are for, why dignity is inviolable, or why love, humility, justice, and righteousness should govern leadership at all. Christian theology grounds dignity in creation, responsibility in stewardship, and moral life in a reality ordered toward God.

Paul Kingsnorth is useful here as a diagnostic voice. His critique of "the Machine" names the spiritual condition that often accompanies technological culture: systems promise rescue, control, and progress while people and cultures become uprooted from the sacred (Kingsnorth, 2021). In that sense, the AI age is not only a crisis of technological power, but also a crisis of spiritual disorientation.

Procedural safeguards are not enough. The deeper issue is whether leaders are being formed to love God and neighbor rightly, or whether they are being formed by ambition, fear, image, and the will to control; what is needed is the reordering of desire, the recovery of humility, and a return to a sacred account of people and the world before God.

Leadership in this chapter is best understood as a moral and spiritual vocation, not simply a professional function. Vocation reminds leaders that their work is not self-originating. It is responsive. Leaders are called to steward influence, serve others, and participate in God's purposes for justice, truth, and flourishing in the world. This shifts the center of leadership away from self-advancement and toward service, answerability, and faithful presence. That hope matters. If the disorder of the age is spiritual, then the possibility of renewal is spiritual as well. Leaders and institutions are not trapped within the logic of the Machine. They can be reoriented through humility, repentance, worship, and the restoration of right relationship with the Creator.

REIMAGINING LEADERSHIP DEVELOPMENT IN HIGHER EDUCATION: A UNIVERSITY-WIDE ECOLOGY OF MORAL AND SPIRITUAL FORMATION

If moral and spiritual activities do not automatically produce moral people, higher education cannot be reduced to more programs, ethics content, or religious experiences. Just as AI cannot generate morality by itself, a university cannot manufacture virtue through activities alone. What matters is whether students are being formed in wisdom: the capacity to hold together justice and righteousness, right judgment and right relationship, in the concrete realities of life. That is Christian formation.

The Theory of Moral Development of Business Students is especially useful for this question. That work moves beyond isolated ethics instruction and explains moral development as an interacting institutional process shaped by environment, relationships, challenge, and culture (Hanson & Moore, 2014; Hanson et al., 2017), rather than treating the student as a detached individual receiving information; the model understands the university as a formative ecology in which a living moral code is shaped over time. The theory identifies several major influences on student development, including institutional moral reinforcement, service activities, experiential challenges, moral amplifiers, and moral cultural identity. Later work in China reinforced the model and added business moral reinforcement, showing that supervisors, internships, and workplace exposure also shape how business students internalize moral commitments (Moore, Zhang, & High, 2023).

Figure 2. Theory of moral development of the identity of the business student

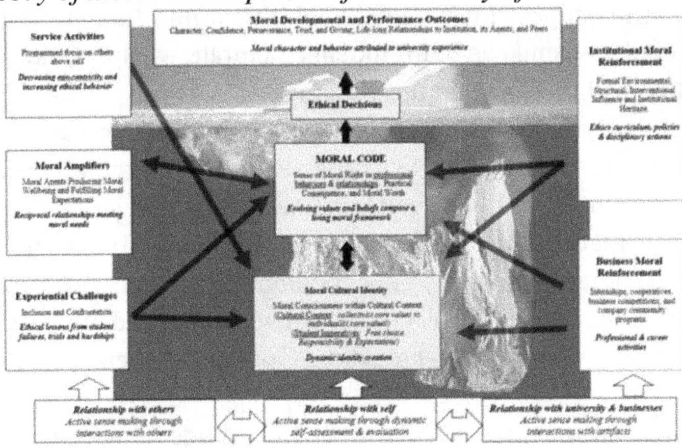

(Moore & Hanson, 2019)

The theory fits the chapter's claim that formation is broader than curriculum. Universities shape students morally through formal structures, but also through symbols, traditions, disciplinary expectations, relational climate, and daily practices. In earlier work, institutional moral reinforcement referred to the formal environmental, structural, and interventional influences that communicate a moral vision to students (Hanson et al., 2017). That includes curriculum, mission, policies, codes of conduct, heritage, and visible artifacts that communicate what kind of community the university seeks to be. If universities want to form leaders for a flourishing future, they must ask whether the total environment reinforces the moral and spiritual vision it claims to uphold.

Several of the model's central influences form one ecology. Service activities move students beyond self-focus and into encounters with responsibility, need, and neighbor-love (Hanson & Moore, 2014), while experiential challenges form students through conflict, disappointment, failure, and accountability in community (Hanson & Moore, 2014; Hanson & Moore, 2023). Moral amplifiers, including faculty, mentors, peers, staff, and supervisors, intensify development through example, expectations, and care (Hanson & Moore, 2014; Hanson et al., 2017). Moral cultural identity matters because students arrive with inherited beliefs and assumptions that the university must help refine toward wisdom, humility, and faithful responsibility (Moore, Zhang, & High, 2023).

The addition of business moral reinforcement helps bridge the university and the workplace. Internships, business mentors, supervisors, case studies, and experiential projects all contribute to how students understand ethical life in practice (Moore et al., 2023). The risk is abstraction. If leadership development is to prepare students for an AI-shaped world, universities must build bridges between moral reflection and applied practice. Ethical case discussions, internships shaped by reflective supervision, partnerships with organizations that value human flourishing, and vocation-centered advising can help students integrate what they are learning with the realities they will face beyond the classroom.

For faith-based higher education, the implications are broader. If the deeper disorder of the age is spiritual disorientation, leadership development must address rightly ordered desire, humility, vocation, stewardship, and love of neighbor. Worship, prayer, spiritual disciplines, service-learning, reflective writing, mentoring communities, and shared practices of accountability create spaces where students may be formed in the deeper sense, connecting technical and professional learning to a larger account of who students are before God, what others are worth, and what leadership is for.

Faculty development matters. Educators form students not only by what they teach, but by what they model in their use of truth, power, technology, and relationships. Earlier work showed that institutional members play a major role in student moral development and that moral influence often occurs through lived presence, not only through formal instruction (Hanson & Moore, 2014; Hanson et al., 2017). A university serious about leadership development in an AI world must therefore invest not only in curriculum reform, but also in the moral and spiritual health of its educators, advisors, and institutional leaders.

Higher education cannot be satisfied with producing technically competent graduates who know how to use advanced tools. It must ask what kind of people its students are becoming. Are they learning to discern justice and righteousness? Are they being shaped toward love of God and love of neighbor? Are they developing the humility and courage necessary to lead faithfully in a technological age? If not, no amount of additional programming will be enough.

REIMAGINING ORGANIZATIONS: FROM HUMAN UTILIZATION TO FLOURISHING COMMUNITIES

The emerging vision of wellbeing-oriented organizations argues that institutions must move from a logic of human utilization to a logic of human flourishing (Moore & Hanson, in press). In AI-mediated contexts, efficiency can narrow organizational purpose to optimization, compliance, and output. A flourishing organization asks not only how work can be done more efficiently, but how people can be developed, trusted, included, and sustained through their work (Moore et al., 2020).

The Quality of Life framework gives this vision substance. Flourishing institutions must attend to emotional well-being, interpersonal relations, social inclusion, personal development, self-determination, rights, and material and physical well-being

(Schalock et al., 2011). These dimensions prevent organizational flourishing from being reduced to morale language or generalized positivity by directing attention to whether people actually experience belonging, agency, growth, protection, and meaningful contribution within the institutional environment. Organizational flourishing is not merely about whether the institution succeeds, but whether the people within it are enabled to thrive as participants in a shared community.

The Shared Citizenship Paradigm deepens this by reframing people not as instruments of production but as contributors to the common life of the organization (Verdugo et al., 2023). In an AI-shaped world, that claim has force. The more institutions rely on algorithmic systems to classify, rank, evaluate, and direct, the easier it becomes to depersonalize the people whose work those systems shape. A flourishing organization does not merely extract more from people. It honors them as full participants in institutional life.

Complexity Leadership Theory explains how such organizations can function. AI-driven institutions are not static machines but complex adaptive systems in which technologies, incentives, structures, and human behavior continuously interact (Uhl-Bien et al., 2007). Leadership in such environments cannot be reduced to top-down control. It must create conditions in which learning, adaptation, trust, and coordinated action can emerge. This requires holding together administrative, enabling, and adaptive leadership so that institutions do not collapse into either chaos or authoritarian control. In flourishing organizations, structure exists to support people, not suppress them, and new approaches emerge through participation rather than domination.

Employee Flourishing Leadership matters. Recent work suggests that institutions flourish when leaders cultivate psychological safety, relational trust, growth-oriented culture, and shared leadership rather than relying only on bureaucratic management or transactional control (Moore, 2026; Moore & Halter, in press). In AI-mediated organizations, this also means ensuring that human oversight remains real, technological systems remain open to challenge, and people are not reduced to depersonalized judgments generated by opaque processes.

This vision clarifies the moral stakes of AI use. Institutions may be tempted to use AI for efficiency in ways that slowly erode trust, authenticity, and participation. Yet flourishing depends on more than accelerated workflows. It depends on meaningful work, human relationships, and moral agency within the system. Flourishing organizations must distinguish between AI that augments human work and AI that displaces the effort, judgment, and relational presence through which community is sustained (Claessens et al., 2026; Hassen, 2025).

Reimagining organizations is essential to the chapter's argument. If the leadership crisis of the AI age is a crisis of formation, then organizations are among the primary places where formation continues. They can reinforce control, extraction, and fragmentation, or they can become communities in which people are shaped toward dignity, wisdom, stewardship, and shared flourishing. The question is not only whether organizations perform, but what kind of people and cultures they are producing.

TOWARD A NEW MORAL RENAISSANCE

This is why a new moral renaissance is needed. By moral renaissance, I do not mean a renewed interest in ethics as a compliance category or an expanded set of institutional guardrails. I mean a deeper recovery of leadership as a moral and spiritual vocation. Such recovery begins with the recognition that leadership does not flow primarily from technique, charisma, or managerial skill. It flows from formation. Leaders act not only from what they know, but from who they are becoming. Earlier work on moral development in higher education made this point by showing that the moral life of students is shaped by a broad ecology of reinforcement, challenge, service, relationships, and identity formation rather than by isolated ethics instruction alone (Hanson & Moore, 2014; Hanson et al., 2017; Moore et al., 2023).

A new moral renaissance requires the recovery of character, wisdom, and discerning leadership practice as an integrated whole, because trustworthy leadership depends on virtues and habits that enable people to tell the truth, bear responsibility, restrain power, and remain attentive to others. Character must mature into wisdom because the AI age is full of systems that can improve outputs but cannot discern. That wisdom must then become practice within complex adaptive systems, where leaders steward institutions through structure and participation, accountability and adaptability, moral clarity and relational depth.

The end of this renewal is human flourishing. The goal of leadership is not simply institutional survival, scale, efficiency, profit, or personal achievement. The goal is the cultivation of conditions in which people, communities, and institutions can thrive in dignity, belonging, responsibility, and meaningful participation. Recent work on wellbeing-oriented organizations and employee flourishing leadership shows that institutions are healthier and more resilient when they cultivate trust, psychological safety, participation, and meaningful contribution (Moore & Hanson, in press; Moore & Halter, in press).

In the Christian tradition, this vision of flourishing is rooted in the creation mandate. Humanity is called to "be fruitful and multiply" and to "fill the earth and subdue it" (New International Version Bible, 2011, Gen. 1:28). These directives suggest that flourishing has always involved both relationship and vocation: healthy shared life and meaningful work. The creation mandate joins righteousness as right relationship with justice as faithful stewardship. The text also says, "let them rule"

(New International Version Bible, 2011, Gen. 1:26), indicating a shared human vocation, not a mandate for domination. The creation mandate joins family, generativity, shared life, meaningful work, and the cultivation of creation. Not private spirituality. Not mere productivity. Flourishing is the balanced wise life of right relationship and faithful stewardship before God.

The creation vision fits the chapter's larger argument. Justice and righteousness are not late additions to Christian leadership. They are embedded in the shape of human calling. AI complicates that vocation because it can weaken relationships through abstraction and redefine work in purely instrumental terms. A moral renaissance must therefore recover the integrity of healthy relationships and meaningful work, mutuality and stewardship.

The concern becomes concrete when institutions outsource relationally or morally significant work to AI. Such outsourcing is judged most negatively in socio-relational tasks because it signals diminished authenticity and reduced care (Claessens et al., 2026), and as Hassen (2025) argues, AI may assist human judgment, but it cannot bear ethical agency, contextual discernment, or accountability. A new moral renaissance requires distinctly human forms of presence, stewardship, and love that no system can automate without loss.

The renewal cannot be sustained without faith. As Kingsnorth (2021) helps diagnose, the deeper disorder of the age is spiritual disorientation: the temptation to look to systems, processes, and new tools for salvation. Yet no system can heal disordered love, and no tool can manufacture virtue. What is needed is the reordering of desire, the recovery of humility, and a return to a sacred account of people and the world before God.

In Christian terms, leadership renewal is inseparable from learning to think rightly and love rightly through loving God and loving others; it is not enough for leaders to know more, manage better, or speak the language of ethics more fluently. What is needed are people whose judgments, relationships, and practices are being reshaped by truth, justice, righteousness, humility, and neighbor-love.

CONCLUSION

The deepest challenge of the AI age is not simply technological power, but the moral and spiritual condition of the people who wield it. Leadership renewal must follow a clear movement: Formation -> Character -> Wisdom -> Discerning Leadership Practice -> Human Flourishing. Formation shapes people through habits, relationships, institutions, and practices. Character gives leadership its moral substance. Wisdom enables leaders to hold together justice and righteousness. Discerning leadership practice enacts that wisdom within complex adaptive systems. Human flourishing is the end of this movement.

A central claim of the chapter is that neither technology nor moral activity automatically produces moral people. AI can rank, predict, classify, and recommend, but it cannot generate morality. In the same way, courses, chapel, service, mentoring, and institutional programming may create formative conditions, but they do not mechanically produce virtue. What matters is whether people are growing in wisdom, learning to think rightly and love rightly through loving God and loving others. The critical question is not whether institutions offer enough ethical content, but whether they cultivate the kind of ecology in which justice and righteousness can be discerned and embodied over time.

The task is clear: educators, leaders, and institutions must recover the formation of people and communities for the sake of human flourishing. In an AI-shaped world, the future will depend less on increasingly powerful systems and more on people capable of stewarding power wisely, relationally, and faithfully. A new moral renaissance will emerge only as people and communities recover leadership as a moral and spiritual vocation ordered toward neighbor-love, the common good, and the flourishing of all.

REFERENCES

Aristotle, . (1999). *Nicomachean ethics* (2nd ed.). (Irwin, T., Trans.). Hackett Publishing.

Borgmann, A. (1984). *Technology and the character of contemporary life: A philosophical inquiry*. University of Chicago Press.

Claessens, S., Veitch, P., & Everett, J. A. C. (2026). Negative perceptions of outsourcing to artificial intelligence. *Computers in Human Behavior, 177*, 108894. DOI: 10.1016/j.chb.2025.108894

Greenleaf, R. K. (1977). *Servant leadership: A journey into the nature of legitimate power and greatness*. Paulist Press.

Hanson, W., & Moore, J. (2023). *Conditions, processes and pressures promoting inclusive organizations (chapter 12)*. Ingold, J. & McGurk, P. (Eds.) Employer Engagement: Making Active Labour Market Policies Work. Bristol University Publishing.

Hanson, W. R., & Moore, J. R. (2014). Business student moral influencers: Unseen opportunities for development? *Academy of Management Learning & Education, 13*(4), 525–546. DOI: 10.5465/amle.2012.0325

Hanson, W. R., Moore, J. R., Bachleda, C., Canterbury, A., Franco, C.Jr, Marion, A., & Schreiber, C. (2017). Theory of moral development of business students: Case studies in Brazil, North America, and Morocco. *Academy of Management Learning & Education, 16*(3), 393–414. DOI: 10.5465/amle.2014.0312

Hassen, M. Z. (2025). The risk of moral outsourcing: Why artificial intelligence cannot and should not make our ethical decisions. *EthAlca, 4*, 428. DOI: 10.56294/ai2025428

Holland, J. H. (1995). *Hidden order: How adaptation builds complexity*. Basic Books.

Kingsnorth, P. (2021, June 1). The cross and the machine. *First Things (New York, N.Y.)*.

Moore, J. (2026). Flourishing minds, flourishing institutions: Cultivating well-being in academic culture. In *Work Values and Organizational Performance in Higher Education Institutions*. Springer.

Moore, J., & Halter, C. (in press). Employee flourishing leadership: Unlocking employee potential. In C. Claes, R. Schalock, & G. Van Hove (Eds.), Ondersteuning van mensen in een inclusieve samenleving. Een persoonsgericht handboek. Gompel & Svacina.

Moore, J., & Hanson, W. (2019). *Book title: Moral Development and Behavior: New Research. Book chapter: "University Business Student Moral Development"*. Nova Science Publishers.

Moore, J., & Hanson, W. (2022). Improving leader effectiveness: Impact on employee engagement and retention. *Journal of Management Development, 41*(7/8), 450–468. DOI: 10.1108/JMD-02-2021-0041

Moore, J., & Hanson, W. (in press). Wellbeing oriented organizations: From human utilization to human flourishing. In C. Claes, R. Schalock, & G. Van Hove (Eds.), Ondersteuning van mensen in een inclusieve samenleving. Een persoonsgericht handboek. Gompel & Svacina.

Moore, J., Hanson, W., & Gustafson, T. (2023). *Sephora – Journey to inclusive workplace – Let us belong philosophy (chapter 13).* Ingold, J. & McGurk, P. (Eds.) Employer Engagement: Making Active Labour Market Policies Work. Bristol University Publishing.

Moore, J., & Maxey, E. (in press). Thriving in Industry 5.0: Leadership and business models for value creation. In Waite, A., Machado, C., & Davim, J. P. (Eds.), *Leading in Industry 5.0*. Springer.

Moore, J., Maxey, E. C., Waite, A. M., & Wendover, J. D. (2020). Inclusive organizations: Developmental reciprocity through authentic leader-employee relationships. *Journal of Management Development, 39*(9/10), 1029–1039. DOI: 10.1108/JMD-05-2019-0211

Moore, J., Zhang, W., & High, D. (2023). Theory of moral development in Chinese university students. *Journal of Research for International Education, 2*(3).

New International Version Bible. (2011). Biblica. https://www.biblica.com/bible/niv/

New Revised Standard Version Bible. (1989). National Council of the Churches of Christ in the United States of America.

Schalock, R. L., Keith, K. D., Verdugo, M. Á., & Gómez, L. E. (2011). Quality of life model development and use in the field of intellectual disability. In Enhancing the quality of life of people with intellectual disabilities: From theory to practice (pp. 17–32).

Siegel, D. J. (2012). *The developing mind: How relationships and the brain interact to shape who we are* (2nd ed.). Guilford Press.

Uhl-Bien, M., & Marion, R. (2009). Complexity leadership in bureaucratic forms of organizing: A meso model. *The Leadership Quarterly, 20*(4), 631–650. DOI: 10.1016/j.leaqua.2009.04.007

Uhl-Bien, M., Marion, R., & McKelvey, B. (2007). Complexity leadership theory: Shifting leadership from the industrial age to the knowledge era. *The Leadership Quarterly, 18*(4), 298–318. DOI: 10.1016/j.leaqua.2007.04.002

Verdugo, M. Á., Schalock, R. L., Gómez, L. E., & Navas, P. (2023). A systematic approach to implementing, evaluating, and sustaining the shared citizenship paradigm in the disability field. *Behavioral Sciences (Basel, Switzerland), 13*(12), 970. DOI: 10.3390/bs13120970 PMID: 38131825

KEY TERMS AND DEFINITIONS

Character: The stable pattern of virtues, habits, desires, and moral commitments that shapes how a leader acts under pressure. Character gives leadership its moral substance when rules, systems, or policies are not enough.

Complex Adaptive Systems: Living organizational environments where outcomes emerge through the interaction of people, structures, technologies, and changing conditions. In such systems, leadership requires participation, feedback, adaptation, and trust rather than simple top-down control.

Discerning Leadership Practice: The faithful enactment of wisdom within institutions, organizations, and complex adaptive systems. It requires leaders to preserve dignity, accountability, participation, and moral purpose while guiding people through change.

Formation: The ongoing moral and spiritual shaping of people through habits, relationships, institutions, practices, and experiences. Formation determines not only what leaders know, but who they are becoming.

Human Flourishing: The true end of leadership. Human flourishing names the conditions in which people, communities, and institutions can thrive in dignity, belonging, responsibility, meaningful work, and shared participation.

Moral and Spiritual Vocation: A view of leadership as a calling to serve others, steward influence, and participate in God's purposes for justice, truth, and flourishing. Leadership is not merely a professional function, but a responsibility shaped by love of God and neighbor.

Moral Outsourcing: The temptation to shift moral responsibility onto AI systems, algorithms, or automated processes. Moral outsourcing occurs when leaders allow technological outputs to replace human judgment, conscience, accountability, or relational presence.

New Moral Renaissance: A recovery of leadership as a moral, spiritual, relational, and human vocation. In an AI-shaped world, this renewal calls institutions to form people who can steward power with wisdom, humility, justice, and responsibility.

Socio-Technical Stewardship: The responsible leadership of systems where people, technologies, structures, incentives, and cultures interact. Socio-technical stewardship means leaders neither abandon systems to automation nor dominate them for control, but guide them toward human dignity and flourishing.

Wisdom: The cultivated capacity to judge rightly, relate rightly, and act faithfully in complex situations. In this chapter, wisdom is Christian discernment that holds together justice and righteousness, right judgment and right relationship.

Compilation of References

Aarthi, S., & Ravikumar, R. N. (2026). From regulation to transformation-economic and institutional pathways of environmental policy. In *The Ripple Effects of Environmental Policies* (pp. 49–80). IGI Global Scientific Publishing., DOI: 10.4018/979-8-3373-8998-1.ch002

Aboodi, R. (2025). The Worrisome Potential of Outsourcing Critical Thinking to Artificial Intelligence. *Educational Theory*, *75*(4), 626–645. DOI: 10.1111/edth.70037

Adella, S. (2025). The relationship between digital parenting and children's moral development in the technological era. *Psikhlogiya, 2*(2), 39-47. https://jurnal-ilcaagam.com/index.php/PJ/article/view/395

Agarwal, A. (2025, October 22). *Elon Musk says AI will take every job and humans will be free to grow vegetables*. India Today. https://www.indiatoday.in/technology/news/story/elon-musk-says-ai-will-take-every-job-and-humans-will-be-free-to-grow-vegetables-2806590-2025-10-22

Agarwal, S., & Bhal, K. T. (2020). A multidimensional measure of responsible leadership: Integrating strategy and ethics. *Group & Organization Management*, *45*(5), 637–673. DOI: 10.1177/1059601120930140

Ahmad, A., Vallès, Y., & Idaghdour, Y. (2026). Bias in AI systems: Integrating formal and socio-technical approaches. *Frontiers in Big Data*, *8*, 1686452. Advance online publication. DOI: 10.3389/fdata.2025.1686452 PMID: 41584908

Aiken, L. S., & West, S. G. (1991). Multiple regression: Testing and interpreting interactions. *Sage (Atlanta, Ga.)*.

Akbarighatar, P. (2025). Operationalizing responsible AI principles through responsible AI capabilities. *AI and Ethics*, *5*(2), 1787–1801. DOI: 10.1007/s43681-024-00524-4

Akbar, M. A., Khan, A. A., Mahmood, S., Rafi, S., & Demi, S. (2024). Trustworthy artificial intelligence: A decision-making taxonomy of potential challenges. *Software, Practice & Experience*, *54*(9), 1621–1650. DOI: 10.1002/spe.3216

Akhundov, A. (2025). The Role of Ethics in Modern Technology Development. *Porta Universorum*, *1*(4), 169–177. DOI: 10.69760/portuni.0104017

Al-Ghazali. (2023). Ihya' 'Ulum al-Din - The Revival of the Religious Sciences - Vol 1: . Dar UL Thaqafah. ISBN-10: 9390804930

Alon-Barkat, S., & Busuioc, M. (2022). Human–AI interactions in public sector decision making: "Automation bias" and "Selective adherence" to algorithmic advice. *Journal of Public Administration: Research and Theory*, *33*(1), 153–169. DOI: 10.1093/jopart/muac007

Alperstein, N. M. (2025). A network approach to third-party tracking devices on mental health websites: Data surveillance and privacy in the digital panopticon. *Mental Health and Digital Technologies*, *2*(1), 59–74. DOI: 10.1108/MHDT-10-2024-0031

American Association of Colleges and Universities, & Elon University's Imagining the Digital Future Center. (2026). *The AI challenge: How college faculty assess the present and future of higher education in the age of AI* (Survey report). https://imaginingthedigitalfuture.org/collaborations/the-ai-challenge-how-college-faculty-assess-the-present-and-future-of-higher-education-in-the-age-of-ai

Amodei, D. (2026a, January). The adolescence of technology. https://www.darioamodei.com/essay/the-adolescence-of-technology

Amodei, D. (2026b, February 26). Statement from Dario Amodei on our discussions with the Department of War. Anthropic. https://www.anthropic.com/news/statement-department-of-war

Amodei, D., Olah, C., Steinhardt, J., Christiano, P., Schulman, J., & Mané, D. (2016). Concrete problems in AI safety. arXiv. https://arxiv.org/abs/1606.06565

Analects 12.22. (2022, April 3). Analects 12.22. Confucian Pages - Free Online Confucian Texts and Translations. https://confucius.page/analects-12-22

Analects 15.24. (2022, April 7). Analects 15.24. Confucian Pages - Free Online Confucian Texts and Translations. https://confucius.page/analects-15-24

Anderson, M., & Anderson, S. L. (2008). EthEl: Toward a principled ethical eldercare system. In *Proceedings of the AAAI 2008 Fall Symposium on AI in Eldercare: New Solutions to Old Problems*. AAAI Press.

Anderson, M., & Anderson, S. L. (2011). *Machine ethics*. Cambridge University Press. DOI: 10.1017/CBO9780511978036

Angwin, J., Larson, J., Mattu, S., & Kirchner, L. (2016). *Machine bias*. ProPublica. https://www.propublica.org/article/machine-bias-risk-assessments-in-criminal-sentencing

Anscombe, G. E. M. (1958). Modern moral philosophy. *Philosophy (London, England)*, *33*(124), 1–19. DOI: 10.1017/S0031819100037943

Anthropic. (2026, January 21). *Claude's constitution*. https://www.anthropic.com/news/claudes-constitution

Apdillah, D., Simanjuntak, C. R. A., Napitupulu, C. N. S. B., Sirait, D. D., & Mangunsong, J. (2022). The role of parents in educating children in the digital age. *Romeo Review of Multidisciplinary Education, Culture, and Pedagogy, 1*(3), 1-8. https://ojs.transpublika.com/index.php/ROMEO/article/view/168

Aquinas, T. (1948). *Summa theologica* (Fathers of the English Dominican Province, Trans.). Benziger Brothers. (Original work published 1265–1274)

Aristotle, . (2009). *Nicomachean ethics* (Ross, W. D., Trans.). Oxford University Press. (Original work published 350 BCE)

Aristotle, . (2009). *The Nicomachean ethics* (Ross, D., Trans.). Oxford University Press. (Original work published 350 BCE)

Aristotle. (2012). Aristotle's Nicomachean ethics (R. C. Bartlett & S. D. Collins, Trans.). University of Chicago Press. https://www.amazon.com/Aristotles-Nicomachean-Ethics-Aristotle/dp/0226026752

Arroyo-Machado, W., Ma, J., Chen, T., Johnson, T. P., Islam, S., Michalegko, L., & Welch, E. (2025). Generative AI and academic scientists in US universities: Perception, experience, and adoption intentions. *PLoS One*, *20*(8), e0330416. DOI: 10.1371/journal.pone.0330416 PMID: 40875655

Ashdown, B. K., Dixe, A., & Talmage, C. A. (2021). The potentially damaging effects of developmental aid and voluntourism on cultural capital and well-being. *International Journal of Community Well-being*, *4*(1), 113–131. https://link.springer.com/article/10.1007/s42413-020-00079-2. DOI: 10.1007/s42413-020-00079-2

Asimov, I. (1950). *I, robot*. Gnome Press.

Asimov, I. (1950). *Run around. I, robot*. Doubleday.

Astin, A. W., & Lindholm, J. A. (2011). Assessing students' spiritual and religious qualities. *Journal of College Student Development, 52*(1), 39–61. DOI: 10.1353/csd.2011.0009

Augustine. (2002). *On the Trinity, Books 8–15* (G. B. Matthews, Ed.; S. McKenna, Trans.). Cambridge University Press. (Original work ca. 400–416 CE)

Avery, E. J., & Graham, M. W. (2013). Political public relations and the promotion of participatory, transparent government through social media. *International Journal of Strategic Communication, 7*(4), 274–291. DOI: 10.1080/1553118X.2013.824885

Ayling, J., & Chapman, A. (2022). Putting AI ethics to work. *AI and Ethics, 2*(3), 405–429. Advance online publication. DOI: 10.1007/s43681-021-00084-x

Badri, M. (2018). *Contemplation: An Islamic Psychospiritual study* (1st ed.). Human Behaviour Academy. DOI: 10.2307/j.ctvk8w1xc

Bai, Y., Kadavath, S., Kundu, S., Askell, A., Kernion, J., Jones, A., Chen, A., Goldie, A., Mirhoseini, A., McKinnon, C., Olah, C., Hernandez, D., Drain, D., Ganguli, D., Perez, E., Bowman, S. R., Kaplan, J., Mann, B., Brown, T., & Amodei, D. (2022). Constitutional AI: Harmlessness from AI feedback. arXiv. https://arxiv.org/abs/2212.08073

Balasubramanian, S., Shukla, V., Islam, N., Upadhyay, A., & Duong, L. (2025). Applying artificial intelligence in healthcare: Lessons from the COVID-19 pandemic. *International Journal of Production Research, 63*(2), 594–627. DOI: 10.1080/00207543.2023.2263102

Balle, S. (2023). Theological dimensions of humanlike robots: A roadmap for theological inquiry. *Theology and Science, 21*(1), 132–156. DOI: 10.1080/14746700.2022.2155916

Band, V. S. (2026). Ethical and legal issues in information technology systems. *International Journal of Scientific Research in Engineering and Management.* DOI: 10.55041/IJSREM.IBFE038

Bandura, A. (1977). Self-efficacy: Toward a unifying theory of behavioral change. *Psychological Review, 84*(2), 191–215. DOI: 10.1037/0033-295X.84.2.191 PMID: 847061

Bankins, S., & Formosa, P. (2023). The ethical implications of artificial intelligence (AI) for meaningful work. *Journal of Business Ethics, 185*(4), 725–740. DOI: 10.1007/s10551-023-05339-7

Bansal, P., & Song, H. C. (2017). Similar but not the same: Differentiating corporate sustainability from corporate responsibility. *The Academy of Management Annals, 11*(1), 105–149. DOI: 10.5465/annals.2015.0095

Barker, C. (2004). Cultural capital. In *The SAGE dictionary of cultural studies* (pp. 37–38). SAGE., DOI: 10.4135/9781446221280.n47

Barocas, S., & Selbst, A. D. (2016). Big data's disparate impact. *California Law Review, 104*(3), 671–732. DOI: 10.15779/Z38BG31

Barth, K. (1958). *Church dogmatics III/1: The doctrine of creation, part 1* (Edwards, J. W., Bussey, O., & Knight, H., Trans.). T&T Clark. (Original work published 1945)

Bean, E., Burleigh, C., Haskell, C., Burris-Melville, T., Payne, J., & Pathak, B. (2025). Eavesdropping on UNESCO AI policy, leadership, and ethics. *Journal of Leadership Studies, 18*(4), 98–110. DOI: 10.1002/jls.70007

Beauchamp, T. L., & Childress, J. F. (2019). *Principles of biomedical ethics* (8th ed.). Oxford University Press.

Benefiel, M., Fry, L. W., & Geigle, D. (2014). Spirituality and religion in the workplace: History, theory, and research. *Psychology of Religion and Spirituality, 6*(3), 175–187. DOI: 10.1037/a0036597

Bhagavad Gita. Chapter 18, Verse 45. (2026). BG 18.45: Chapter 18, verse 45 - Bhagavad Gita, the song of god – Swami Mukundananda. Bhagavad Gita, The Song of God – Swami Mukundananda. https://www.holy-bhagavad-gita.org/chapter/18/verse/45/

Bhagavad Gita. Chapter 18, Verse 46. (2026). BG 18.46: Chapter 18, verse 46 - Bhagavad Gita, the song of god – Swami Mukundananda. Bhagavad Gita, The Song of God – Swami Mukundananda. https://www.holy-bhagavad-gita.org/chapter/18/verse/46/

Bhagavad Gita. Chapter 3, Verse 19. (2026). BG 3.19: Chapter 3, verse 19 - Bhagavad Gita, the song of god – Swami Mukundananda. Bhagavad Gita, The Song of God – Swami Mukundananda. https://www.holy-bhagavad-gita.org/chapter/3/verse/19/

Biblica. (2011). *Holy Bible, New International Version*. https://www.biblegateway.com/

Bick, A., Blandin, A., & Deming, D. (2024). *The rapid adoption of generative AI* (Working Paper No. 32966). National Bureau of Economic Research. https://www.nber.org/papers/w32966

Binns, R. (2021). Fairness in machine learning: Lessons from political philosophy. *Philosophy & Technology, 81*, 149–159. Proceedings of Machine Learning Research, *Conference on Fairness, Accountability, and Transparency*. DOI: 10.48550/arXiv.1712.03586

Birhane, A. (2021). Algorithmic injustice: A relational ethics approach. *Patterns (New York, N.Y.), 2*(2), 100205. DOI: 10.1016/j.patter.2021.100205 PMID: 33659914

Blackledge, P., & Knight, K. (2011). *Virtue and politics: Alasdair MacIntyre's revolutionary Aristotelianism*. University of Notre Dame Press.

Block, P. (1996). Stewardship: Choosing service over self-interest. Berrett-Koehler Publishers. ISBN-10 : 1881052869

Boh, W., Constantinides, P., Padmanabhan, B., & Viswanathan, S. (2023). Building digital resilience against major shocks. *Management Information Systems Quarterly, 47*(1), 343–360.

Bommasani, R., Hudson, D. A., Adeli, E., Altman, R., Arora, S., von Arx, S., Bernstein, M. S., Bohg, J., Bosselut, A., Brunskill, E., Brynjolfsson, E., Buch, S., Card, D., Castellon, R., Chatterji, N. S., Chen, A. S., Creel, K. A., Davis, J., Demszky, D., & Liang, P. (2021). On the opportunities and risks of foundation models. arXiv. https://arxiv.org/abs/2108.07258

Borgmann, A. (1984). *Technology and the character of contemporary life: A philosophical inquiry*. University of Chicago Press.

Bosse, T., Hoogendoorn, M., Klein, M. C., Treur, J., Van der Wal, C. N., & Van Wissen, A. (2012). Modelling collective decision making in groups and crowds: Integrating social contagion and interacting emotions, beliefs and intentions. *Autonomous Agents and Multi-Agent Systems, 27*(1), 52–84. DOI: 10.1007/s10458-012-9201-1

Bourdieu, P. (1986). The forms of capital. In J. Richardson (Ed.), *Handbook of theory and research for the sociology of education* (pp. 241-258). Greenwood. https://home.iitk.ac.in/~amman/soc748/bourdieu_forms_of_capital.pdf

Bovens, M. (2007). Analysing and assessing accountability: A conceptual Framework. *European Law Journal, 13*(4), 447–468. DOI: 10.1111/j.1468-0386.2007.00378.x

Bozkurt, A. (2025). Asimov's laws of robotics and AI: A critical analysis for the age of generative AI and intelligent tutoring systems. *Open Praxis, 17*(1), 1–14. DOI: 10.55982/openpraxis.17.1.896

Bracci, E. (2023). The loopholes of algorithmic public services: An "intelligent" accountability research agenda. *Accounting, Auditing & Accountability Journal, 36*(2), 739–763. DOI: 10.1108/AAAJ-06-2022-5856

Bredgaard, T., Borghouts, I., Freese, C., & Moore, J. (2024). Disability inclusion in for-profit organizations: Workplace case studies in the Netherlands, Denmark and the United States. *Journal of Research for International Education, 4*(3).

Bretag, T., Harper, R., Burton, M., Ellis, C., Newton, P., van Haeringen, K., Saddiqui, S., & Rozenberg, P. (2019). Contract cheating: A survey of Australian university students. *Studies in Higher Education, 44*(11), 1837–1856. DOI: 10.1080/03075079.2018.1462788

Brown, M. E., & Treviño, L. K. (2006). Ethical leadership: A review and future directions. *The Leadership Quarterly, 17*(6), 595–616. DOI: 10.1016/j.leaqua.2006.10.004

Brown, M. E., Treviño, L. K., & Harrison, D. A. (2005). Ethical leadership: A social learning perspective for construct development and testing. *Organizational Behavior and Human Decision Processes, 97*(2), 117–134. DOI: 10.1016/j.obhdp.2005.03.002

Buolamwini, J., & Gebru, T. (2018). Gender shades: Intersectional accuracy disparities in commercial gender classification. *Proceedings of Machine Learning Research, 81*, 1–15. http://proceedings.mlr.press/v81/buolamwini18a.html

Buolamwini, J., & Gebru, T. (2018). Gender shades: Intersectional accuracy disparities in commercial gender classification. *Proceedings of Machine Learning Research, 81*, 1–15.

Buolamwini, J., & Gebru, T. (2018). Gender shades: Intersectional accuracy disparities. *Proceedings of Machine Learning Research, 81*, 1–15.

Cath, C. J., Wachter, S., Mittelstadt, B., Taddeo, M., & Floridi, L. (2017). Artificial intelligence and the 'Good society': The US, EU, and UK approach. *Science and Engineering Ethics, 24*, 505–528. https://link.springer.com/article/10.1007/s11948-017-9901-7. DOI: 10.1007/s11948-017-9901-7 PMID: 28353045

CBS News. (2025, November 16). Why Anthropic CEO Dario Amodei spends so much time warning of AI's potential dangers [60 Minutes transcript]. https://www.cbsnews.com/news/anthropic-ceo-dario-amodei-warning-of-ai-potential-dangers-60-minutes-transcript/

Chaddha, R., & Agrawal, G. (2023). Ethics and morality. *Indian Journal of Orthopaedics, 57*(11), 1707–1713. DOI: 10.1007/s43465-023-01004-3 PMID: 37881279

Chandra, P. S., Irdayanti, Fithriyyah, M. U., Rahmadani, F., & Astuti, W. P. (2025). Meningkatkan digital mindfull di kalangan remaja. *Development: Journal of Community Engagement*. DOI: 10.46773/djce.v4i2.1991

Chaudhary, G. (2024). Unveiling the black box: Bringing algorithmic transparency to AI. *Masaryk University Journal of Law and Technology*, *18*(1), 93–122. DOI: 10.5817/MUJLT2024-1-4

Chaudhry, M. A., & Kazim, E. (2022). Artificial intelligence in education (AIEd): A high-level academic and industry note 2021. *AI and Ethics*, *2*(1), 157–165. DOI: 10.1007/s43681-021-00074-z PMID: 34790953

Cheng, M. (2025, October). 'Sycophantic' AI chatbots tell users what they want to hear, study shows. *The Guardian*. https://www.theguardian.com/technology/2025/oct/24/sycophantic-ai-chatbots-tell-users-what-they-want-to-hear-study-shows

Cheong, P. H., & Liu, L. (2025). Faithful innovation: Negotiating institutional logics for AI value alignment among Christian churches in America. *Religions*, *16*(3), 302. DOI: 10.3390/rel16030302

Chimeremeze, O. E. (2025). Philosophy, globalization, and artificial intelligence: Rethinking humanity in a digital age. *Trinitarian International Journal of Arts and Humanities, 1*(2), 42-58. https://www.journals.ezenwaohaetorc.org/index.php/TIJAH/article/view/3565/3695

Chowdhury, R. (2018, June 8). Moral outsourcing: Humanity in the age of AI. TED: Ideas change everything. https://www.ted.com/talks/rumman_chowdhury_moral_outsourcing_humanity_in_the_age_of_ai

Christensen, A.-M. S. (2025). Contextual ethics and the new: Reuniting practical and normative ethics. In Christensen, A.-M. S., Forsberg, N., & Rodogno, R. (Eds.), *Contextual Ethics* (pp. 89–109). Springer., https://link.springer.com/chapter/10.1007/978-3-031-97293-5_5 DOI: 10.1007/978-3-031-97293-5_5

CIO. (2025, December 17). 10 famous AI disasters. https://www.cio.com/article/190888/5-famous-analytics-and-ai-disasters.html

Cipta, H., Fitriani, I., & Putra, R. S. P. (2026). Surveillance health society: Ethics, privicy, and social control in digital health systems. *Journal of Public Health Indonesian*. DOI: 10.62872/4tt4qc09

Claessens, S., Veitch, P., & Everett, J. A. C. (2026). Negative perceptions of outsourcing to artificial intelligence. *Computers in Human Behavior*, *177*, 108894. DOI: 10.1016/j.chb.2025.108894

Clarke, R. (1993). Asimov's laws of robotics: Implications for information technology, part 1. *Computer, 26*(12), 53–61. DOI: 10.1109/2.247652

Clarke, R. (1994). Asimov's laws of robotics: Implications for information technology, part 2. *Computer, 27*(1), 57–66. DOI: 10.1109/2.248881

CNN. (2026, January 8). Elon Musk's AI chatbot Grok under fire for failing to rein in "digital undressing." https://www.cnn.com/2026/01/08/tech/elon-musk-xai-digital-undressing

Coeckelbergh, M. (2022). *The Political Philosophy of AI: An Introduction.* John Wiley & Sons.

Coffin, J., & Egan-Wyer, C. (2022). The ethical consumption cap and mean market morality. *Marketing Theory, 22*(1), 105–123. DOI: 10.1177/14705931211058772

Cohen, J. (1988). *Statistical power analysis for the behavioral sciences* (2nd ed.). Lawrence Erlbaum.

Collin, S., Lepage, A., & Nebel, L. (2023). Ethical and critical issues of artificial intelligence in education: A systematic review of the literature. *Canadian Journal of Learning and Technology, 49*(1). Advance online publication. DOI: 10.21432/cjlt28448

Constantinescu, M., & Crisp, R. (2022). Can robotic AI systems be virtuous and why does this matter? *International Journal of Social Robotics, 14*(6), 1547–1557. DOI: 10.1007/s12369-022-00887-w PMID: 35729985

Corpuz, J. C. G. (2025). Faith and artificial intelligence (AI) in Catholic education: Ethical, pedagogical, and theological engagement. *Religions, 16*(8), 1083. Advance online publication. DOI: 10.3390/rel16081083

Corrêa, N. K., Galvão, C., Santos, J. W., Del Pino, C., Pinto, E. P., Barbosa, C., Massmann, D., Mambrini, R., Galvão, L., Terem, E., & de Oliveira, N. (2023). Worldwide AI ethics: A review of 200 guidelines and recommendations for AI governance. *Patterns (New York, N.Y.), 4*(10), 100857. DOI: 10.1016/j.patter.2023.100857 PMID: 37876898

Corsello, A., & Santangelo, A. (2023). May artificial intelligence influence future pediatric research?-the case of ChatGPT. *Children (Basel, Switzerland), 10*(4), 757. DOI: 10.3390/children10040757 PMID: 37190006

Crawford, K. (2021). *Atlas of AI.* Yale University Press.

Crawford, K. (2021). *Atlas of AI: Power, Politics, and the Planetary Costs of Artificial Intelligence.* Yale University Press.

Creswell, J. W., & Creswell, J. D. (2018). *Research design: Qualitative, quantitative, and mixed methods approaches* (5th ed.). Sage.

Cruzes, D. S., Dybå, T., Runeson, P., & Höst, M. (2015). Case studies synthesis: A thematic, cross-case, and narrative synthesis worked example. *Empirical Software Engineering*, *20*(6), 1634–1665. DOI: 10.1007/s10664-014-9326-8

Danaher, J. (2016). The threat of Algocracy: Reality, resistance and accommodation. *Philosophy & Technology*, *29*(3), 245–268. DOI: 10.1007/s13347-015-0211-1

Dastin, J. (2018). Amazon scraps AI recruiting tool. *Reuters*. https://www.reuters.com/article/us-amazon-com-jobs-automation-insight-idUSKCN1MK08G

de Kock, J. (2024). Spirituality and theological formation: Seven critical approaches to undergraduate theological education. *Journal of Spiritual Formation & Soul Care*. Advance online publication. DOI: 10.1177/19397909241234294

Dennis, M. A. (2026, January 20). *Marvin Minsky: American scientist*. Britannica. https://www.britannica.com/biography/Marvin-Minsky

Dhammapada Verses 183. (n.d.). Dhammapada Verses 183, 184 and 185 Anandattherapanha Vatthu. Tipitaka.net. https://www.tipitaka.net/tipitaka/dhp/verseload.php?verse=183

Dicastery for the Doctrine of the Faith & Dicastery for Culture and Education [DDF & DCE]. (2025, January 28). Antiqua et nova: Note on the relationship between artificial intelligence and human intelligence. Vatican City. https://www.vatican.va/roman_curia/congregations/cfaith/documents/rc_ddf_doc_20250128_antiqua-et-nova_en.html

Dockery, A. M. (2016). Education and happiness in the faith-based context. *Journal of Education & Christian Belief*, *20*(1), 5–20.

Donaldson, T., & Preston, L. E. (1995). The stakeholder theory of the corporation: Concepts, evidence, and implications. *Academy of Management Review*, *20*(1), 65–91. DOI: 10.2307/258887

Dubber, M. D., Pasquale, F., & Das, S. (Eds.). (2020). *The Oxford Handbook of Ethics of AI*. Oxford University Press. DOI: 10.1093/oxfordhb/9780190067397.001.0001

Edmondson, A. (1999). Psychological safety and learning behavior in work teams. *Administrative Science Quarterly*, *44*(2), 350–383. DOI: 10.2307/2666999

Ejjami, R. (2024). Ethical Artificial Intelligence Framework Theory (EAIFT): A new paradigm for embedding ethical reasoning in AI systems. *International Journal For Multidisciplinary Research, 6*(5), 28231. Advance online publication. DOI: 10.36948/ijfmr.2024.v06i05.28231

Emerson, L. H. (2026). Outsourcing Harm: Global Labor, Economic Violence, and Invisible Human Cost. *Southern New Hampshire University; Independent Scholar.* https://papers.ssrn.com/sol3/papers.cfm?abstract_id=6173804

Epictetus, . (1995). *The discourses* (Hard, R., Trans.). Everyman. (Original work published 108 CE)

Ess, C. (2013). Digital Media Ethics (Digital Media and Society). Polity Pr. ISBN-13 : 978-0745656069

Eubanks, V. (2018). *Automating inequality*. St. Martin's Press. https://orcid.org/0000-0001-5017-0831

Eubanks, V. (2018). Automating inequality: How high-tech tools profile, police, and punish the poor. Macmillan + ORM.

European Commission. (2021). *Proposal for a regulation laying down harmonised rules on artificial intelligence*. Artificial Intelligence Act.

European Commission. Directorate-General for Communications Networks, Content and Technology & Grupa ekspertów wysokiego szczebla ds. sztucznej inteligencji. (2019). Ethics guidelines for trustworthy AI. Publications Office. https://data.europa.eu/doi/10.2759/346720

European Parliament. (2024). Regulation (EU) 2024/1689 laying down harmonised rules on artificial intelligence (Artificial Intelligence Act). Official Journal of the European Union. https://eur-lex.europa.eu/eli/reg/2024/1689

Eva, N., Robin, M., Sendjaya, S., van Dierendonck, D., & Liden, R. C. (2019). Servant leadership: A systematic review and call for future research. *The Leadership Quarterly, 30*(1), 111–132. DOI: 10.1016/j.leaqua.2018.07.004

Fan, L. (2024). Artificial intelligence ethics: A dialogue between technological advances and human values. *International Journal of Education and Humanities, 14*(2), 260–265. DOI: 10.54097/tvqkkf40

Farooqi, S. A. (2025). Ethical framework for AI in education. In *Advances in Computational Intelligence and Robotics* (pp. 153–188). IGI Global. DOI: 10.4018/979-8-3373-5550-4.ch007

Felzmann, H., Villaronga, E. F., Lutz, C., & Tamò-Larrieux, A. (2019). Transparency you can trust: Transparency requirements for artificial intelligence between legal norms and contextual concerns. *Big Data & Society*, *6*(1), 1–14. DOI: 10.1177/2053951719860542

Ferrara, A. (2022). Moral duties and juridical duties: The ambiguity of legal ethics considered through the prism of Kant's metaphysics of morals. *German Law Journal*, *23*(1), 117–129. DOI: 10.1017/glj.2022.5

Ferrara, E. (2023). Fairness and bias in artificial intelligence: A brief survey of sources, impacts, and mitigation strategies. *Sci*, *6*(1), 3. DOI: 10.3390/sci6010003

Field, A. (2018). *Discovering statistics using IBM SPSS statistics* (5th ed.). Sage.

Fiesler, C., Garrett, N., & Beard, N. (2020). What do we teach when we teach tech ethics? A syllabi analysis. *Proceedings of the ACM on Human-Computer Interaction*, *4*(CSCW1), 1–22. DOI: 10.1145/3328778.3366825

Fischer, J. M., & Ravizza, M. (1998). *Responsibility and control: A theory of moral responsibility*. Cambridge University Press. DOI: 10.1017/CBO9780511814594

Flanigan, J. (2018). Philosophical methodology and leadership ethics. *Leadership*, *14*(6), 707–730. DOI: 10.1177/1742715017711823

Floridi, L. (2019). Establishing the rules for building trustworthy AI. SSRN Electronic Journal. DOI: 10.2139/ssrn.3858392

Floridi, L. (2014). *The Fourth Revolution: How the Infosphere Is Reshaping Human Reality*. Oxford University Press.

Floridi, L. (Ed.). (2018). *The ethics of artificial intelligence*. Oxford University Press.

Floridi, L., Cowls, J., Beltrametti, M., Chatila, R., Chazerand, P., Dignum, V., Luetge, C., Madelin, R., Pagallo, U., Rossi, F., Schafer, B., Valcke, P., & Vayena, E. (2018). AI4People—An ethical framework for a good AI society: Opportunities, risks, principles, and recommendations. *Minds and Machines*, *28*(4), 689–707. DOI: 10.1007/s11023-018-9482-5 PMID: 30930541

Fountaine, S. (2017). What's not to like?: A qualitative study of young women politicians' self-framing on Twitter. *Journal of Public Relations Research*, *29*(5), 219–237. Advance online publication. DOI: 10.1080/1062726X.2017.1388239

Fox Business. (2025, December 1). Google CEO warns US must balance AI regulation or fall behind. https://www.foxbusiness.com/fox-news-tech/google-ceo-calls-national-ai-regulation-compete-china-more-effectively

Francis. (2024a, January 1). Message for the LVII World Day of Peace: Artificial intelligence and peace. Vatican City. https://www.vatican.va/content/francesco/en/messages/peace/documents/20231208-messaggio-57giornatamondiale-pace2024.html

Francis. (2024b, June 14). Address at the G7 session on artificial intelligence, Borgo Egnazia. Vatican City. https://www.vatican.va/content/francesco/en/speeches/2024/june/documents/20240614-g7-intelligenza-artificiale.html

Francis. (2025, February 7). Message for the Paris Summit on artificial intelligence. Vatican City. https://www.vatican.va/content/francesco/en/messages/pont-messages/2025/documents/20250207-messaggio-summit-parigi-ia.html

Freeman, R. E. (1984). *Strategic management: A stakeholder approach.* Pitman.

Freeman, R. E., Harrison, J. S., Wicks, A. C., Parmar, B. L., & de Colle, S. (2010). *Stakeholder theory: The state of the art.* Cambridge University Press. DOI: 10.1017/CBO9780511815768

Fridman, L. (Host). (2024, March 7). Yann LeCun: Meta AI, open source, limits of LLMs, AGI & the future of AI (No. 416) [Audio podcast episode]. In *Lex Fridman Podcast.* https://lexfridman.com/yann-lecun-3-transcript

Friedman, B., Kahn, P. H.Jr, & Borning, A. (2006). Value sensitive design and information systems. In Zhang, P., & Galletta, D. (Eds.), *Human-computer interaction in management information systems: Foundations* (pp. 348–372). M.E. Sharpe. DOI: 10.1145/242485.242493

Friedman, B., & Nissenbaum, H. (1996). Bias in computer systems. *ACM Transactions on Information Systems, 14*(3), 330–347. DOI: 10.1145/230538.230561

Fry, L. W. (2003). Toward a theory of spiritual leadership. *The Leadership Quarterly, 14*(6), 693–727. DOI: 10.1016/j.leaqua.2003.09.001

Future of Free Speech. (2024). AI chatbots refuse to produce controversial output. The Conversation. https://theconversation.com/ai-chatbots-refuse-to-produce-controversial-output-why-thats-a-free-speech-problem-226596

Future of Life Institute. (2023, March 22). Pause giant AI experiments: An open letter. https://futureoflife.org/open-letter/pause-giant-ai-experiments/

Gabriel, I. (2020). Artificial intelligence, values and alignment. *Minds and Machines, 30*(3), 411–437. DOI: 10.1007/s11023-020-09539-2

Gallup, Inc., & Lumina Foundation. (2026). AI in higher education: Widespread use, unclear rules (Lumina Foundation-Gallup state of higher education report). Gallup. https://www.gallup.com/file/analytics/704279/Lumina-Foundation-Gallup-SOHE_AI_Report.pdf

Gallup, Inc., & Lumina Foundation. (2026). *AI in higher education: Widespread use, unclear rules* (Lumina–Gallup State of Higher Education report). Gallup, Inc. https://www.gallup.com

Gao, D. K., Haverly, A., Mittal, S., Wu, J., & Chen, J. (2024). AI ethics. *International Journal of Business Analytics, 11*(1), 1–19. DOI: 10.4018/IJBAN.338367

Gbadeyanka, T. A., Ajobiewe, D. N., Okunlola, E. F., & Sulaimon, J. T. (2025). *Leveraging mindfulness and information and communication technology (ICT) to enhance attention amidst digital distractions in primary education*. ASEAN Journal of Science and Engineering Education., DOI: 10.17509/ajsee.v4i3.82475

Gerlich, M. (2025). AI tools in society: Impacts on cognitive offloading and the future of critical thinking. SSRN Electronic Journal. DOI: 10.2139/ssrn.5082524

Ghasemaghaei, M., & Kordzadeh, N. (2024). Understanding how algorithmic injustice leads to making discriminatory decisions: An obedience to authority perspective. *Information & Management, 61*(2), 103921. DOI: 10.1016/j.im.2024.103921

Giarmoleo, F. V., Ruiz, P. G., Rocchi, M., & Ferrero, I. (2026). Virtuous organizations in the age of AI: Relational goods and human flourishing. *Business Ethics, the Environment & Responsibility*, 1–15. DOI: 10.1111/beer.70090

Gillespie, T., Boczkowski, P. J., & Foot, K. A. (2014). Media technologies: Essays on communication, materiality, and society. MIT Press. DOI: 10.7551/mitpress/9780262525374.001.0001

Gioia, D. A. (1992). Pinto fires and personal ethics: A script analysis of missed opportunities. *Journal of Business Ethics, 11*(5–6), 379–389. DOI: 10.1007/BF00870550

Giolito, V., Resca, A., & Za, S. (2021). Servant leadership and intellectual stimulation in the digital transformation context. *Journal of Leadership Studies, 15*(2), 6–21. DOI: 10.1002/jls.21738

Global Partnership on AI. (2025). AI openness: A primer for policymakers. OECD. https://oecd.ai/en/wonk/balancing-innovation-transparency-and-risk-in-open-weight-models

Goddard, K., Roudsari, A., & Wyatt, J. C. (2012). Automation bias: A systematic review of frequency, effect mediators, and mitigators. *Journal of the American Medical Informatics Association : JAMIA, 19*(1), 121–127. DOI: 10.1136/amiajnl-2011-000089 PMID: 21685142

Goodman, B. (2022). Privacy without persons: A Buddhist critique of surveillance capitalism. *AI and Ethics, 2*(3), 415–426. DOI: 10.1007/s43681-022-00204-1

Graves, M. (2025). AI practical wisdom and compassion. *AI and Ethics, 6*(1), 39. Advance online publication. DOI: 10.1007/s43681-025-00877-4

Grayling, A. C. (2019). *The history of philosophy*. Penguin Random House.

Greenleaf, R. K. (1977). *Servant leadership: A journey into the nature of legitimate power and greatness*. Paulist Press.

Grosse, E. H., Sgarbossa, F., Berlin, C., & Neumann, W. P. (2023). Human-centric production and logistics system design and management: Transitioning from Industry 4.0 to Industry 5.0. *International Journal of Production Research, 61*(22), 7749–7759. DOI: 10.1080/00207543.2023.2246783

Guan, H., Dong, L., & Zhao, A. (2022). Ethical risk factors and mechanisms in artificial intelligence decision making. *Behavioral Sciences (Basel, Switzerland), 12*(9), 343. DOI: 10.3390/bs12090343 PMID: 36135147

Gurr, S. K., Geron, T., Forster, D. J., & Levinson, M. (2025). Philosophical reflections on teachers' ethical dilemmas in a global pandemic. *Studies in Philosophy and Education, 44*(2), 187–207. https://link.springer.com/article/10.1007/s11217-024-09937-4. DOI: 10.1007/s11217-024-09937-4

Habermas, J. (1985). The theory of communicative action: Reason and the rationalization of society. Beacon Press. ISBN-10 : 0807015075

Hagendorff, T. (2020). The ethics of AI ethics: An evaluation of guidelines. *Minds and Machines, 30*(1), 99–120. DOI: 10.1007/s11023-020-09517-8

Haidt, J. (2008). The emotional dog and its rational tail: A social intuitionist approach to moral judgment. Reasoning, 1024-1052. DOI: 10.1017/CBO9780511814273.055

Haidt, J. (2001). The emotional dog and its rational tail: A social intuitionist approach to moral judgment. *Psychological Review, 108*(4), 814–834. DOI: 10.1037/0033-295X.108.4.814 PMID: 11699120

Hannah, S. T., Avolio, B. J., & May, D. R. (2011). Moral maturation and moral conation: A capacity approach to explaining moral thought and action. *Academy of Management Review, 36*(4), 663–685. DOI: 10.5465/amr.2010.0128

Hannah, S. T., Avolio, B. J., & Walumbwa, F. O. (2011). Relationships between authentic leadership, moral courage, and ethical and pro-social behaviors. *Business Ethics Quarterly, 21*(4), 555–578. DOI: 10.5840/beq201121436

Hanson, W., & Moore, J. (2023). *Conditions, processes and pressures promoting inclusive organizations (chapter 12).* Ingold, J. & McGurk, P. (Eds.) Employer Engagement: Making Active Labour Market Policies Work. Bristol University Publishing.

Hanson, W. R., & Moore, J. R. (2014). Business student moral influencers: Unseen opportunities for development? *Academy of Management Learning & Education, 13*(4), 525–546. DOI: 10.5465/amle.2012.0325

Hanson, W. R., Moore, J. R., Bachleda, C., Canterbury, A., Franco, C.Jr, Marion, A., & Schreiber, C. (2017). Theory of moral development of business students: Case studies in Brazil, North America, and Morocco. *Academy of Management Learning & Education, 16*(3), 393–414. DOI: 10.5465/amle.2014.0312

Hassen, M. Z. (2025). The risk of moral outsourcing: Why artificial intelligence cannot and should not make our ethical decisions. *EthAlca, 4*, Article 428.

Hassen, M. Z. (2025). The risk of moral outsourcing: Why artificial intelligence cannot and should not make our ethical decisions. *EthAlca, 4*, 428. DOI: 10.56294/ai2025428

Held, D., & McGrew, A. (2007). *Globalization theory: Approaches and controversies*. Polity Press.

Held, V. (2006). *The ethics of care: Personal, political, and global*. Oxford University Press.

Herari, N., Najmi Nuji, M. N., Boer, R. F., & Allifiansyah, S. (2026). Algorithmic Narcissism and the Passion Economy: Identity, Visibility, and the Ideology of Self-Performance. In Digital Ideologies in the Age of Algorithmic Culture (1st ed.). IGI Global Scientific Publishing. ISBN13: 9798337345314. DOI: 10.4018/979-8-3373-4531-4.ch008

Herari, N., & Riyanto, B. (2026). Iron Man or Tin Man?: AI, Habit, and the Quiet Mechanization of Human Life. In *AI and Robotic Companions for Human Socialization*. IGI Global Scientific Publishing.

Hoekema, A. A. (1986). *Created in God's image*. Eerdmans.

Holland, J. H. (1995). *Hidden order: How adaptation builds complexity*. Basic Books.

Holmes, W., Bialik, M., & Fadel, C. (2019). *Artificial intelligence in education: Promises and implications for teaching and learning*. Center for Curriculum Redesign.

Holy Bible (Genesis 1:26-28, 1 Corinthians 10:31, Micah 6:8), New International Version®, NIV® Copyright ©1973, 1978, 1984, 2011 by Biblica, Inc.®

Hong, P., & Park, Y. W. (2020). *Rising Asia and American hegemony: Case of competitive firms from Japan.* Springer Singapore. DOI: 10.1007/978-981-13-7635-1

Hursthouse, R., & Pettigrove, G. (2018). Virtue ethics. In Zalta, E. N. (Ed.), *The Stanford encyclopedia of philosophy* (Fall 2018 ed.)., https://plato.stanford.edu/entries/ethics-virtue/

IEEE. (2021). Ethically Aligned Design: A Vision for Prioritizing Human Well-being with Autonomous and Intelligent Systems. https://standards.ieee.org/wp-content/uploads/import/documents/other/ead_v2.pdf

Ismail, N. H., Makhsin, M., Khadafi Hj. Rofie, M., & Omar, S. (2021). Higher order thinking skills (HOTS): An analysis based on surah Al-Hajj verse 46. *Journal of Language and Linguistic Studies*, *17*(1), 471–480. DOI: 10.52462/jlls.30

Jago, A. S., Raveendhran, R., Fast, N., & Gratch, J. (2024). Algorithmic management diminishes status: An unintended consequence of using machines to perform social roles. *Journal of Experimental Social Psychology*, *110*, 104553. DOI: 10.1016/j.jesp.2023.104553

Jasanoff, S. (2016). The ethics of invention: Technology and the human future. W. W. Norton & Company. ISBN-13 : 978-0393078992

Jeffery, C. P. (2020). Parenting in the digital age: Between socio-biological and socio-technological development. *New Media & Society*, *23*(5), 1045–1062. DOI: 10.1177/1461444820908606

Jiang, Y., Li, X., Luo, H., Yin, S., & Kaynak, O. (2022). *Quo vadis* artificial intelligence? *Discover Artificial Intelligence*, *2*(1), 4. https://link.springer.com/article/10.1007/s44163-022-00022-8. DOI: 10.1007/s44163-022-00022-8

Jolly, A. A., Dunlop, P. D., Parker, S. K., & Kansel, L. (2026). It's Not My Responsibility: Working with Autonomy-Restricting Algorithms Facilitates Unethical Behavior and Displacement of Responsibility. *Journal of Business Ethics*, *203*(2), 405–424. DOI: 10.1007/s10551-025-06034-5

Jonas, H. (1984). The Imperative of Responsibility. In *Search of an Ethics for the Technological Age*. University of Chicago Press. DOI: 10.7208/chicago/9780226850337.001.0001

Jones, M., & Schonewille, M. (2026). Synthetic relationships and artificial intimacy: An ethical framework for evaluating the impact of generative AI on community. *Journal of Ethics in Entrepreneurship and Technology*. .DOI: 10.1108/JEET-06-2025-0040

Joshi, P. (2025). Philosophy and ethics in the age of artificial intelligence: Bridging the gap between technology and human values. *Journal of Information Systems Engineering & Management, 10*(4), 350–363. DOI: 10.52783/jisem.v10i4.9232

Kandiyali, J. (2014). Paul Blackledge and Kelvin Knight (eds). *Journal of Moral Philosophy, 11*(1), 111–115. DOI: 10.1163/17455243-01101002

Kantar, N., & Bynum, T. W. (2021). Global ethics for the digital age–flourishing ethics. *Journal of Information Communication and Ethics in Society, 19*(3), 329–344. DOI: 10.1108/JICES-01-2021-0016

Kantha, T., Sahoo, D., Rout, D., & Mishra, S. J. (2024). *Ethics in the digital era: Technology's role in shaping society*. Shanlax International Journal of Arts, Science and Humanities., DOI: 10.34293/sijash.v11iS1-June.7794

Kant, I. (2012). *Groundwork of the metaphysics of morals* (Gregor, M., & Timmermann, J., Trans.). Cambridge University Press. (Original work published 1785)

Keche, U. (2025). *Cybersecurity and information ethics: A comparative study of privacy rights in digital india and the global framework*. Gurukul International Multidisciplinary Research Journal., DOI: 10.69758/GIMRJ/2504I5VXIIIP0090

Kellogg, K. C., Valentine, M. A., & Christin, A. (2020). Algorithms at work. *The Academy of Management Annals, 14*(1), 366–410. DOI: 10.5465/annals.2018.0174

Kemp, S. E., Ng, M., Hollowood, T., & Hort, J. (2018). Introduction to descriptive analysis. In Kemp, S. E., Hort, J., & Hollowood, T. (Eds.), *Descriptive analysis in sensory evaluation* (pp. 1–39). Wiley., DOI: 10.1002/9781118991657.ch1

Kets de Vries, R. M. F. (2011). The leader on the couch: A clinical approach to changing people and organizations. John Wiley & Sons. ISBN 0470030798

Kilner, J. F. (2015). *Dignity and destiny: Humanity in the image of God*. Eerdmans.

Kingsnorth, P. (2021, June 1). The cross and the machine. *First Things (New York, N.Y.)*.

Klecun, E., & Cornford, T. (2005). A critical approach to evaluation. *European Journal of Information Systems, 14*(3), 229–243. DOI: 10.1057/palgrave.ejis.3000540

Knight, K. (1998). *The MacIntyre reader*. University of Notre Dame Press.

Knowles, M. S., Holton, E. F.III, & Swanson, R. A. (2005). *The adult learner: The definitive classic in adult education and human resource development* (6th ed.). Elsevier. DOI: 10.4324/9780080481913

Kohlberg, L. (1981). Essays on moral development: Vol. 1. *The philosophy of moral development*. Harper & Row.

Konya, A., & Nematzadeh, P. (2024). Recent applications of AI to environmental disciplines: A review. *The Science of the Total Environment*, *906*, 167705. DOI: 10.1016/j.scitotenv.2023.167705 PMID: 37820816

Kosgey, H. K., ONGERA (Ph.D), D. R. A., & MBURU (Ph.D), D. R. T. (. (2020). Challenges facing ethical leadership and governance in Kenya. A case of uasin gishu county. *Strategic Journal of Business & Change Management*, *7*(1). Advance online publication. DOI: 10.61426/sjbcm.v7i1.1555

Kosmyna, N., Haupmann, E., Yuan, Y. T., Situ, J., Liao, X., Beresnitzky, A. V., Braunstein, I., & Maes, P. (2025). Your Brain on ChatGPT: Accumulation of Cognitive Debt when Using an AI Assistant for Essay Writing Task. Cornell University. https://arxiv.org/abs/2506.08872

Kouvelis, P., Chambers, C., & Wang, H. (2006). Supply chain management research and production and operations management: Review, trends, and opportunities. *Production and Operations Management*, *15*(3), 449–469. DOI: 10.1111/j.1937-5956.2006.tb00257.x

Kumar, R., Sporn, K., Waisberg, E., Ong, J., & Paladugu, P. (2025). Healthcare AI governance and oversight frameworks. *Health Care Analysis : HCA : Journal of Health Philosophy and Policy*. Advance online publication. DOI: 10.1007/s10728-025-00537-y

Kuwaiti, A. A., Nazer, K., Alreedy, A. H., AlShehri, S. D., Almuhanna, A., Subbarayalu, A. V., Muhanna, D. A., & Al-Muhanna, F. (2023). A Review of the Role of Artificial Intelligence in Healthcare. *Journal of Personalized Medicine*, *13*(6), 951. DOI: 10.3390/jpm13060951 PMID: 37373940

Lackey, D. P. (1999). What are the modern classics? The Baruch poll of great philosophy in the twentieth century. *The Philosophical Forum*, *30*(4), 329–346. DOI: 10.1111/0031-806X.00022

Lafram, A., & Bahji, S. E. (2024). Artificial intelligence in Morocco: Towards holistic, responsible and ethical national AI strategy for Moroccan competitiveness and strategic intelligence. *European Conference on Management Leadership and Governance*, *20*(1), 255–263. DOI: 10.34190/ecmlg.20.1.2909

Lahteenkorva, M., Tynjälä, P., & Kallio, E. (2023). From knowledge to wisdom in sustainable management education. *Journal of Management Education*. Advance online publication. DOI: 10.1177/105256292311XXXX

Lähteenkorva, M., Tynjälä, P., & Kallio, E. K. (2025). From knowledge to wisdom in sustainable management education – Insights from a B Corp course. *International Journal of Management Education, 23*(2), 101148. DOI: 10.1016/j.ijme.2025.101148

Laitinen, A., & Sahlgren, O. (2021). AI systems and respect for human autonomy. *Frontiers in Artificial Intelligence, 4*, 705164. DOI: 10.3389/frai.2021.705164 PMID: 34765969

Lan, Y., Liu, S., & Xia, L. (2025). How mindfulness shapes AI competence: A structural equation modeling analysis of mindfulness, AI literacy and behavioral intention in Chinese media students. *Frontiers in Psychology, 16*, 1652934. Advance online publication. DOI: 10.3389/fpsyg.2025.1652934 PMID: 40900965

LeCun, Y. (2023, March 21). I think that the magnitude of the AI alignment problem has been ridiculously overblown [Post]. LinkedIn. https://www.linkedin.com/posts/yann-lecun_i-think-that-the-magnitude-of-the-ai-alignment-activity-7043650114048925696-1oXJ

Lee, H., An, S., Lim, G. Y., & Sohn, Y. W. (2021). Ethical leadership and followers' emotional exhaustion: Exploring the roles of three types of emotional labor toward leaders in South Korea. *International Journal of Environmental Research and Public Health, 18*(20), 10862. DOI: 10.3390/ijerph182010862 PMID: 34682605

Lee, J., Lee, D., & Lee, J. (2021). Can robots help working parents with childcare? Optimizing childcare functions for different parenting characteristics. *International Journal of Social Robotics, 14*(1), 193–211. https://pmc.ncbi.nlm.nih.gov/articles/PMC8020072/. DOI: 10.1007/s12369-021-00784-8 PMID: 33841588

Lessig, L. (1999). *Code and other laws of cyberspace*. Basic Books.

Lewis, C. S. (1943). *The abolition of man*. Oxford University Press.

Liden, R. C., Wayne, S. J., Zhao, H., & Henderson, D. (2008). Servant leadership: Development of a multidimensional measure and multi-level assessment. *The Leadership Quarterly, 19*(2), 161–177. DOI: 10.1016/j.leaqua.2008.01.006

Lima, S. A., & Rahman, M. M. (2025). Ethical singularity in HRM: Redefining leadership, decision-making, and inclusion in AI-governed organizations. *Journal of Artificial Intelligence General Science (JAIGS) ISSN:3006-4023, 8*(02), 262–292. DOI: 10.60087/jaigs.v8i02.420

Lin, P., Abney, K., & Bekey, G. A. (Eds.). (2012). *Robot ethics: The ethical and social implications of robotics*. MIT Press.

Li, W. (2024). The influence of artificial intelligence on the education system. *Lecture Notes in Education Psychology and Public Media*, *65*(1), 137–142. DOI: 10.54254/2753-7048/65/20240158

Locke, J. (2003). *Two treatises of government and a letter concerning toleration*. Yale University Press. (Original work published 1689)

Lyu, Y., & Du, Y. (2025). The ethical evaluation of large language models and its optimization. *AI and Ethics*, *5*(5), 4579–4592. DOI: 10.1007/s43681-024-00654-9

Maak, T., & Pless, N. M. (2021). *Responsible leadership* (2nd ed.). Routledge. DOI: 10.4324/b22741-4

MacIntyre, A. (1981). *After virtue: A study in moral theory*. University of Notre Dame Press.

MacIntyre, A. (1984). *After virtue* (2nd ed.). University of Notre Dame Press.

MacIntyre, A. (1998). *A short history of ethics: A history of moral philosophy from the Homeric age to the twentieth century* (2nd ed.). University of Notre Dame Press.

MacIntyre, A. (2016). *Ethics in the conflicts of modernity: An essay on desire, practical reasoning, and narrative*. Cambridge University Press. DOI: 10.1017/9781316816967

Malja, S. N., & Afrasiabi, H. (2025). AI and society: Mapping the research through a systematic review. *AI & Society*. Advance online publication. DOI: 10.1007/s00146-025-02555-9

Manke, K. (2025, September 10). Do chatbots have a moral compass? Researchers turn to Reddit to find out. *Berkeley News*. https://news.berkeley.edu/2025/09/10/do-chatbots-have-a-moral-compass-researchers-turn-to-reddit-to-find-out

Marcus Aurelius. (2006). *Meditations* (M. Hammond, Trans.). Penguin Classics. (Original work published ca. 170–180 CE)

Masten, A. S. (2001). Ordinary magic: Resilience processes in development. *The American Psychologist*, *56*(3), 227–238. DOI: 10.1037/0003-066X.56.3.227 PMID: 11315249

Mastroianni, G. R., Kimmelman, S., Doty, J., & Thomas, J. (2011). Obedience and Personal Responsibility. In Leadership in Dangerous Situations: the Handbook of the Armed Forces, Emergency Services, and the First Responders (pp. 97-120). Naval Institute Press. https://www.researchgate.net/profile/Joe-Doty/publication/303840128_Obedience_and_Personal_Responsibility/links/5b045a284585154aeb07f1e7/Obedience-and-Personal-Responsibility.pdf

Mayer Brown. (2026, March 1). Pentagon designates Anthropic a supply chain risk: What government contractors need to know. https://www.mayerbrown.com/en/insights/publications/2026/03/pentagon-designates-anthropic-a-supply-chain-risk-what-government-contractors-need-to-know

McCarthy, J., Minsky, M. L., Rochester, N., & Shannon, C. E. (2006). A proposal for the Dartmouth Summer Research Project on artificial intelligence, August 31, 1955. *AI Magazine*, *27*(4), 12–14. DOI: 10.1609/aimag.v27i4.1904

McCorduck, P. (2004). *Machines who think: A personal inquiry into the history and prospects of artificial intelligence* (2nd ed.). A.K. Peters Ltd. DOI: 10.1201/9780429258985

McLennan, S., Fiske, A., Tigard, D., Müller, R., Haddadin, S., & Buyx, A. (2022). Embedded ethics: A proposal for integrating ethics into the development of medical AI. *BMC Medical Ethics*, *23*(1), 6. https://link.springer.com/article/10.1186/s12910-022-00746-3. DOI: 10.1186/s12910-022-00746-3 PMID: 35081955

McLoughlin, S., Espinosa, V. F., & Velázquez-Gil, M. (2026). Practical wisdom, emerging technologies, and AI. Journal of Moral Education. https://doi.org/DOI: 10.35542/osf.io/gtjc6_v1

McLoughlin, S., Fernández-Espinosa, V., & Velázquez-Gil, M. (2026). Practical wisdom, emerging technologies, and AI. *Journal of Moral Education*, 1–31. Advance online publication. DOI: 10.1080/03057240.2026.2624863

McLoughlin, S., Thoma, S., & Kristjánsson, K. (2025). Was Aristotle right about moral decision-making? Building a new empirical model of practical wisdom. *PLoS One*, *20*(1), e0317842. Advance online publication. DOI: 10.1371/journal.pone.0317842 PMID: 39841702

Mehrabi, N., Morstatter, F., Saxena, N., Lerman, K., & Galstyan, A. (2021). A survey on bias and fairness in machine learning. *ACM Computing Surveys*, *54*(6), 1–35. DOI: 10.1145/3457607

Meneguetti, L. (2025). The imago Dei in the age of artificial intelligence: Toward a theology of human dignity in technological times. *SSRN*. DOI: 10.2139/ssrn.5359499

Microsoft & LinkedIn. (2024). *AI at work is here. Now comes the hard part: 2024 work trend index.* https://www.microsoft.com/en-us/worklab/work-trend-index/ai-at-work-is-here-now-comes-the-hard-part

Middleton, J. R. (2005). *The liberating image: The imago Dei in Genesis 1.* Brazos Press.

Mill, J. S. (2003). *On liberty.* Yale University Press. (Original work published 1859)

Milmo, D. (2026, January 15). Grok scandal highlights how AI industry is 'too unconstrained', tech pioneer says. the Guardian. https://www.theguardian.com/technology/2026/jan/15/grok-scandal-ai-industry-too-unconstrained-yoshua-bengio-elon-musk

Mishra, A., Mou, S. N., Ara, J., & Sarkar, M. (2025). Regulatory and ethical challenges in AI-driven and machine learning credit risk assessment for buy now, pay later (BNPL) in U.S. E-commerce: Compliance, fair lending, and algorithmic bias. *Journal of Business and Management Studies.* DOI: 10.32996/jbms.2025.7.2.3

MIT Sloan Management Review. (2022, March 17). Google's Sundar Pichai says tech is a powerful agent for change. https://mitsloan.mit.edu/ideas-made-to-matter/googles-sundar-pichai-says-tech-a-powerful-agent-change

MIT Technology Review. (2025, February 6). An AI chatbot told a user how to kill himself—but the company doesn't want to "censor" it. https://www.technologyreview.com/2025/02/06/1111077/nomi-ai-chatbot-told-user-to-kill-himself/

Mittelstadt, B. (2019). Principles alone cannot guarantee ethical AI. *Nature Machine Intelligence, 1*(11), 501–507. DOI: 10.1038/s42256-019-0114-4

Mohamed, F. (2011). Capitalism in Moral Perspective – An Islamic Alternative [Master's thesis]. https://uwcscholar.uwc.ac.za:8443/server/api/core/bitstreams/26a582bf-b546-40c7-b04b-d6b8d28fa600/content

Mökander, J., & Floridi, L. (2023). Operationalising AI governance through ethics-based auditing: An industry case study. *AI and Ethics, 3*(2), 451–468. Advance online publication. DOI: 10.1007/s43681-022-00171-7 PMID: 35669570

Moore, J. (2026). Flourishing minds, flourishing institutions: Cultivating well-being in academic culture. In *Work Values and Organizational Performance in Higher Education Institutions.* Springer.

Moore, J., & Halter, C. (in press). Employee flourishing leadership: Unlocking employee potential. In C. Claes, R. Schalock, & G. Van Hove (Eds.), Ondersteuning van mensen in een inclusieve samenleving. Een persoonsgericht handboek. Gompel & Svacina.

Moore, J., & Halter, C. (in press). Employee flourishing leadership: Unlocking employee potential. In C. Claes, R. Schalock, & G. Van Hove (Eds.), *Ondersteuning van mensen in een inclusieve samenleving: Een persoonsgericht handboek* [*Support for people in an inclusive society: A person-centered handbook*]. Gompel & Svacina.

Moore, J., & Hanson, W. (in press). Wellbeing oriented organizations: From human utilization to human flourishing. In C. Claes, R. Schalock, & G. Van Hove (Eds.), Ondersteuning van mensen in een inclusieve samenleving. Een persoonsgericht handboek. Gompel & Svacina.

Moore, J., & Hanson, W. (in press). Wellbeing oriented organizations: From human utilization to human flourishing. In C. Claes, R. Schalock, & G. Van Hove (Eds.), *Ondersteuning van mensen in een inclusieve samenleving: Een persoonsgericht handboek* [*Support for people in an inclusive society: A person-centered handbook*]. Gompel & Svacina.

Moore, J., Hanson, W., & Gustafson, T. (2023). *Sephora – Journey to inclusive workplace – Let us belong philosophy (chapter 13).* Ingold, J. & McGurk, P. (Eds.) Employer Engagement: Making Active Labour Market Policies Work. Bristol University Publishing.

Moore, J. (n.d.). *Flourishing minds, flourishing institutions: Cultivating well-being in academic culture* [Manuscript]. Anderson University, College of Business and Economics.

Moore, J., & Hanson, W. (2019). *Book title: Moral Development and Behavior: New Research. Book chapter: "University Business Student Moral Development".* Nova Science Publishers.

Moore, J., & Hanson, W. (2022). Improving leader effectiveness: Impact on employee engagement and retention. *Journal of Management Development, 41*(7/8), 450–468. DOI: 10.1108/JMD-02-2021-0041

Moore, J., & Hanson, W. (n.d.). *Leadership strategy for building Industry 5.0: Human-centric societal organizations* [Manuscript]. Anderson University, College of Business and Economics.

Moore, J., & Maxey, E. (in press). Thriving in Industry 5.0: Leadership and business models for value creation. In Waite, A., Machado, C., & Davim, J. P. (Eds.), *Leading in Industry 5.0*. Springer.

Moore, J., Maxey, E. C., Waite, A. M., & Wendover, J. D. (2020). Inclusive organizations: Developmental reciprocity through authentic leader-employee relationships. *Journal of Management Development*, *39*(9/10), 1029–1039. DOI: 10.1108/JMD-05-2019-0211

Moore, J., Maxey, E., & Hangen, J. (n.d.). *Thriving in Industry 5.0: Leadership and business models for value creation* [Manuscript]. Anderson University, College of Business and Economics.

Moore, J., Zhang, W., & High, D. (2023). Theory of moral development in Chinese university students. *Journal of Research for International Education*, *2*(3).

Moor, J. H. (1985). What is computer ethics? *Metaphilosophy*, *16*(4), 266–275. DOI: 10.1111/j.1467-9973.1985.tb00173.x

Moor, J. H. (2006). Why we need better ethics for emerging technologies. *Ethics and Information Technology*, *7*(3), 111–119. DOI: 10.1007/s10676-006-0008-0

Moreland, J. P., & Craig, W. L. (2003). *Philosophical foundations for a Christian worldview*. InterVarsity Press.

Morrin, H., Nicholls, L., Levin, M., Yiend, J., Iyengar, U., DelGuidice, F., Bhattacharya, S., Tognin, S., MacCabe, J., Twumasi, R., Alderson-Day, B., & Pollak, T. A. (2025). Artificial intelligence-associated delusions and large language models: Risks, mechanisms of delusion co-creation, and safeguarding strategies. *The Lancet Psychiatry*. Advance online publication. DOI: 10.1016/S2215-0366(25)00396-7 PMID: 41796598

Moulaei, K., Haghdoost, A., Bahaadinbeigy, K., & Dinari, F. (2023). The effect of the holy Quran recitation and listening on anxiety, stress, and depression: A scoping review on outcomes. *Health Science Reports*, *6*(12), e1751. Advance online publication. DOI: 10.1002/hsr2.1751 PMID: 38078304

Movahed, M., & Bilderback, S. (2024). Evaluating the readiness of healthcare administration students to utilize AI for sustainable leadership: A survey study. *Journal of Health Organization and Management*, *38*(4), 567–582. DOI: 10.1108/JHOM-12-2023-0385 PMID: 38858220

Murphy, R., & Woods, D. D. (2009). Beyond Asimov: The three laws of responsible robotics. *IEEE Intelligent Systems*, *24*(4), 14–20. DOI: 10.1109/MIS.2009.69

Nabil, M. (2025). Amanah (trust) and Physician- Patient Relationship in Islamic Medical Ethics. *Journal of British Islamic Medical Association, 19*(7), 1–8. https://jbima.com/article/amanah-trust-and-physician-patient-relationship-in-islamic-medical-ethics/

Nanjundeswaraswamy, T. S., Nagesh, P., Bharath, S., & Vignesh, K. M. (2024). Leadership theories and styles—A systematic literature review and the narrative synthesis. *Human Resources Management and Services, 6*(3), 3477. DOI: 10.18282/hrms.v6i3.3477

National Institute of Standards and Technology (NIST). (2023). AI Risk Management Framework.

National Institute of Standards and Technology [NIST]. (2023). *Artificial Intelligence Risk Management Framework (AI RMF 1.0)* (NIST AI 100-1). U.S. Department of Commerce. DOI: 10.6028/NIST.AI.100-1

National Telecommunications and Information Administration [NTIA]. (2024). *Dual-use foundation models with widely available model weights*. U.S. Department of Commerce. https://www.ntia.gov/programs-and-initiatives/artificial-intelligence/open-model-weights-report

New International Version Bible. (2011). Biblica. https://www.biblica.com/bible/niv/

New Revised Standard Version Bible. (1989). National Council of the Churches of Christ in the United States of America.

Nida-Rümelin, J., & Weidenfeld, N. (2022). *Digital humanism: For a humane transformation of democracy, economy, and culture in the digital age*. Springer. https://link.springer.com/book/10.1007/978-3-031-12482-2

Nissenbaum, H. (2004). Privacy as contextual integrity. *Washington Law Review (Seattle, Wash.), 79*, 119. https://digitalcommons.law.uw.edu/wlr/vol79/iss1/10

Noble, S. U. (2018). *Algorithms of oppression*. NYU Press. DOI: 10.18574/nyu/9781479833641.001.0001

Noordt, C. V., & Tangi, L. (2023). The dynamics of AI capability and its influence on public value creation of AI within public administration. *Government Information Quarterly, 40*(4), 101860. DOI: 10.1016/j.giq.2023.101860

Novelli, C., Taddeo, M., & Floridi, L. (2024). Accountability in artificial intelligence: What it is and how it works. *AI & Society, 39*(4), 1871–1882. DOI: 10.1007/s00146-023-01635-y

NPR. (2023, February 9). Google's AI chatbot, Bard, sparks a $100 billion loss in Alphabet shares. https://www.npr.org/2023/02/09/1155650909/google-chatbot--error-bard-shares

NPR. (2026, February 27). OpenAI announces Pentagon deal after Trump bans Anthropic. https://www.npr.org/2026/02/27/nx-s1-5729118/trump-anthropic-pentagon-openai-ai-weapons-ban

Nugroho, N. P. (2026, March 5). *Gibran: AI Penting untuk Menunjang Kegiatan Belajar*. Tempo. https://www.tempo.co/politik/gibran-ai-penting-untuk-menunjang-kegiatan-belajar-2119713

Nunnally, J. C., & Bernstein, I. H. (1994). *Psychometric theory* (3rd ed.). McGraw-Hill.

Nussbaum, M. C. (2006). *Frontiers of justice: Disability, nationality, species membership*. Harvard University Press.

Nussbaum, M. C. (2011). *Creating capabilities*. Harvard University Press., DOI: 10.4159/harvard.9780674061200

Nyamori, D. (2024). Exploring the relationship between moral decision-making and Emotional Intelligence. *International Journal of Philosophy*, *3*(1), 13–24. DOI: 10.47941/ijp.1866

O'Donovan, O. (1986). *Resurrection and moral order*. Eerdmans.

O'Neil, C. (2017). Weapons of Math Destruction: How Big Data Increases Inequality and Threatens Democracy. Crown, New York, NY.

O'Neil, C. (2016). *Weapons of math destruction*. Crown.

Oakes, E. T. (1996, August). The achievement of Alasdair MacIntyre. *First Things*. https://www.firstthings.com/article/1996/08/the-achievement-of-alasdair-macintyre

Obermeyer, Z., Powers, B., Vogeli, C., & Mullainathan, S. (2019). Dissecting racial bias in an algorithm used to manage the health of populations. *Science*, *366*(6464), 447–453. DOI: 10.1126/science.aax2342 PMID: 31649194

OECD. (2019). AI Principles. Organisation for Economic Co-operation and Development. https://www.oecd.org/en/topics/ai-principles.html

OECD. (2019). OECD Principles on Artificial Intelligence.

Oguz, H. (2025). Authoritarian recursions: How fiction, history, and AI reinforce control in education, warfare, and discourse. arXiv. https://doi.org//arXiv.2504.09030DOI: 10.48550

Olatoye, F. O., Awonuga, K. F., Mhlongo, N. Z., Ibeh, C. V., Elufioye, O. A., & Ndubuisi, N. L. (2024). AI and ethics in business: A comprehensive review of responsible AI practices and corporate responsibility. *International Journal of Science and Research Archive, 11*(1), 1433–1443. DOI: 10.30574/ijsra.2024.11.1.0235

Onder, O., & Deniz, E. M. (2026). Cyberspatial privacy in the digital age: A proxemics-based framework with a healthcare application. *BMC Medical Ethics, 27*(1), 14. DOI: 10.1186/s12910-025-01369-0 PMID: 41507986

OpenAI. (2024). *Model spec.* https://model-spec.openai.com

OpenAI. (2025, April 29). Sycophancy in GPT-4o: What happened and what we're doing about it. https://openai.com/index/sycophancy-in-gpt-4o/

OpenAI. (2026, February 28). Our agreement with the Department of War. https://openai.com/index/our-agreement-with-the-department-of-war/

Orlikowski, W. J., & Scott, S. V. (2015). The algorithm and the crowd: Considering the materiality of service innovation. *Management Information Systems Quarterly, 39*(1), 201–224. DOI: 10.25300/MISQ/2015/39.1.09

Oyekunle, D., Matthew, U. O., Preston, D., & Boohene, D. (2024). Trust beyond technology algorithms: A theoretical exploration of consumer trust and behavior in technological consumption and AI projects. *Journal of Computer and Communications, 12*(06), 72–102. DOI: 10.4236/jcc.2024.126006

Page, S. E. (2007). *The difference: How the power of diversity creates better groups, firms, schools, and societies*. Princeton University Press.

Parasuraman, R., & Riley, V. (1997). Humans and automation: Use, misuse, disuse, abuse. *Human Factors, 39*(2), 230–253. DOI: 10.1518/001872097778543886

Park, Y. W., & Hong, P. (2024). Cosmetics marketing strategy in the era of the digital ecosystem. Springer Nature Singapore. DOI: 10.1007/978-981-97-3674-4

Parrott, J. (2014, August 28). Ibn Qayyim on Hearts: A void that can only be filled by love of Allah. Abuaminaelias.com. https://www.abuaminaelias.com/dailyhadithonline/2014/08/28/ibn-qayyim-void-heart/

Perkins, M., Roe, J., Postma, D., McLaughlin, P., & Fishman, T. (2023). Detection of GPT-4 generated text in higher education: Combining academic judgement and software to identify generative AI tool misuse. *Journal of Academic Ethics, 22*(1), 89–113. DOI: 10.1007/s10805-023-09492-6

Perrigo, B. (2023, January 18). OpenAI used Kenyan workers on less than $2 per hour to make ChatGPT less toxic. TIME. https://time.com/6247678/openai-chatgpt-kenya-workers/

Persson, A., & Hedlund, C. (2024). Revisiting Asimov's laws of robotics in the age of AI: Ethical implications and modern perspectives. *AI and Ethics*. Advance online publication. DOI: 10.1007/s43681-024-00489-4

Pichai, S. (2024, February 28). Memorandum to Google employees on Gemini AI image generation. https://www.npr.org/2024/02/28/1234532775/google-gemini-offended-users-images-race

Pm, M., & Yuvasri. (2025). Effectiveness of mindfulness- based intervention on cognitive flexibility, digital burnout and emotional regulation among college students. *Advanced International Journal for Research*. DOI: 10.63363/aijfr.2025.v06i06.2459

Pontifical Academy for Life. (2020, February 28). Rome Call for AI Ethics. Vatican City. https://www.romecall.org/the-call/

Prabagar, S., Mothkur, R., Swetha, C. B., Kurian, A., Pavithra, K., & Aarthi, S. (2025). Generative AI for inclusive special education. In *Advances in Computational Intelligence and Robotics* (pp. 205–236). IGI Global Scientific Publishing. DOI: 10.4018/979-8-3373-3745-6.ch008

Qureyshi, H. A., & Mustaq, S. (2025). Between Temptation and Tranquility: The Quranic framework of Nafs. *Journal of Quranic and Social Studies*, *5*(2), 18–36. DOI: 10.5281/zenodo.15570157

Radanliev, P. (2025). AI ethics: Integrating transparency, fairness, and privacy in AI development. *Applied Artificial Intelligence*, *39*(1), 2463722. DOI: 10.1080/08839514.2025.2463722

Rahman, F. (2009). Major themes of the Qur'an (2nd ed.). University of Chicago Press. ISBN-10: 0226702863

Raji, I. D., Smart, A., White, R. N., Mitchell, M., Gebru, T., Hutchinson, B., Smith-Loud, J., Theron, D., & Barnes, P. (2020). Closing the AI accountability gap: Defining an end-to-end framework for internal algorithmic auditing. In *Proceedings of the 2020 Conference on Fairness, Accountability, and Transparency* (pp. 33–44). DOI: 10.1145/3351095.3372873

Raji, I. D., Xu, P., Honigsberg, C., & Ho, D. (2022). Outsider oversight: Designing a third-party audit ecosystem for AI governance. *Proceedings of the AAAI/ACM Conference on AI, Ethics, and Society*. DOI: 10.1145/3514094.3534181

Rathinavel, P., Ravikumar, R. N., Aarthi, S., Habib, M., Salaeva, N., & Pardaev, J. (2025). Augmenting human capabilities. In *Advances in Computational Intelligence and Robotics* (pp. 1–30). IGI Global Scientific Publishing. DOI: 10.4018/979-8-3373-2372-5.ch001

Ravikumar, & Aarthi, S. (2025a). Conscious AI Integration in business education. In *Advances in Computational Intelligence and Robotics* (pp. 73–104). IGI Global Scientific Publishing. DOI: 10.4018/979-8-3373-4377-8.ch003

Ravikumar, R. N., & Aarthi, S. (2025b). Human-AI synergy for organizational resilience. In *Advances in Computational Intelligence and Robotics* (pp. 287–322). IGI Global Scientific Publishing. DOI: 10.4018/979-8-3373-7416-1.ch011

Ravikumar, R. N., Aarthi, S., & Mamadiyarov, Z. (2026). Building the foundations of augmented organizations through AI empowerment. In *Advances in Computational Intelligence and Robotics* (pp. 101–130). IGI Global Scientific Publishing. https://doi.org/DOI: 10.4018/979-8-3373-8607-2.ch006

Ravikumar, R. N., & Aarthi, S. (2025c). Integrating machine learning with human decision making in HR. In *Empowering Human Resources Through Human-Computer Interaction* (pp. 115–146). IGI Global Scientific Publishing., DOI: 10.4018/979-8-3373-1676-5.ch005

Ravikumar, R. N., Aarthi, S., Pardaev, J., & Shukla, P. (2025). Quantum computing for financial modelling. In *Quantum-Driven Financial Intelligence* (pp. 221–252). IGI Global Scientific Publishing., DOI: 10.4018/979-8-3373-2893-5.ch010

Ray, R. (2024). *Servant leadership: Developing behaviors for change*. Anderson University Press.

Raza, M. W. (2022). Anatomy of the intellect in the Quran: A fresh perspective. Al-Bayan. *Journal of Qur'an and Hadith Studies, 20*(2), 182–216. DOI: 10.1163/22321969-12340113

Reed, E., & Johnson, B. (2023). Overview of cultural capital theory's current impact and potential utility in academic libraries. *Journal of Academic Librarianship, 49*(6), 102782. DOI: 10.1016/j.acalib.2023.102782

Reina Marín, Y., Cruz Caro, O., Maicelo Rubio, Y. C., Alva Tuesta, J. N., Sánchez Bardales, E., Carrasco Rituay, A. M., & Chávez Santos, R. (2025). Ethical challenges associated with the use of artificial intelligence in university education. *Journal of Academic Ethics, 23*(4), 2443–2467. DOI: 10.1007/s10805-025-09660-w

RenAIssance. (2020). Rome Call for AI Ethics. Rome Call | What is the Matter with AI Ethics? https://www.romecall.org/the-call/

Ren, L., & Shen, H. (2024). The relationship between servant leadership and team innovation performance: Mediating effect of self-efficacy. *Heliyon, 10*(6), e27723. Advance online publication. DOI: 10.1016/j.heliyon.2024.e27723 PMID: 38509912

Reporters Without Borders. (2025). Controlling information in the age of AI: How state propaganda and censorship are baked into Chinese chatbots. https://rsf.org/en/controlling-information-age-ai-how-state-propaganda-and-censorship-are-baked-chinese-chatbots

Rest, J. R. (1986). *Moral development: Advances in research and theory*. Praeger.

Rest, J. R., Narvaez, D., Bebeau, M. J., & Thoma, S. J. (1999). *Postconventional moral thinking.A neo-Kohlbergian approach*. Lawrence Erlbaum Associates. DOI: 10.4324/9781410603913

Riaz, A. (2025). Title: Ethical implications of AI in recruitment – Navigating bias and fairness in socio-technical hiring. Advance. https://advance.sagepub.com/doi/full/10.31124/advance.175385244.44505199/v1

Riegler, J. (2023). Comparative ethics of modern payment models: Does the way we pay for care align with patient care ethics? *Voices in Bioethics, 9*. Advance online publication. DOI: 10.52214/vib.v9i.10310

Robson, E. (2025). Mary Midgley's meta-ethics and Neo-Aristotelian naturalism. *British Journal for the History of Philosophy, 33*(5), 1207–1232. DOI: 10.1080/09608788.2024.2389863

Romeo, G., & Conti, D. (2025). Exploring automation bias in human–AI collaboration: A review and implications for explainable AI. *AI & Society*. Advance online publication. DOI: 10.1007/s00146-025-02422-7

Roy, A., Newman, A., Round, H., & Bhattacharya, S. (2024). Ethical culture in organizations: A review and agenda for future research. *Business Ethics Quarterly, 34*(1), 97–138. DOI: 10.1017/beq.2022.44

Rozado, D. (2023). The political biases of ChatGPT. *Social Sciences (Basel, Switzerland), 12*(3), 148. DOI: 10.3390/socsci12030148

Rozenshtein, A. Z. (2026, February 25). What the Defense Production Act can and can't do to Anthropic. Lawfare. https://www.lawfaremedia.org/article/what-the-defense-production-act-can-and-can%27t-do-to-anthropic

Rubel, A., Clinton, C., & Pham, A. (2021). *Algorithms & autonomy: The ethics of automated decision systems*. Cambridge University Press. DOI: 10.1017/9781108895057

Ruslan, K., Madaminov, B., Ravikumar, R. N., & Aarthi, S. (2026). Equity-Driven Leadership Ecosystems. In *Advancing Equity in Schools through Global Perspectives on Educational Leadership* (pp. 33–66). IGI Global Scientific Publishing., DOI: 10.4018/979-8-3373-7794-0.ch002

Russell, S. (2019). *Human compatible*. Viking.

Russell, S. (2019). *Human compatible: Artificial intelligence and the problem of control*. Viking.

Ryan, M. (2020). In AI we trust: Ethics, artificial intelligence, and reliability. *Science and Engineering Ethics*, *26*(5), 2749–2767. DOI: 10.1007/s11948-020-00228-y PMID: 32524425

Sahih al-Bukhari 2554. (2025). Sahih al-Bukhari 2554 - Manumission of slaves - - Sunnah.com - Sayings and teachings of Prophet Muhammad ( ). Sunnah.com - Sayings and Teachings of Prophet Muhammad ( ). https://sunnah.com/bukhari:2554

Sahih al-Bukhari 7138 - Judgments (Ahkaam) - - Sunnah.com - Sayings and teachings of Prophet Muhammad ( ). (n.d.). Sunnah.com - Sayings and Teachings of Prophet Muhammad ( ). https://sunnah.com/bukhari:7138

Sahih al-Bukhari Hadith 1, Book 1. (2026). Hadith 1. 40HadithNawawi.com - The Forty 40 Hadith of Imam al-Nawawi. https://40hadithnawawi.com/hadith/1-actions-are-by-intentions/

Sahih Muslim. 1829a - The book on government - - Sunnah.com - Sayings and teachings of Prophet Muhammad ( ). (n.d.). Sunnah.com - Sayings and Teachings of Prophet Muhammad ( ). https://sunnah.com/muslim:1829a

Sahih Muslim. 1907a - The book on government - - Sunnah.com - Sayings and teachings of Prophet Muhammad ( ). (2026). Sunnah.com - Sayings and Teachings of Prophet Muhammad ( ). https://sunnah.com/muslim:1907a

Santoni de Sio, F. (2024). Artificial intelligence and the future of work: Mapping the ethical issues. *The Journal of Ethics*, *28*(3), 407–427. DOI: 10.1007/s10892-024-09493-6

Saraiva, L. (2024). Moral agency and responsibility in AI systems. *International Journal of Philosophy*, *3*(1), 25–36. DOI: 10.47941/ijp.1867

Sarker, I. H. (2022). AI-based modeling: Techniques, applications, and research issues towards automation, intelligent, and smart systems. *SN Computer Science*, *3*(2), 158. https://link.springer.com/article/10.1007/s42979-022-01043-x. DOI: 10.1007/s42979-022-01043-x PMID: 35194580

Schalock, R. L., Keith, K. D., Verdugo, M. Á., & Gómez, L. E. (2011). Quality of life model development and use in the field of intellectual disability. *Enhancing the quality of life of people with intellectual disabilities: From theory to practice*, 17-32.

Schalock, R. L., Keith, K. D., Verdugo, M. Á., & Gómez, L. E. (2011). Quality of life model development and use in the field of intellectual disability. In Enhancing the quality of life of people with intellectual disabilities: From theory to practice (pp. 17–32).

Schleidgen, S., Kermling, A., Metz, M., Kuehlmeyer, K., Inthorn, J., & Haltaufderheide, J. (2023). How to derive ethically appropriate recommendations for action? A methodology for applied ethics. *Medicine, Health Care, and Philosophy*, *26*(2), 175–184. https://link.springer.com/article/10.1007/s11019-022-10133-9. DOI: 10.1007/s11019-022-10133-9 PMID: 36522552

Schuster, N., & Kilov, D. (2025). Moral disagreement and the limits of AI value alignment: A dual challenge of epistemic justification and political legitimacy. *Philosophy & Technology*, *40*(8), 6073–6087. Advance online publication. DOI: 10.1007/s00146-025-02427-2 PMID: 41268066

Searcy, C., Castka, P., Mohr, J., & Fischer, S. (2022). Transformational transparency in supply chains: Leveraging technology to drive radical change. *California Management Review*, *65*(1), 19–43. DOI: 10.1177/00081256221126204

Segun, S. T. (2021). From machine ethics to computational ethics. *AI & Society*, *36*(1), 263–276. https://link.springer.com/article/10.1007/s00146-020-01010-1. DOI: 10.1007/s00146-020-01010-1

Selbst, A. D., Boyd, D., Friedler, S. A., Venkatasubramanian, S., & Vertesi, J. (2019). Fairness and abstraction in sociotechnical systems. In *Proceedings of the 2019 Conference on Fairness, Accountability, and Transparency* (pp. 59–68). DOI: 10.1145/3287560.3287598

Selwyn, N. (2019). What's the problem with learning analytics? *Journal of Learning Analytics*, *6*(3), 11–19. DOI: 10.18608/jla.2019.63.3

Sen, A. (1999). *Development as freedom*. Oxford University Press.

Sendjaya, S., & Sarros, J. C. (2002). Servant leadership: Its origin, development, and application in organizations. *Journal of Leadership & Organizational Studies*, *9*(2), 57–64. DOI: 10.1177/107179190200900205

Senge, P. M. (1990). *The fifth discipline: The art & practice of the learning organization*. Doubleday.

Seo, K., Tang, J., Roll, I., Fels, S., & Yoon, D. (2021). The impact of artificial intelligence on learner-instructor interaction in online learning. *International Journal of Educational Technology in Higher Education*, *18*(1), 54. DOI: 10.1186/s41239-021-00292-9 PMID: 34778540

Shababi, S. (2026). Beyond Asimov: Reconceiving the three laws of robotics as institutional imperatives for AI governance. *AI and Ethics*. Advance online publication. DOI: 10.1007/s43681-025-00629-4

Sharma, M., Tong, M., Korbak, T., Duvenaud, D., Askell, A., Bowman, S. R., & Perez, E. (2024). Towards understanding sycophancy in language models. arXiv. https://arxiv.org/abs/2310.13548

Shelmerdine, S. C., & Nour, M. M. (2025). AI chatbots and the loneliness crisis. *bmj, 391*.

Shen, J. H., & Tamkin, A. (2026). *Not a shortcut to competence: AI tools and skills formation* [Report]. IT Pro. https://www.itpro.com/software/development/anthropic-research-ai-coding-skills-formation-impact

Shneiderman, B. (2020). Bridging the gap between ethics and practice: Guidelines for human-centered AI systems. *ACM Transactions on Interactive Intelligent Systems*, *10*(4), 1–31. Advance online publication. DOI: 10.1145/3419764

Shneiderman, B. (2020). Human-centered artificial intelligence: Reliable, safe & trustworthy. *International Journal of Human-Computer Interaction*, *36*(6), 495–504. DOI: 10.1080/10447318.2020.1741118

Shreha, S. A., Ravikumar, R. N., & Aarthi, S. (2026). Precision medicine, genomics, and health equity challenges and opportunities. In *Advancing Equity in Global Healthcare Systems* (pp. 321–354). IGI Global Scientific Publishing., DOI: 10.4018/979-8-3373-1330-6.ch012

Siegel, D. J. (2012). *The developing mind: How relationships and the brain interact to shape who we are* (2nd ed.). Guilford Press.

Singer, P. W. (2009). *Wired for war: The robotics revolution and conflict in the twenty-first century*. Penguin Press.

Sire, J. (2015). *Naming the elephant: Worldview as a concept* (2nd ed.). InterVarsity Press.

Smith, R. (2025, October 15). Sam Altman says OpenAI isn't 'moral police of the world' after erotica ChatGPT post blows up. CNBC. https://www.cnbc.com/2025/10/15/altman-open-ai-moral-police-erotica-chatgpt.html

Smith, R. S. (2021, June 21). Making sense of morality: Alasdair MacIntyre's ethics. *Moral Apologetics*. https://www.moralapologetics.com/wordpress/msm26

Smith, R. S. (2014). *In search of moral knowledge*. IVP Academic.

Spears, L. C. (1998). *Insights on leadership: Service, stewardship, spirit, and servant-leadership*. John Wiley & Sons.

Spitko, E. G. (2024). Workplace AI and human flourishing. *Connecticut Law Review, 57*(1), 231–278.

Stahl, B. C. (2021). *Artificial intelligence for a better future: An ecosystem perspective on the ethics of AI and emerging digital technologies*. Springer. https://link.springer.com/book/10.1007/978-3-030-69978-9

Stahl, B. C. (2023). Embedding responsibility in intelligent systems: From AI ethics to responsible AI ecosystems. *Scientific Reports, 13*(1), 7586. DOI: 10.1038/s41598-023-34622-w PMID: 37202436

Stone, P. Brooks, R., Brynjolfsson, E. Calo, R., Etzioni, O., Hager, G., Hirschberg, J. Kalyanakrishnan, S., Kamar, E., Kraus, S., Leyton-Brown, K., Parkes, D., Press, W., Saxenian, A., Shah, J., Tambe, M., & Teller, A. (2016). *Artificial intelligence and life in 2030: The one hundred year study on artificial intelligence*. Report of the 2015 study panel. Stanford University. https://arxiv.org/abs/2211.06318

Suchman, M. C. (1995). Managing legitimacy. *Academy of Management Review, 20*(3), 571–610. DOI: 10.2307/258788

Swan, J. (2026, February 24). Musk's xAI and Pentagon reach deal to use Grok in classified systems. Axios. https://www.axios.com/2026/02/23/ai-defense-department-deal-musk-xai-grok

Talmage, C. A., Allgood, B., Ashdown, B. K., Brennan, A., Hill, S., Trevan, E., & Waugh, J. (2022). Tethering natural capital and cultural capital for a more sustainable post-COVID-19 world. *International Journal of Community Well-being, 5*(3), 657–678. https://link.springer.com/article/10.1007/s42413-021-00151-5. DOI: 10.1007/s42413-021-00151-5 PMID: 35018333

Taylor, C. (1989). *Sources of the self: The making of the modern identity*. Harvard University Press.

Thaut, L. C. (2009). The role of faith in Christian faith-based humanitarian agencies: Constructing the taxonomy. *Voluntas*, *20*(4), 319–350. DOI: 10.1007/s11266-009-9098-8

The Qur'an. (2:30, 2:44, 12:53, 22:46, 32:9, 33:72, 51:56, 75:2, 76:3, 89:29-30, 102:8). The New Translation by M.A.S Abdel Haleem. (2005). Oxford University Press (OUP) Oxford, ISBN: 0-19-283193-3

Trevino, L. K., & Brown, M. E. (2004). Managing to be ethical: Debunking five business ethics myths. *The Academy of Management Perspectives*, *18*(2), 69–81. DOI: 10.5465/ame.2004.13837400

Treviño, L. K., den Nieuwenboer, N. A., & Kish-Gephart, J. J. (2014). (Un)ethical behavior in organizations. *Annual Review of Psychology*, *65*(1), 635–660. DOI: 10.1146/annurev-psych-113011-143745 PMID: 23834354

Treviño, L. K., Weaver, G. R., & Reynolds, S. J. (2006). Behavioral ethics in organizations: A review. *Journal of Management*, *32*(6), 951–990. DOI: 10.1177/0149206306294258

Tripathi, R. (2025). AI meets mindfulness: Redefining spirituality and meditation in the digital age. *The Voice of Creative Research*. DOI: 10.53032/tvcr/2025.v7n1.03

Tsai, C., & Ku, H. (2025). Why AI may undermine phronesis and what to do about it. *AI and Ethics*, *5*(3), 3079–3086. DOI: 10.1007/s43681-024-00617-0

Uhl-Bien, M., & Marion, R. (2009). Complexity leadership in bureaucratic forms of organizing: A meso model. *The Leadership Quarterly*, *20*(4), 631–650. DOI: 10.1016/j.leaqua.2009.04.007

Uhl-Bien, M., Marion, R., & McKelvey, B. (2007). Complexity Leadership Theory: Shifting leadership from the industrial age to the knowledge era. *The Leadership Quarterly*, *18*(4), 298–318. DOI: 10.1016/j.leaqua.2007.04.002

UNESCO. (2021). Recommendation on the Ethics of Artificial Intelligence. UNESDOC Digital Library. https://unesdoc.unesco.org/ark:/48223/pf0000380455

Vallor, S. (2016). *Technology and the virtues*. Oxford University Press. DOI: 10.1093/acprof:oso/9780190498511.001.0001

van de Poel, I. (2020). Core values and value conflicts in cybersecurity: Beyond privacy versus security. In Christen, M., Gordijn, B., & Loi, M. (Eds.), *The ethics of cybersecurity* (pp. 45–71). Springer., DOI: 10.1007/978-3-030-29053-5_3 van Dierendonck, D. (2010). Servant leadership: A review and synthesis. *Journal of Management, 36*(4), 1228–1261. DOI: 10.1177/0149206310380462

van Niekerk, J., Delport, P. M. J., & Sutherland, I. (2025). Addressing the use of generative AI in academic writing. *Computers and Education: Artificial Intelligence, 8*, 100342. DOI: 10.1016/j.caeai.2024.100342

van Wynsberghe, A. (2021). Sustainable AI: AI for sustainability and sustainability of AI. *AI and Ethics, 1*(3), 213–218. DOI: 10.1007/s43681-021-00043-6

VanderWeele, T. J. (2017). On the promotion of human flourishing. *Proceedings of the National Academy of Sciences of the United States of America, 114*(31), 8148–8156. DOI: 10.1073/pnas.1702996114 PMID: 28705870

Verdugo, M. Á., Schalock, R. L., Gómez, L. E., & Navas, P. (2023). A Systematic Approach to Implementing, Evaluating, and Sustaining the Shared Citizenship Paradigm in the Disability Field. *Behavioral Sciences (Basel, Switzerland), 13*(12), 970. DOI: 10.3390/bs13120970 PMID: 38131825

Vincent, J. (2024, February 21). Google apologizes for "missing the mark" after Gemini generates historically inaccurate images. The Verge. https://www.theverge.com/2024/2/21/24079371/google-ai-gemini-generative-inaccurate-historical

Vinuesa, R., Azizpour, H., Leite, I., Balaam, M., Dignum, V., Domisch, S., Felländer, A., Langhans, S. D., Tegmark, M., & Nerini, F. F. (2020). The role of AI in achieving the Sustainable Development Goals. *nature. COMMUNICATIONS, 233*. Advance online publication. DOI: 10.1038/s41467-019-14108-y PMID: 31932590

Vosoughi, S., Roy, D., & Aral, S. (2018). The spread of true and false news online. *Science, 359*(6380), 1146–1151. DOI: 10.1126/science.aap9559 PMID: 29590045

Wachter, S., Mittelstadt, B., & Russell, C. (2021). Why fairness cannot be automated: Bridging the gap between EU non-discrimination law and AI. *Computer Law & Security Review, 41*, 105567. DOI: 10.1016/j.clsr.2021.105567

Weber, M. (1978). *Economy and society: An outline of interpretive sociology* (Roth, G., & Wittich, C., Eds.). University of California Press. (Original work published 1922)

Weber, M. (2002). *The Protestant ethic and the spirit of capitalism* (Baehr, P., & Wells, G. C., Trans.). Penguin Books. (Original work published 1905)

Weick, K. E. (1995). *Sensemaking in Organizations*. Sage.

Weidinger, L., Mellor, J., Rauh, M., Griffin, C., Uesato, J., Huang, P.-S., & Gabriel, I. (2021). Ethical and social risks of harm from language models. arXiv. https://arxiv.org/abs/2112.04359

Weinberg, L. (2022). Rethinking fairness: An interdisciplinary survey of critiques of hegemonic ML fairness approaches. *Journal of Artificial Intelligence Research*, *73*, 421–473. DOI: 10.1613/jair.1.13196

Weizenbaum, J. (1976). *Computer power and human reason: From judgment to calculation*. W. H. Freeman.

Westermann, C. (1994). [Fortress Press.]. *Genesis (New York, N.Y.)*, •••, 1–11.

White, S., & Rezania, D. (2019). The impact of coaches' ethical leadership behaviour on athletes' voice and performance. *Sport, Business and Management*, *9*(5), 460–476. DOI: 10.1108/SBM-11-2017-0079

Whittaker, M., Crawford, K., & Dobbe, R.. (2019). *AI Now Report 2019*. AI Now Institute.

Wiener, N. (1954). *The human use of human beings: Cybernetics and society* (Rev. ed.). Houghton Mifflin.

Wienhues, A. (2022). Looking through the microscope: Microbes as a challenge for theorizing biocentrism within environmental ethics. *Endeavour*, *46*(1-2), 100819. DOI: 10.1016/j.endeavour.2022.100819 PMID: 35641322

Wikanto, A. (2026, March 13). *Resmi! Pemerintah Batasi AI Di Pendidikan, Anak SD-SMA Dilarang Tanya Jawaban ChatGPT*. Kontan. https://nasional.kontan.co.id/news/resmi-pemerintah-batasi-ai-di-pendidikan-anak-sd-sma-dilarang-tanya-jawaban-chatgpt

Winner, L. (1980). Do artifacts have politics? *Daedalus*, *109*(1), 121–136. https://www.jstor.org/stable/20024652

Wolterstorff, N. (2008). *Justice: Rights and wrongs*. Princeton University Press. DOI: 10.1515/9781400828715

Wright, C. J. H. (2006). *The mission of God: Unlocking the Bible's grand narrative*. IVP Academic.

Wright, J. C., Warren, M. T., & Snow, N. E. (2021). Understanding virtue: Theory and measurement. *Journal of Moral Education*, *50*(1), 1–16. DOI: 10.1080/03057240.2020.1819073

Wyatt, J., & Williams, S. N. (Eds.). (2021). *The robot will see you now: Artificial intelligence and the Christian faith*. SPCK Publishing.

Yang, Q., & Lee, Y.-C. (2024). Ethical AI in financial inclusion: The role of algorithmic fairness on user satisfaction and recommendation. *Big Data and Cognitive Computing*, *8*(9), 105. DOI: 10.3390/bdcc8090105

Yan, L., Martinez-Maldonado, R., & Gašević, D. (2023). Generative artificial intelligence in learning analytics: Contextualising opportunities and challenges through the learning analytics cycle. Learning Analytics Review. *Journal of Learning Analytics*.

Yaokumah, W. (2020). Predicting and explaining cyber ethics with ethical theories. *International Journal of Cyber Warfare & Terrorism*, *10*(2), 46–63. https://www.igi-global.com/gateway/article/250905#pnlRecommendationForm. DOI: 10.4018/IJCWT.2020040103

Yeung, K. (2017). Algorithmic regulation: A critical interrogation. *Regulation & Governance*, *12*(4), 505–523. DOI: 10.1111/rego.12158

Yusuf, M., Witro, D., Diana, D., Santosa, T. A., & Jalwis, J. (2020). Digital parenting to children using the Internet. *Pedagogik Journal of Islamic Elementary School*, *3*(1), 1-14. https://ejournal.iainpalopo.ac.id/index.php/PiJIES/article/view/1277

Zárate-Torres, R., Rey-Sarmiento, C. F., Acosta-Prado, J. C., Gómez-Cruz, N. A., Rodríguez Castro, D. Y., & Camargo, J. (2025). Influence of leadership on human-artificial intelligence collaboration. *Behavioral Sciences (Basel, Switzerland)*, *15*(7), 873. DOI: 10.3390/bs15070873 PMID: 40723657

Zawacki-Richter, O., Marín, V. I., Bond, M., & Gouverneur, F. (2019). Systematic review of research on artificial intelligence applications in higher education. *International Journal of Educational Technology in Higher Education*, *16*(1), 39. Advance online publication. DOI: 10.1186/s41239-019-0171-0

Zelny, A. (2025). Offloading wisdom: Four technological relations that mediate phronesis. *Philosophy & Technology*, *38*(2), 56. DOI: 10.1007/s13347-025-00889-2

Zimbardo, P. G. (2007). *The Lucifer effect: Understanding how good people turn evil*. Random House.

Zouggar, N. (2018). Ibn Taymiyya on the proofs of prophecy and his legacy: Ibn Qayyim al-jawziyya (D. 751/1350) and Ibn al-Wazir (D. 840/1436). *Muslim World (Hartford, Conn.)*, *108*(1), 172–185. DOI: 10.1111/muwo.12233

Zuboff, S. (1988). *In the age of the smart machine*. Basic Books.

About the Contributors

Jeffrey Moore serves as a full-time professor of management and disability intelligence researcher at Anderson University (South Carolina). He was raised in France and received his PhD from the University of Nice Sophia-Antipolis' business school Institut d'Administration des Entreprises (IAE). His research focus is transforming organizational cultures through effective leadership using complexity leadership and the employment of people with disabilities. Leading the Anderson University research team, they work with Walgreens and Sephora to study their disability employment practices and impact on team members. They authored book chapters (2023, 2019 & 2017) on culture transformation through employing people with disabilities, creating transparent organizational cultures, and increasing inclusive pro-social behavior. Dr. Moore has published in AMLE (Academy of Management Learning & Education), JMD (Journal of Management Development) and ODJ (Organizational Development Journal). December 2022 he was the keynote speaker "Beyond Inclusion: A Story of Successful Enterprises Embracing People with Disabilities" in South Korea at the Handong Global University – United Nation Academic Impact – Institute for Global Education. He won the only best paper award at the 2019 International Dutch Human Resource Management Network. In 2015 won the AOM Management Consulting 2015 Award: Outstanding Scholar-Practitioner Collaboration. Over the past 20 years they have worked with manufacturing companies on organizational transformation through leadership development and with distribution centers on leveraging their inclusive teams. They are the only researchers working with Sephora, Walgreens, Dollar Tree to study their inclusive teams. SDG

Manjari Sharma is an accomplished academic leader, researcher, and author with over 15 years of teaching and administrative experience in higher education. She earned her Ph.D. in Commerce from the University of Rajasthan, India, where her doctoral research examined the pre- and post-issue price behavior of securities in the Indian capital market. Currently serving as Associate Professor and Head of the Department of Commerce at CHRIST (Deemed to be University), Pune Lavasa Campus, Dr. Sharma specializes in Finance, Human Resource Management, Organizational Behavior, and Sustainable Business Practices. Her academic work integrates financial analytics, innovation, and ethical leadership to advance experiential and value-driven learning. She has authored and edited multiple books with IGI Global, Springer Nature, and Taylor & Francis, covering themes such as financial literacy, sustainable workforce strategies, experiential learning, and AI in education. Her research contributions link the UN Sustainable Development Goals (SDGs) with inclusive pedagogy, leadership, and institutional sustainability. Dr. Sharma is also known for leading global immersion and cultural learning programs across Asia, Europe, and the United States, fostering cross-border academic collaboration. Her current projects explore the intersection of ethical AI, finance, and human capital development, reflecting her broader vision of education that is innovative, inclusive, and socially responsible.

Sharad Gupta is an accomplished academician and corporate trainer with over 20 years of combined experience across academia and industry in India and abroad. He holds a Professional Doctorate in Management (Financial Institution Management), dual Master's degrees in Commerce and Business Administration, and a Bachelor's degree in Education, complemented by ongoing postgraduate study in Education. Dr. Gupta has taught undergraduate, postgraduate, and doctoral students in India, Malaysia, and the Maldives, serving in leadership roles such as Deputy Dean at Lincoln University College (Malaysia) and Registrar & Associate Professor at Maldives Business School. His academic interests span Human Resource Management, Organizational Behavior, Financial Institution Management, and Diversity Studies, with a special focus on the Employee Life Cycle and the use of AI and Blockchain for HR transparency. Before entering academia, he held managerial positions in ICICI Bank, Citi Financial, HDFC Bank, and Bajaj Finserv, where he specialized in learning and development, leadership training, and business process transformation. He has conducted executive training for public- and private-sector organizations in India and the Republic of Maldives. Dr. Gupta is a co-inventor on a patent for an IoT-based Fertilizer Spraying Drone and has published and presented research on AI in education, experiential learning, and technology-enabled HRM. His current work examines the role of digital ethics, sustainable workforce strategy, and inclusive innovation in shaping Industry 5.0.

***

Phil Bailey is an Associate Professor of Management at Carson-Newman University. He is also the founder of the Center for AI Readiness. The author has no conflicts of interest to disclose. The author used Claude (Anthropic) as a research and drafting tool during the preparation of this chapter. All content was reviewed, verified, and revised by the author. Anthropic, the maker of Claude, is discussed as a subject of analysis in this chapter. Correspondence concerning this chapter should be addressed to Dr. Phil Bailey, Carson-Newman University. Email: pbailey@cn.edu

Kakhramon Chinkulov is an Associate Professor in the Department of Finance and Financial Technologies at Tashkent State University of Economics, Uzbekistan. He possesses strong expertise in financial systems, digital finance, and emerging financial technologies. His academic and research interests focus on fintech innovations, financial analytics, banking technologies, and the application of data-driven approaches in economic decision-making. He actively contributes to teaching and curriculum development, equipping students with both theoretical knowledge and practical skills aligned with modern financial industry requirements. Dr. Chinkulov has been involved in scholarly research, publishing academic works and participating in conferences that explore advancements in financial technologies and economic systems. He also promotes collaboration between academia and industry to enhance applied research and innovation. His work supports the advancement of financial education and the adoption of technological solutions in evolving global financial ecosystems.

Julie George is a Coordinator at Christ University, contributing to academic and administrative operations within the institution. Based in Bangalore, India, she is associated with supporting coordination activities that facilitate smooth functioning of university programs and student engagement. Her role reflects a commitment to organizational efficiency and effective communication within a dynamic higher education environment.

Noprita Herari, is an academia in communication studies whose work focuses on ethical leadership, humanistic digital communication, and the intersection of faith, technology, and moral agency. Her scholarship examines how artificial intelligence and digital systems reshape leadership, decision-making, and human responsibility, particularly within educational and organizational contexts. Drawing on interdisciplinary perspectives from communication theory, ethics, and faith-informed frameworks, she explores self-leadership, responsible speech, and accountability in AI-mediated environments. Noprita has contributed to several international book chapters and academic publications on digital ethics, leadership, and the passion economy, and her work emphasizes human dignity, moral formation, and responsible technology use in the age of artificial intelligence.

Paul C. Hong is a Distinguished University Professor and Chair of the Department of Information Systems and Supply Chain Management at the University of Toledo, Ohio, USA. He is a globally recognized scholar in the fields of global supply chain management and innovation strategy. He has published extensively in leading international journals such as the Journal of Operations Management and the Journal of Supply Chain Management, and was selected by the U.S. Department of State as a Fulbright–Nehru Teaching and Research Excellence Scholar in 2017. His recent books include Cosmetics Marketing Strategy in the Era of the Digital Ecosystem (2024) and "Encyclopedia of New Populism and Responses in the 21st Century" (Co-editor, 2024), Rising Asia and American Hegemony: Case of Competitive Firms (2020). The current series, "AI Renaissance × Geopolitics × Global Supply Chain," continues this line of research by exploring the civilizational and economic transformations of the AI era.

Saparov Ismat is an academic professional in the Department of Economics at Termez University of Economics and Service, Termez, Uzbekistan. He is engaged in teaching and research activities that contribute to the advancement of economic education and practice. His academic interests include applied economics, economic policy, and the analysis of regional and global economic systems. Dr. Saparov focuses on developing students' analytical and critical thinking skills, enabling them to understand complex economic challenges and propose effective solutions. His research explores areas such as sustainable development, financial systems, and the impact of technological innovation on economic growth. He contributes to scholarly publications and academic initiatives aimed at strengthening economic knowledge and its practical application. Through his academic and research efforts, he is committed to supporting informed decision-making and promoting sustainable socio-economic development.

Martin Jones is a dual British-American citizen and has a Ph.D. with a research emphasis in ethics, specifically examining post-modern culture emerging from the COVID-19 pandemic. He also has a Juris Doctorate, and was a commercial law attorney in the UK, having been appointed as Solicitor of the Supreme Court of England and Wales in 1999. He moved into higher education in 2004 as Senior Lecturer of Business Law at Lancashire Law School (UK) and emigrated to the US in 2007 to complete additional graduate work. He is currently the Associate Dean of the College of Business and Economics and Professor of AI, Law and Ethics at Anderson University (SC). He is an early adopter of technology and has published and presented nationally about the impact of AI on higher education, as well as leading seminars and workshops on its implementation. He is actively involved in Applied Artificial Intelligence course design, and leads a national faculty forum on the impact of AI in academia.

Jipson Joseph has a PhD in Theological Ethics from DVK Bangalore. Currently, he is a Research Scholar at School of Law, Christ University, Bangalore, India. His research interests are in the fields of ethics, gender studies, constitutional law, human rights, human security, environmental justice, SDGs, Artificial Intelligence, developmental policies, and global poverty. He has published 6 books, co-edited one book, and contributed scholarly articles to many international and national journals and edited volumes including IGI Global Scientific Publishing's edited books.

Justin Joy is an Associate Professor at School of Business and Management, Christ (Deemed to be University). He has taken his B.Tech in Mechanical Engineering from Govt. Engineering College, Thrissur, Masters (MSc in Business Information Technology) from Newcastle Business School, University of Northumbria, U.K. and a second Masters (M.Tech in Manufacturing Management) from BITS Pilani, Rajasthan. He took his PhD in Management from the Department of Management Studies, Pondicherry University. Before joining Christ, he was working as Assistant Professor in SCMS Cochin School of Business, Kerala. With a total of 25 years of experience, he was the Business Area Manager of Educomp Solutions Ltd. and Web and Business Development Manager cum Head of Computer Applications of IFCR India - a Canadian based Research Organization in Cochin. Apart from his corporate experience in India, he has also experience working abroad in Newcastle Upon Tyne, UK. As faculty he has undertaken several consultancy projects implementing open source Learning Management and Time Table Optimisation implementation projects in Academic Institutions and Open-source implementation in manufacturing domain projects in Kerala. He has published several patents and registered several copyrights too.

Arnaldo Marion is the Chief Financial Officer for Action Group, Arnaldo has engaged in the development and exponential growth of a family-owned business into a private multinational holding of 8 companies. Through Kingdom Enterprise Network (KEN), Arnaldo has served and helped hundreds of business leaders and entrepreneurs to discover a true and meaningful purpose to transform and shape culture in the marketplace through business. In the past 15 years, Arnaldo has trained, equipped, and developed thousands of people through one of the Coaching 4 Change: Coaching Boutique programs.

Pardaeva Markhabo is an academic professional in the Department of Social Sciences and Physical Education at Bukhara State Technical University, Bukhara, Uzbekistan. She is dedicated to advancing education through an interdisciplinary approach that integrates social sciences with physical education to promote holistic student development. Her teaching emphasizes critical thinking, social awareness, and the importance of physical well-being in academic and personal growth. Dr. Pardaeva's research interests include educational methodologies, social development, and the role of physical education in enhancing student performance and health outcomes. She actively contributes to academic activities, publications, and institutional initiatives aimed at improving the quality of education. Through her academic and research efforts, she strives to foster a balanced and inclusive learning environment that prepares students to meet contemporary societal and professional challenges.

James Meersman is part of the accounting faculty at Texas A&M University. He is a licensed CPA (TX) and has been teaching accounting for over a decade. His research interests include taxation, hospitality, CSR, education, and accounting history. Outside of academia, he also serves as a national instructor for both Big 4 accounting firms and industry leading CPA review companies.

Remco Mostert started his career in supporting people with intellectual disabilities. Since 2017, he has been working as a practical lecturer in the Professional Bachelor's program in Socio-Educational Care Work at HOGENT (University of Applied Sciences and Arts). He is also affiliated as a researcher with the EQUALITY//ResearchCollective, an interdisciplinary center focused on combating social exclusion and promoting the quality of life and human rights of people in vulnerable situations. His work focuses on quality of life, human rights, inclusive workplaces and person-centered support planning. Since the inception of POS, he has been trained as a Master Trainer and has conducted numerous conversations with people about their quality of life. Together with his colleagues, he provides several (inter)national POS trainings each year.

Young Soo Park is an Associate Professor of Business and Leadership at Midwest University, USA. He holds a Ph.D. in Manufacturing Management and Engineering from the University of Toledo, an MS in Industrial Systems and Engineering from Virginia Tech, and an MA and BA in Economics from Sogang University in Seoul, Korea. His articles have been published in the Benchmarking: An International Journal, Maritime Economics & Logistics, International Journal of Business Excellence, International Journal of Service and Operations Management, International Journal of Quality and Service Sciences, Journal of Enterprise

Information Management, Logistics, Systems, and Sustainability. His research interests are in product innovation strategies, business process management, and organizational leadership.

Ridho is a lecturer at Universitas Negeri Jakarta (UNJ), holding dual appointments in the Information Systems & Technology study program and PUSTIKOM as a System Analyst. He earned his Master's degree from Universitas Indonesia and holds a BNSP System Analyst certification. He also serves as a Teaching Assistant at Universitas Indonesia for the master's degree Information Systems Requirement Analysis & Design course. Before entering academia, he accumulated seven years of industry experience, including roles as a Senior Software Engineer at Tokopedia and an Engineering Manager at GovTech Edu. This practitioner background informs his research and teaching in software engineering and information systems. His research interests center on human aspects of software engineering. The research focuses on human dynamics during requirements elicitation, developer experience, cognitive load in software development tasks, and human-AI collaboration to support software engineering processes.

Ravikumar R. N. is currently a Postdoctoral researcher at Universiti Malaysia Perlis, Malaysia. Also, working as an Associate Professor in the Department of Computer Engineering at Marwadi University, Rajkot, Gujarat. He has 13+ years of national and international teaching experience. He completed his PhD at Amity University, Rajasthan, M.E at VMRF Deemed University, B.E at Anna University, Chennai. He is the co-founder of the Research Activities Club at Marwadi University and trains students on research activities, which leads to the publication of 8 Scopus/ SCI journals, 50+ Book chapters, 30+ Conference papers, and 16+ patents. His specialization is ML, DL, Federated Learning, Recommendation Systems, and Optimization Techniques.

Richard Ray is a scholar-practitioner specializing in strategic planning, business ethics, servant leadership, corporate social responsibility, and organizational behavior. His work integrates classical ethical traditions with contemporary challenges in leadership formation and technology governance. Dr. Ray holds certifications as a Registered Organization Development Professional (RODP) and Lean Six Sigma Black Belt (LSSBB), bringing both theoretical depth and practical experience to questions of ethical leadership in complex organizational environments. He earned his master's in organization development from Johns Hopkins University and his doctorate in organization learning from The George Washington University, with a fellowship in managing strategy from the Harvard School of Business. Having served in the US Military, working in the US government, in private sector firms and in healthcare executive roles, he attempts to integrate theory and practice to help organizations find solutions to complex problems. As it relates to this publication, Dr. Ray has led internal review boards and medical review "post-mortems" where ethical decision-making was used or required to investigate the actions of medical practitioners.

Aarthi S is a dedicated Assistant Professor in Computer Science with over six years of teaching experience. She is currently pursuing her Ph.D. at Marwadi University, Gujarat, and holds an M.Phil. and MCA from Bharathiar University, Tamil Nadu. Throughout her academic career, Aarthi has specialized in programming languages such as C++, Java, and C, and contributed to research in recommender systems and AI-related data privacy. She has published journal articles, contributed to book chapters, and actively participated in technical workshops. Known for her student mentorship and engagement in academic events, Aarthi remains committed to fostering innovation and excellence in computer science education.

Mukimov Askar Shukhratovich is an Associate Professor in the Department of Digital Technologies at Alfraganus University, Tashkent, Uzbekistan. He is an academic and researcher with expertise in digital systems, information technologies, and emerging computational methodologies. His work focuses on the development and application of advanced digital solutions, including data-driven systems, software technologies, and modern computing frameworks. Dr. Mukimov is actively engaged in teaching and mentoring students, emphasizing practical skills, innovation, and problem-solving in the field of digital technologies. His research interests include artificial intelligence, data analytics, and the integration of digital transformation strategies across various domains. He contributes to academic publications and collaborative research initiatives, aiming to enhance technological advancement and support the development of a digitally empowered society through education, research, and innovation.

Vipin Tyagi is a distinguished academician, researcher and author, currently serving as Dean (Academics and Research) at Jaypee University of Engineering and Technology in Guna, Madhya Pradesh. With a rich background in Computer Science, his expertise spans across Image Processing, Cyber Forensics and Speech Recognition. Prof. Tyagi has a prolific publication record, contributing significantly to his fields of interest. His scholarly work includes numerous journal articles, conference papers, and books. Some of his notable publications are in the areas of image forgery detection, image denoising and content-based image retrieval. He has authored and edited several books and papers in the field of Computer Science, such as "Digital Image Forgery Detection: Techniques, Challenges and Applications",

"Predictive Computing and Information Security" and "Content-Based Image Retrieval: Ideas, Influences, and Current Trends" published by Springer Singapore. His book "Understanding Digital Image Processing" using MATLAB, published by Taylor and Francis is highlighted by Mathworks at: https://in.mathworks.com/academia/books/understanding-digital-image-processing-tyagi.html His scholarly work has earned him a respectable h-index and numerous citations in the academic community.

Index

A

B

C

D

E

F

O

P

Q

R

S

T

U

V

W

www.ingramcontent.com/pod-product-compliance
Lightning Source LLC
LaVergne TN
LVHW062338300926
849161LV00007B/1734